Lecture Notes in Computer Science

Lecture Notes in Bioinformatics 14548

Series Editors

Sorin Istrail, *Brown University, Providence, USA*
Pavel Pevzner, *University of California, San Diego, USA*
Michael Waterman, *University of Southern California, Los Angeles, USA*

Editorial Board Members

Søren Brunak, *Technical University of Denmark, Kongens Lyngby, Denmark*
Mikhail S. Gelfand, *IITP, Research and Training Center on Bioinformatics, Moscow, Russia*
Thomas Lengauer, *Max Planck Institute for Informatics, Saarbrücken, Germany*
Satoru Miyano, *University of Tokyo, Tokyo, Japan*
Eugene Myers, *Max Planck Institute of Molecular Cell Biology and Genetics, Dresden, Germany*
Marie-France Sagot, *Université Lyon 1, Villeurbanne, France*
David Sankoff, *University of Ottawa, Ottawa, ON, Canada*
Ron Shamir, *Tel Aviv University, Ramat Aviv, Israel*
Terry Speed, *Walter and Eliza Hall Institute of Medical Research, Melbourne, Australia*
Martin Vingron, *Max Planck Institute for Molecular Genetics, Berlin, Germany*
W. Eric Wong, *University of Texas at Dallas, Richardson, USA*

The series Lecture Notes in Bioinformatics (LNBI) was established in 2003 as a topical subseries of LNCS devoted to bioinformatics and computational biology.

The series publishes state-of-the-art research results at a high level. As with the LNCS mother series, the mission of the series is to serve the international R & D community by providing an invaluable service, mainly focused on the publication of conference and workshop proceedings and postproceedings.

Mukul S. Bansal · Wei Chen · Yury Khudyakov ·
Ion I. Măndoiu · Marmar R. Moussa ·
Murray Patterson · Sanguthevar Rajasekaran ·
Pavel Skums · Sharma V. Thankachan ·
Alexander Zelikovsky
Editors

Computational Advances in Bio and Medical Sciences

12th International Conference, ICCABS 2023
Norman, OK, USA, December 11–13, 2023
Revised Selected Papers

Editors
Mukul S. Bansal
University of Connecticut
Storrs, CT, USA

Wei Chen
University of Oklahoma
Norman, OK, USA

Yury Khudyakov
Centers for Disease Control
Atlanta, GA, USA

Ion I. Măndoiu
University of Connecticut
Storrs, CT, USA

Marmar R. Moussa
University of Oklahoma
Norman, OK, USA

Murray Patterson
Georgia State University
Atlanta, GA, USA

Sanguthevar Rajasekaran
University of Connecticut
Storrs, CT, USA

Pavel Skums
University of Connecticut
Storrs, CT, USA

Sharma V. Thankachan
NC State University
Raleigh, NC, USA

Alexander Zelikovsky
Georgia State University
Atlanta, GA, USA

ISSN 0302-9743　　　　　　　　ISSN 1611-3349　(electronic)
Lecture Notes in Bioinformatics
ISBN 978-3-031-82767-9　　　ISBN 978-3-031-82768-6　(eBook)
https://doi.org/10.1007/978-3-031-82768-6

LNCS Sublibrary: SL8 – Bioinformatics

© The Editor(s) (if applicable) and The Author(s), under exclusive license
to Springer Nature Switzerland AG 2025, corrected publication 2025
Chapter "An Explainable Deep Learning Framework for Mandibular Canal Segmentation from Cone Beam Computed Tomography Volumes" is licensed under the terms of the Creative Commons Attribution 4.0 International License (http://creativecommons.org/licenses/by/4.0/). For further details see license information in the chapter.

This work is subject to copyright. All rights are solely and exclusively licensed by the Publisher, whether the whole or part of the material is concerned, specifically the rights of translation, reprinting, reuse of illustrations, recitation, broadcasting, reproduction on microfilms or in any other physical way, and transmission or information storage and retrieval, electronic adaptation, computer software, or by similar or dissimilar methodology now known or hereafter developed.
The use of general descriptive names, registered names, trademarks, service marks, etc. in this publication does not imply, even in the absence of a specific statement, that such names are exempt from the relevant protective laws and regulations and therefore free for general use.
The publisher, the authors and the editors are safe to assume that the advice and information in this book are believed to be true and accurate at the date of publication. Neither the publisher nor the authors or the editors give a warranty, expressed or implied, with respect to the material contained herein or for any errors or omissions that may have been made. The publisher remains neutral with regard to jurisdictional claims in published maps and institutional affiliations.

This Springer imprint is published by the registered company Springer Nature Switzerland AG
The registered company address is: Gewerbestrasse 11, 6330 Cham, Switzerland

If disposing of this product, please recycle the paper.

Preface

This volume contains the papers presented at ICCABS 2023, the 12th International Conference on Computational Advances in Bio and Medical Sciences, held on December 11–13, 2023 in Norman, Oklahoma, USA. ICCABS 2023 was hosted by the School of Computer Science at the University of Oklahoma with Marmar R. Moussa as General Chair and Pavel Skums and Sharma V. Thankachan as Program Committee Chairs. ICCABS has the goal of bringing together researchers, scientists, and students from academia, laboratories, and industry to discuss recent advances in computational techniques and applications in the areas of biology, medicine, and drug discovery.

There were 65 paper submissions. Following a rigorous review process in which each submission was reviewed by at least 2 reviewers from the Program Committee, the committee decided to accept 23 papers for oral presentation and publication in the post-proceedings volume. The program also includes a poster session with 15 accepted posters and several invited talks presented in the 4 satellite workshops hosted by the conference: 10 invited talks presented at the 12th Workshop on Computational Advances for Next-Generation Sequencing (CANGS 2023), 10 invited talks presented at the 11th Workshop on Computational Advances in Molecular Epidemiology (CAME 2023), 7 invited talks presented at the 5th Workshop on Computational Advances for Single-Cell Omics Data Analysis (CASCODA 2023), and 7 invited talks presented at the 2nd Workshop on Advances in Systems Immunology (ASI 2023). Workshop speakers were invited to submit 12-page extended abstracts that followed the same review process used for the main conference for publication in the post-proceedings volume. All submissions included in the volume have been revised to address reviewers' comments.

The technical program of ICCABS 2023 also featured keynote talks by three distinguished speakers: Courtney Montgomery from the Oklahoma Medical Research Foundation (OMRF) gave a talk on "Growing Opportunities for Computationa Advancements in Biomedical Research", Karen Jonscher from the Harold Hamm Diabetes Center gave a talk on "Unraveling the Early-Life Origins of Nonalcoholic Fatty Liver Disease (NAFLD) Using Multi-Omics Approaches", and Jie Chen from Augusta University gave a talk on "Computational and modeling aspects of analyzing high-throughput genomic data".

We would like to thank all keynote speakers and authors for presenting their work at the conference. We would also like to thank the Program Committee members and external reviewers for volunteering their time to review and discuss the submissions. Additionally, we would like to extend special thanks to the Steering Committee members for their continued leadership, and to the webmaster, and the publicity chairs. Last but not least, we would like to thank our sponsors, especially the University of Oklahoma and the National Science Foundation (NSF) for their support in making ICCABS 2023 a successful event.

October 2024

Mukul S. Bansal
Wei Chen
Yury Khudyakov
Ion I. Măndoiu
Marmar R. Moussa
Murray Patterson
Sanguthevar Rajasekaran
Pavel Skums
Sharma V. Thankachan
Alexander Zelikovsky

Organization

Steering Committee

Srinivas Aluru	Georgia Institute of Technology, USA
Reda A. Ammar	University of Connecticut, USA
Tao Jiang	University of California, Riverside, USA
Vipin Kumar	University of Minnesota, USA
Ming Li	University of Waterloo, Canada
Sanguthevar Rajasekaran (Chair)	University of Connecticut, USA
John Reif	Duke University, USA
Sartaj Sahni	University of Florida, USA

General Chair

Marmar R. Moussa — University of Oklahoma, USA

Program Chairs

Pavel Skums	University of Connecticut, USA
Sharma V. Thankachan	North Carolina State University, USA

Workshop Chairs

Mukul S. Bansal	University of Connecticut, USA
Wei Chen	University of Oklahoma, USA
Yury Khudyakov	Centers for Disease Control and Prevention, USA
Ion Măndoiu	University of Connecticut, USA
Marmar Moussa	University of Oklahoma, USA
Murray Patterson	Georgia State University, USA
Pavel Skums	Georgia State University, USA
Alex Zelikovsky	Georgia State University, USA

Publicity Chair

Olga Glebova   University of Connecticut, USA

Webmaster

Rany Kamel   University of Connecticut, USA

Program Committee

Derek Aguiar	University of Connecticut, USA
Max Alekseyev	George Washington University, USA
Sahar Al Seesi	Southern Connecticut State University, USA
Marmar Moussa	University of Oklahoma, USA
Mukul Bansal	University of Connecticut, USA
Jaime Davila	Mayo Clinic, USA
Jorge Duitama	Universidad de los Andes, Colombia
Richard Edwards	University of New South Wales, Australia
Oliver Eulenstein	Iowa State University, USA
Daniel Gibney	University of Texas at Dallas, USA
Danny Krizanc	Wesleyan University, USA
M. Oğuzhan Külekci	Istanbul Technical University, Turkey
Manuel Lafond	Université de Montréal, Canada
Ion Măndoiu	University of Connecticut, USA
Pavel Skums (Chair)	University of Connecticut, USA
Yanni Sun	Michigan State University, USA
Sing-Hoi Sze	Texas A&M University, USA
Sharma V. Thankachan (Chair)	North Carolina State University, USA
Balaji Venkatachalam	Google, USA
Jianxin Wang	Central South University, China
Fang Xiang Wu	University of Saskatchewan, Canada
Wei Zhang	University of Central Florida, USA
Shaojie Zhang	University of Central Florida, USA
Cuncong Zhong	University of Kansas, USA
Alexander Zelikovsky	Georgia State University, USA
Shibu Yooseph	University of Central Florida, USA
Maria Poptsova	Moscow State University, Russia
Ugo Vaccaro	University of Salerno, Italy
Yufeng Wu	University of Connecticut, USA

Additional Reviewers

Abdelnaby, Mohamed
Abedin, Paniz
Adeniyi, Ezekiel A.
Avdeyev, Pavel
Chubet, Oliver
Das, Arghya Kusum
Ganguly, Arnab
Khan, Nabila Shahnaz
Kuzmin, Kiril
Liu, Ben
Nemira, Alina
Nguyen, Tuan
Nihalani, Rahul
Oladapo, Olajumoke
Pranjal, Smriti
Rahaman, Md Mahfuzur
Thippabhotla, Sirisha
Vijendran, Sriram
Wagle, Sanket
Weiner, Samson
Xuan, Hao

Contents

An Explainable Deep Learning Framework for Mandibular Canal
Segmentation from Cone Beam Computed Tomography Volumes 1
   Konstantinos Barzas, Shereen Fouad, Gainer Jasa, and Gabriel Landini

Identification of Chimeric RNAs: A Novel Machine Learning Perspective 14
   Paola Bonizzoni, Clelia De Felice, Yuri Pirola, Raffaella Rizzi,
   Rocco Zaccagnino, and Rosalba Zizza

PartialFibers: An Efficient Method for Predicting Drug-Drug Interactions 27
   Aysegul Bumin, Kejun Huang, and Tamer Kahveci

Optimizing Deep Learning for Biomedical Imaging 40
   Ayush Chaturvedi, Guohua Cao, and Wu-chun Feng

Exploring a Solution Curve in the Phase Plane for Extreme Firing Rates
in the Izhikevich Model 53
   Chu-Yu Cheng and Chung-Chin Lu

Cancer and Tissue Prediction Using Mutational Signatures in Highly
Mutated Cancers 65
   Julia Cordes and Jaime Davila

On the Hardness of Wildcard Pattern Matching on de Bruijn Graphs 75
   Arnab Ganguly, Daniel Gibney, Arghya Kusum Das,
   and Sharma V. Thankachan

Plastic: An Easy to Use and Modular Tool for Benchmarking Tumor
Phylogeny Reconstruction Pipelines 82
   Akshay Juyal, Zahra Tayebi, Alexander Zelikovsky,
   Mauricio Soto-Gomez, Simone Ciccolella, Gianluca Della Vedova,
   and Murray Patterson

A 3D Deep Learning Architecture for Denoising Low-Dose CT Scans 94
   Armen Kasparian, Guohua Cao, and Wu-chun Feng

A Simple and Interpretable Deep Learning Model for Diagnosing
Pneumonia from Chest X-Ray Images 107
   Lucas Otavio Leme Silva, Karine Marques Hara,
   Pedro Henrique Mendes de Paula, Alexandre Rossi Paschoal,
   and Fabricio Martins Lopes

FedDP: Secure Federated Learning with Differential Privacy for Disease Prediction 119
  Bin Li, Hongchang Gao, and Xinghua Shi

Computational Tumor Progression Analysis via Seriation Based Trajectory Inference 132
  Marmar R. Moussa, Charles H. Street, and Sriram Boddeda

Multilayer Network Analysis of Brain Signals for Detecting Alzheimer's Disease 145
  Sean M. Nguyen, Mohammad Amin Basiri, and Sina Khanmohammadi

DNA Methylation Based Subtype Classification of Breast Cancer 154
  Sri Lakshmi Bhavani Pagolu, S. Suba, and Nita Parekh

Repeated Measures Latent Dirichlet Allocation for Longitudinal Microbiome Analysis 166
  Namitha Viona Pais, Nalini Ravishanker, Sanguthevar Rajasekaran, and George Weinstock

Improving Disease Comorbidity Prediction with Biologically Supervised Graph Embedding 178
  Xihan Qin and Li Liao

Lightweight and Generalizable Model for COVID-19 Detection Using Chest Xray Images 191
  Suba Suseela and Nita Parekh

Decoding Heterogeneity in Quadruple-Negative Breast Cancer: A Data-driven Clustering Approach 203
  Bikram Sahoo, Nikita Jinna, Padmashree Rida, Zandra Pinnix, and Alex Zelikovsky

Determining Temporal Linkages in Dynamic Epidemiological Networks Using the Earth Mover's Distance 218
  Rahul Singh and Jiadong Yu

Functional Connectivity Disruptions in Alzheimer's Disease: A Maximum Flow Perspective 229
  Emma T. Stubby, Seyed Majid Razavi, and Sina Khanmohammadi

On Multi-phase Metagenomics Reads Binning 238
  Francesco Tomasella and Cinzia Pizzi

A Unified Machine Learning Framework for Multi-subtype Tumour
Classification Across Diverse Datasets 251
   Ankur Yadav and Ovidiu Daescu

AFA: Abstract Functional Analysis Identifies New Microglial Subtypes
at Single Cell Level in Alzheimer's Disease 262
   *Chenyu Zhang, Honglin Wang, Seung-Hyun Hong, Riqiang Yan,
and Dong-Guk Shin*

Correction to: Identification of Chimeric RNAs: A Novel Machine
Learning Perspective C1
   *Paola Bonizzoni, Clelia De Felice, Yuri Pirola, Raffaella Rizzi,
Rocco Zaccagnino, and Rosalba Zizza*

Author Index 277

An Explainable Deep Learning Framework for Mandibular Canal Segmentation from Cone Beam Computed Tomography Volumes

Konstantinos Barzas[1], Shereen Fouad[1(✉)], Gainer Jasa[2], and Gabriel Landini[3]

[1] School of Computer Science and Digital Technologies, Aston University, Birmingham, UK
s.fouad@aston.ac.uk
[2] Division of Oral Radiology, Faculty of Dentistry, Republic University, Montevideo, Uruguay
[3] Oral Pathology Unit, Dentistry, School of Health Sciences, University of Birmingham, Birmingham, UK

Abstract. Cone Beam Computed Tomography (CBCT) is an indispensable imaging modality in oral radiology, offering comprehensive dental anatomical information. Accurate detection of the mandibular canal (MC), a crucial anatomical structure in the lower jaw, within CBCT volumes is essential to support clinical dentistry workflows, including diagnosis, preoperative treatment planning, and postoperative evaluation. In this study, we present a deep learning-based (DL) approach for MC segmentation using 3D U-Net and 3D Attention U-Net networks. We collected a unique dataset of CBCT scans from 20 anonymous hemisected mandibular bones, which were further processed for analysis. The samples were scanned using a CBCT scanner after inserting a wire through the whole length of the MC to identify its location in space (as a gold standard). Our experimental results demonstrate that the 3D Attention U-Net outperforms the standard 3D U-Net in detecting the MC's location, with Dice similarity score, Precision, and Recall values of 0.65, 0.75, and 0.60, respectively. Unlike current DL-enabled methods for MC segmentation, which face deployment and trust challenges due to their "black-box" nature, our approach incorporates a post-hoc visual explainability feature through the Grad-CAM++ (Gradient-weighted Class Activation Mapping) algorithm. This tool highlights important regions within the CBCT volumes that influence the model's predictions, providing valuable insights into the segmentation process, and bridging the gap between cutting-edge DL technology and clinical practice.

Keywords: Dental Cone Beam Computed Tomography · mandibular canal segmentation · U-Net deep learning model · explainable artificial intelligence · Grad-CAM

S. Fouad—Equal Contribution.

1 Introduction and Background

The mandible is the largest and strongest bone in the human orofacial region, giving hard tissue to the lower jaw and part of the mouth. From an anatomical perspective, the mandibular canal (MC) runs along part of the mandible, containing the inferior alveolar nerve and blood vessels. In dental surgery, the accurate detection of MC is crucial to a wide range of dental procedures including dental extractions, implant placement, thirds molar surgeries and jaw alignment surgery to avoid damage to the inferior alveolar nerve and vessels [2]. Advancements in imaging technologies, like Cone Beam Computed Tomography (CBCT), have enhanced the ability to precisely identify the MC. CBCT is a three-dimensional (3D) imaging modality that captures images from various angles with a cone-shaped beam to reconstruct volumetric information as a series of axial images. MC segmentation of volumetric CBCT images is a challenging task due to the complexity of dental anatomical structures that feature variability in shape and position. Segmenting the MC in CBCT images manually is a time-intensive process and susceptible to both intra- and inter-observer variability. Earlier segmentation approaches have used traditional segmentation methods, such as image thresholding, mathematical morphology and statistical approaches [1,11]. However, those methods suffered often resulted in low segmentation accuracy, long processing time and poor reliability.

More recently, Deep Learning-based (DL) (a subset of machine learning) has shown remarkable performance in segmenting MC from CBCT volumes. In the context of image segmentation, machine learning-based algorithms aim at labelling each image pixel with a specific category [7,8]. Many attempts to perform MC segmentation in CBCT using DL have primarily employed 3D convolutional neural networks (CNN). For example, Jaskari et al. [9] demonstrated that 3D CNN models produced better MC segmentation quality than statistical shape models, but the segmentation results were able to achive a Dice similarity score of just 0.57. Kwak et al. [12] used 2D U-Net network [6], 2D SegNet and 3D U-Net [5], to automatically localise the MC in 2D and 3D CBCT images. A thresholding-based tooth segmentation technique was applied initially to remove non-mandibular bone areas from the scans. 3D U-Net outperformed the 2D approach, however, it was unable to detect the canal when the surrounding cortical bone layers were not clearly defined and the pre-processing thresholding step was difficult to initialise. Lahoud et al. [13] proposed a two step segmentation framework using 3D CNN networks to firstly preform voxel-wise segmentations to adjust for variations in MC shape and width and then train another 3D CNN model to produce a full-resolution segmentation output. They achieved a Dice similarity score of 0.77 using a dataset of 235 CBCT scans. Jeoun et al. [10] introduced the Canal-Net approach, based on the 3D U-Net architecture, enhanced with bidirectional convolutional long short-term memory units within a multi-task learning framework. This approach outperformed 2D U-Net, SegNet and 3D U-Net, but has large memory requirements. Recently, a two-stage 3D U-Net method for MC segmentation was proposed in [14] which yielded a Dice similarity score of 0.95 on a different dataset that is significantly larger than our studied images.

Despite remarkable results achieved by DL-methods in MC segmentation tasks, these methods suffer from the "black-box" nature, i.e. while they can produce accurate results, it is often challenging to understand and interpret why the model arrived to a particular decision. This lack of transparency can be problematic in clinical settings where trust and interpretability are paramount. This lack of explainability, common in DL-based segmentation tools, hinders the their adoption and use in critical clinical settings. Additionally, the "black-box" nature of DL-based models can easily pose challenges in meeting the requirements of regulatory bodies. Recently, the concept of Explainable AI (XAI) has emerged, introducing a range of techniques that aim to strike a balance between explainability and robust detection and prediction performance [18]. Post-hoc visual XAI in medical image segmentation provides visual explanations for the decisions made by DL models after they have performed the image segmentation. This is achieved by highlighting image regions that exert the most significant influence on the model's decision. This is a crucial step for understanding and validating the segmentation results and ensuring that its trustworthy and accurate. Various post-hoc visual XAI methods have been proposed in the literature, however, the application of XAI has not been well explored in the context of MC segmentation in 3D CBCT.

To close this gap, we propose an explainable DL framework for MC segmentation from CBCT volumes using U-Net architectures and Grad-CAM++ (Gradient-weighted Class Activation Mapping) algorithms [4]. In particular, we study the performance of 3D U-Net [5] and 3D Attention U-Net [15] networks (a modified version of the U-Net). 3D U-Net has shown remarkable results in segmenting volumetric biomedical images [10,12,14], however, 3D Attention U-Net has not been fully explored in the context of detecting MC in 3D CBCT images. The 'attention' mechanism seems particularly effective when handling medical images characterized by high levels of noise or when the region of interest occupies only a small portion of the complete image, which applies to our CBCT images. Furthermore, the proposed framework addresses explainability by incorporating Grad-CAM++ [4] (an extended version of the original Grad-Cam [17]) into the segmentation pipeline, thereby providing clinicians with an additional and deeper understanding of the decision-making process.

2 Dataset Description and Preprocessing

The sample set consisted of 20 anonymous dry hemisected mandibular bones, obtained from the Anatomy Museum collection at the Faculty of Dentistry, Universidad de la República, Uruguay. The ethical approvals for collecting and using the data were obtained prior to conducting the study. To create gold standard images, each half mandible was scanned with a wire inside the MC to precisely locate it. The wire was then removed without moving the sample, and the half mandibles were scanned using the same field of view and the same exposure time, but at varying tube voltages and currents, using a combination of kilovoltage (kV) and milliamperage (mA) values. The bones were scanned, submerged in water to simulate soft tissue absorption and positioned so that the MC was parallel to the axial plane. Data volumes captured with and without the inserted wire were spatially registered (see Fig. 1). Digital files were exported in the digital imaging and communication in medicine (DICOM) format for further analysis with Python software. Notably, the dimensions of the CBCT voxels varied within the range of $(543, 543, 80)$ and $(543, 543, 190)$ across the X, Y, and Z axes respectively, with the associated Hounsfield units spanning from -1000 to 2000.

The ground truth segmentation masks of the MC were hand traced guided by the inserted wire (see Fig. 1) using Slicer imaging software (version 5.2.2). Volumes were cropped to a resolution of $(120, 240, 512)$, denoting the designated Region of Interest (ROI). Thereafter, a 3D spherical paint brush instrument with a diameter of 2% of the ROI was used to mark the location of the wire and thus create the binary mask (ground truth).

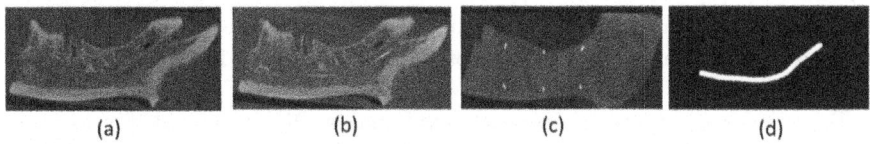

Fig. 1. (a) mandibular canal without wire inserted, and (b) with wire. (c) original image after applying maximum projection, and (d) binary mask image

Data augmentation is a key data pre-processing technique that is commonly used to improve the robustness and generalization of DL models, particularly where annotated training data is small in size, which applies to this study. It involves applying various geometric transformations to the original image to create additional training data, including random rotation, scaling, and flipping. The addition of Gaussian noise to the original images also helps the model to become more robust to real-life image artifacts. In the context of rotational transformations, an initial large angle was selected, followed by gradual reductions until the model exhibited convergence, which was assessed by monitoring the decline of the Dice loss over increasing training epochs. The model was trained for a total of 100 epochs and where divergence was noted, the angle of rotation was decreased incrementally, specifically by 5% with each divergence occurrence. Subsequently, once convergence was achieved, a further reduction of the angle by 30% was implemented to establish the final range for that particular parameter. This strategy was also applied for the range of the added random Gaussian noise. After experimentation, the values of mean and standard deviation of Gaussian noise are 0 and 0.1, respectively. While the angle range for the rotational transformation is 0.2 radians in x, y, and z coordinates. The data augmentation was performed by the MONAI library transforms [3].

3 The Segmentation Model

In this study, we propose a DL-based approach for MC segmentation. We investigated the performance of 3D U-Net [5] and 3D Attention U-Net [15] networks. Both methods have demonstrated exceptional performance in similar semantic segmentation tasks [10,12,14].

3.1 3D U-Net Architecture

3D U-Net [5] is an adapted version of the conventional 2D U-Net [16] configured to process 3D images. It preserves the core structure of the initial 2D U-Net model, but it substitutes all 2D operations with their 3D counterparts, such as 3D convolutions, 3D max pooling, and 3D up-convolutions, resulting in a segmented image in three dimensions. A graphical illustration of the applied 3D U-Net model is presented in Fig. 2.

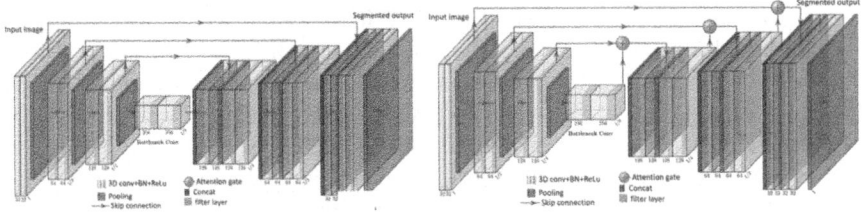

Fig. 2. 3D U-Net (left) and 3D Attention U-Net architectures (right)

The 3D U-Net architecture contains two paths, the contracting path (or encoder) captures the high-level semantic features of the image, while the expanding path (or decoder) restores the high-level semantic feature map back to the original image resolution. The layers in the encoder part are skip connected and concatenated with layers in the decoder part (highlighted in Fig. 2). This enables 3D U-Nets to utilize the finely-detailed information acquired during the encoder phase to generate an image within the decoder phase.

In the contracting path, the network starts with a 3D convolutional layer featuring 32 filters, followed by instance normalization and dropout layers and a ReLU activation function to introduce non-linearity. This phase follows a consistent pattern of convolution layers, doubling the number of filters at each subsequent stage, ultimately reaching 256 filters. This incremental approach enhances feature representation while progressively reducing the output spatial dimensions, resulting in a systematic down-sampling. In the expanding path, the network uses up-convolution layers to gradually increase the spatial dimensions of the feature maps. It starts with a layer containing 128 filters and incorporates skip connections from corresponding layers in the contracting phase. Feature channels are halved at each step, aiming to reduce feature channels during up-sampling. This process includes two 3D convolutions followed by a ReLU activation function to refine features. Finally, a single-filter layer is used to convert feature vectors into the segmented output.

3.2 3D Attention U-Net Architecture

Skip connection in original 3D U-Net combines spatial information from the contracting path with the expanding path to retain good spatial information. But this process brings along some irrelevant features from the initial layers. Attention in U-Net is a recent extension of the basic U-Net method, and it aims at highlighting only important/relevant parts of the image while ignoring unnecessary areas during training. The 3D Attention U-Net architecture, illustrated in Fig. 2, utilises an attention mechanism by adding attention gates at each level of the expansive path. These gates re-calibrate the feature channels by taking input feature maps from both the contracting and expanding path and merges them to generate a set of attention coefficients which are then used to adjust the feature maps from the contracting path before they are concatenated with the corresponding feature maps in the expanding path. The resultant attention-gated feature maps are subsequently passed through up-convolution layers and other successive layers in the expanding path, mirroring the original 3D U-Net structure. 3D Attention U-Net also adopts the ReLU activation function, and it maintains the use of instance normalization and dropout layers to reduce over-fitting and ensuring a stable training process.

4 Post-hoc Visual Explainability - Grad-CAM++

Post-hoc visual explainability refers to the process of providing visual explanation and interpretation of the decisions made by the DL models [18]. It helps DL users understand why a model has generated a specific result, which seems particularly important in medical imaging applications, where trust and interpretability are crucial.

Gradient-weighted Class Activation Mapping (Grad-CAM) [17] is a XAI technique for generating visual explanations of decisions made by CNNs. It generates a 'heatmap' that highlights the regions of the input image most responsible for a particular output. The heatmap is constructed by leveraging the gradient information from the last convolutional layer in the network, which is activated for different channels with respect to the class. Specifically, for each feature map A_{ij}^k at this convolutional layer, an importance weight α_k^c is computed with respect to the kth feature map and class c. This is done through global-average-pooling over pixel location (i, j) for the gradient $\frac{\partial y^c}{\partial A_{ij}^k}$ of respective classification score y^c of class c. Mathematically, the weights can be estimated as:

$$\alpha_k^c = \frac{1}{Z} \sum_i \sum_j \frac{\partial y^c}{\partial A_{ij}^k}$$

where Z is the number of pixels in the activation map. Once these importance weights are obtained for each feature map, the next step is to calculate the heatmap, denoted here as L^c. This is done by taking a weighted sum of the feature maps and applying a ReLU function to keep only the positive contributions: $L^c = \text{ReLU}\left(\sum_k \alpha_k^c A^k\right)$. The resulting heatmap L^c therefore shows the areas of the input image that were most influential in producing the class c, thus providing a visualization of the model's focus and decision-making process.

Grad-CAM++ [4] is an extension of the original Grad-CAM technique, which uses second order gradients. It offers a more fine-grained explanation of model decisions by accounting for the importance of individual pixels within the feature maps. On the other hand, the pixel gradients that have no impact on the prediction will be scaled down. In particular, while Grad-CAM computes a global importance weight α_k^c for each feature map A_k through average-pooling, Grad-CAM++ goes a step further and calculates the following weight

$$\alpha_{ij}^{kc} = \frac{\frac{\partial^2 y^c}{\partial (A_{ij}^k)^2}}{2\frac{\partial^2 y^c}{\partial (A_{ij}^k)^2} + \sum_a \sum_b A_{ab}^k \frac{\partial^3 y^c}{\partial (A_{ij}^k)^3}}$$

where α_{ij}^{kc} is the value of α at pixel location (i,j) for the k-th feature map corresponding to the output class c. Here, (a,b) are iterators over the same activation map A_k and are used to avoid confusion. The heatmap L^c can be expressed as follows: $L^c = \sum_i \sum_i \alpha_{ij}^{ck}.\text{ReLU}(\frac{\partial y^c}{\partial A_{i,j}^k})$. The equation considers both the positive and negative contributions of each pixel in determining the final output. This allows Grad-CAM++ to produce more nuanced heatmaps compared to its predecessor Grad-CAM.

5 Model Training and Hyperparameters Fine-Tuning

The dataset was partitioned into training, validation, and testing subsets, adhering to a ratio of 60:20:20, respectively. The Hyperparameters of the U-Net model were fine-tuned utilizing the three parameters,

(a) Dice Loss. and it is estimated as, $\frac{2\times\sum_{i=1}^{N} p_i g_i}{\sum_{i=1}^{N} p_i^2 + g_i^2}$. Where p_i and g_i represent pairs of corresponding pixel values of predicted binary segmentation volume P and ground truth binary volume G,

(b) Mean Intersection over Union (mIoU), and it is estimated as, $\frac{1}{C}\sum_c \text{IoU}_c$, where C is the number of classes, $\text{IoU}_c = \frac{TP_c}{TP_c + FP_c + FN_c}$, and TP_c, FP_c, and FN_c denotes true positives, false positives, and false negatives for class c, respectively.

(c) Generalized Dice Focal Loss (GDFL), and it is estimated as $-\frac{\sum_{i=1}^{N} 2 \cdot TP_i \cdot (1-p_i)^\gamma}{\sum_{i=1}^{N} TP_i + \sum_{i=1}^{N}(1-p_i)^\gamma \cdot (TP_i + FN_i) + \epsilon}$, where N is the number of classes, p_i is the predicted probability for class i, TP_i is the True Positives, FN_i is the False Negatives, γ is the Focal Loss hyperparameter that modulates the focusing effect, and ϵ is a constant value.

As shown in Table 1, a range of Hyperparameters have been adjusted to optimize both the computational efficiency and the predictive performance of the two studied models. A random search was conducted with these parameters; the search was run for 60 iterations for 3D U-Net and 20 for 3D Attention U-Net. Both models were trained with patches from randomly rotated and translated CBCT volumes using the loss objectives.

Table 1. Hyperparameters to be tuned

Hyperparameters	Model	
	3D U-Net	3D Attention U-Net
Loss function	Dice Loss, GDFL, Mean IoU	Dice Loss, GDFL
Batch Size	1, 2, 3, 4	1, 2
Channels	[8, 16, 32, 64], [16, 32, 64, 128], [32, 64, 128, 256], [64, 128, 256, 512]	[8, 16, 32, 64], [16, 32, 64, 128]
Dropout	0, 0.2, 0.3, 0.4, 0.6	0, 0.1, 0.2
Learning Rate	3e-3, 5e-4, 5e-5	3e-3, 5e-4, 2e-5
Early Stopping	25	25
Max Epochs	150	150
Val Interval	5	5

6 Results

6.1 Optimization Results

Figure 3 reports the hyperparameter optimization results from the random search for the 3D U-Net and 3D Attention U-Net models. For the 3D U-Net model, the optimal values for dropout rate, channel sequence and batch size were identified as 0,[32, 64, 128, 256], and 1, respectively. However, further investigation of the learning rates along with loss functions, revealed that the learning rates of 2×10^{-4}, along with GDFL as loss functions, yielded the best results on the validation set. For the 3D Attention U-Net model, the optimal settings for the loss function, learning rate, dropout rate, and batch size, were GDFL, 0.003, 0.1, and 1, respectively. Further investigation to determine the optimal values for the channel sequence (between [8,16,32,64] and [16,32,64,128]) revealed that the lower number channel sequence converged faster, whereas the higher channel sequence converged slower but provided slightly more accurate results on the evaluation set. Eventually, the lower sequence, [8,16,32,64], was selected as the optimal value as it required less number of trainable parameters, making the model more accessible for practical use.

6.2 Evaluation Metrics

To evaluate the optimal models generated through the random search the following metrics were used:

(a) Dice similarity score, measures the similarity between two image samples, and it is calculated as: $\frac{2 \times TP}{2 \times TP + FP + FN}$,

(b) Precision, measures the proportion of true positive predictions among all the positive predictions, and it is calculated as: $\frac{TP}{TP+FP}$

(c) Recall, measures the model's ability to identify all relevant instances, and it is calculated as: $\frac{TP}{TP+FN}$

(d) Hausdorff Distance, measures the maximum perpendicular distance between the automatic and manual segmentation, and it is defined as:
$\max\left(\max_{a \in A}\left(\min_{b \in B} d(a,b)\right), \max_{b \in B}\left(\min_{a \in A} d(a,b)\right)\right)$, where d(.,.) represents Euclidean distance between two sets of points denoted as a and b.

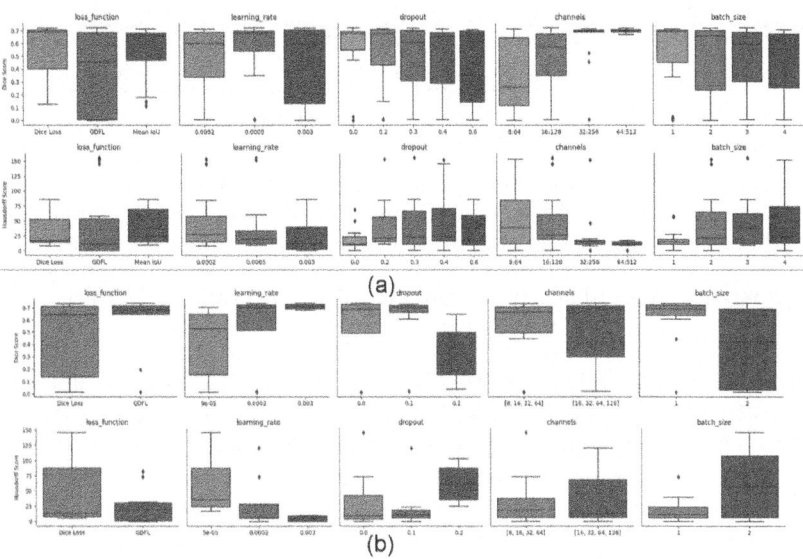

Fig. 3. Hyperparameter results from the random search of 3D U-Net (a) and 3D Attention U-Net (b). Each box contains the first and third quartile of data, and medians are indicated as lines inside each box.

6.3 Segmentation Results Obtained from the Optimised Models

Segmentation quantitative results obtained from the optimised models are shown in Table 2 and qualitative results are presented in Fig. 4. The 3D Attention U-Net outperformed the standard 3D U-Net, specially in handling high-noise scans and in detecting the central portion of the canal. On the other hand, the 3D U-Net demonstrated better performance in identifying the extremes of the canal, as denoted by a more pronounced presence of red markers at the canal ends.

Table 2. Segmentation results obtained from the optimised U-net and Attention U-Net models.

Dataset	Model Name	Dice similarity score	Hausdorff	Precision	Recall
Train	3D U-Net	0.76	15.52	0.81	0.72
	3D Attention U-Net	**0.79**	**6.06**	**0.82**	**0.76**
Validation	3D U-Net	**0.72**	16.67	0.77	**0.67**
	3D Attention U-Net	0.72	**15.81**	**0.79**	0.66
Test	3D U-Net	0.62	47.54	0.74	0.57
	3D Attention U-Net	**0.65**	**35.04**	**0.75**	**0.60**

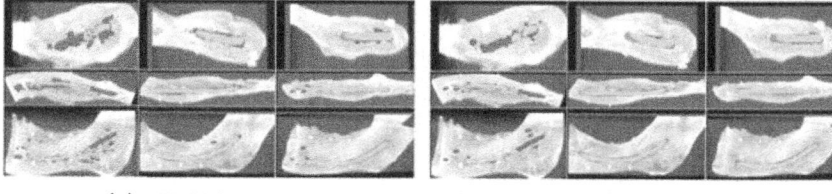

(a) 3D U-Net   (b) 3D Attention U-Net

Fig. 4. The segmentation results on Coronal, Axial and Sagittal view of three example CBCT test scans. Green: manual segmentation, red: automatic segmentation, yellow: overlap between automatic and manual segmentation. (Color figure online)

6.4 Post-hoc Visual Explanation Results via Grad-CAM++

After the segmentation output of 3D U-Net and 3D Attention U-Net, was obtained, the Grad-CAM++ method was applied. The M3d-CAM library was used to streamline the process of visualizing attention maps utilizing the Grad-CAM++. The library provides a simple method for incorporating the required components into the pre-existing PyTorch model. Contrary to the common practice of focusing on the last convolutional layer for attention visualization, exploratory analyses revealed that the first convolutional layer provided more insightful attention maps for both 3D Attention U-Net and 3D U-Net. This notable divergence from standard practices offers an insight into the networks' focus during the image segmentation process. The M3d-CAM library streamlined this exploratory process, enabling to load the best-performing models post-training and swiftly visualize attention at various layers. Based on this, the first convolutional layer was ultimately selected for in-depth attention map analyses. Features in this layer contained some high-level semantics and simultaneously should preserve some spatial information.

The output of the Grad-CAM++ for 3D U-Net and 3D Attention U-Net is shown in Fig. 5. The 3D Attention U-Net heatmap showed better region localizing ability when compared to the 3D U-Net model, which was acting as indicator of the model's confidence rather than providing a coherent heatmap representation. The Grad-CAM++ heatmap highlights the features (ROI) that the model considered before making the segmentation decision, enhancing the transparency

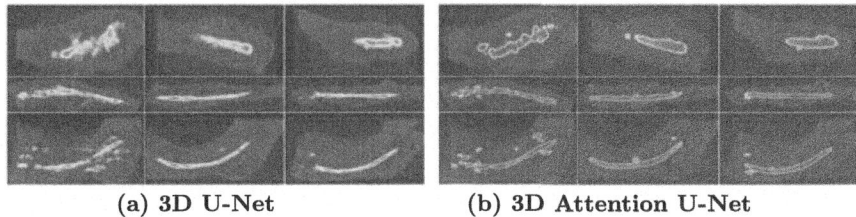

Fig. 5. Grad-CAM++ output/heatmap on Coronal, Axial and Sagittal view of three example CBCT test scans.

and trustworthiness of the segmentation model. This could help clinicians to verify and validate the model's decisions, ultimately improving trust and deployment of DL-based techniques in dental surgical procedures.

7 Conclusion

Mandibular canal (MC) segmentation is a challenging task in maxillofacial radiology. This study introduces an explainable framework for segmentation of CBCT volumes using 3D U-Net and 3D Attention U-Net models, complemented by Grad-CAM++ visualization. We analysed a set of 20 CBCT scans, followed by a series of data preprocessing steps and optimization procedures to identify the optimal training parameters for MC segmentation. The results revealed that the 3D Attention U-Net system outperformed the standard 3D U-Net in terms of both segmentation quality and post-hoc visualization. The segmentation results were relatively lower than those of some previously DL-based methods, possibly due to the limited number of training scans available. However, the primary focus here centred in improving the explainability of MC automatic segmentation over the perceived "black box" approach of other DL segmentations. The post-hoc heatmaps using Grad-CAM++ highlighted significant regions in the images that contributed to the model's decision therefore enhancing explainability and trust, thereby facilitating the deployment of DL-enabled segmentation tools within clinical imaging workflows.

References

1. Abdolali, F., Zoroofi, R.A., Abdolali, M., Yokota, F., Otake, Y., Sato, Y.: Automatic segmentation of mandibular canal in cone beam ct images using conditional statistical shape model and fast marching. Int. J. Comput. Assist. Radiol. Surg. **12**, 581–593 (2017)
2. Batstone, M.D., Scott, B., Lowe, D., Rogers, S.N.: Marginal mandibular nerve injury during neck dissection and its impact on patient perception of appearance. Head Neck: J. Sci. Special. Head Neck **31**(5), 673–678 (2009)
3. Cardoso, M.J., et al.: Monai: An open-source framework for deep learning in healthcare. arXiv preprint arXiv:2211.02701 (2022)

4. Chattopadhay, A., Sarkar, A., Howlader, P., Balasubramanian, V.N.: Gradcam++: Generalized gradient-based visual explanations for deep convolutional networks. In: 2018 IEEE Winter Conference on Applications Of Computer Vision (WACV), pp. 839–847. IEEE (2018)
5. Çiçek, Ö., Abdulkadir, A., Lienkamp, S.S., Brox, T., Ronneberger, O.: 3D U-Net: Learning Dense Volumetric Segmentation from Sparse Annotation. In: Ourselin, S., Joskowicz, L., Sabuncu, M.R., Unal, G., Wells, W. (eds.) MICCAI 2016. LNCS, vol. 9901, pp. 424–432. Springer, Cham (2016)
6. Falk, T., et al.: U-net: deep learning for cell counting, detection, and morphometry. Nat. Methods **16**(1), 67–70 (2019)
7. Fouad, S., Landini, G., Robinson, M., Song, T.H., Mehanna, H.: Human papilloma virus detection in oropharyngeal carcinomas with in situ hybridisation using hand crafted morphological features and deep central attention residual networks. Comput. Med. Imag. Graph. **88**, 101853 (2021)
8. Fouad, S., Randell, D., Galton, A., Mehanna, H., Landini, G.: Unsupervised morphological segmentation of tissue compartments in histopathological images. PLoS ONE **12**(11), e0188717 (2017)
9. Jaskari, J., et al.: Deep learning method for mandibular canal segmentation in dental cone beam computed tomography volumes. Sci. Rep. **10**, 5842 (2020)
10. Jeoun, B.S., et al.: Canal-net for automatic and robust 3d segmentation of mandibular canals in cbct images using a continuity-aware contextual network. Sci. Rep. **12**(1), 13460 (2022)
11. Kainmueller, D., Lamecker, H., Seim, H., Zinser, M., Zachow, S.: Automatic extraction of mandibular nerve and bone from cone-beam ct data. In: Medical Image Computing and Computer-Assisted Intervention–MICCAI 2009: 12th International Conference, London, UK, September 20-24, 2009, Proceedings, Part II 12, pp. 76–83. Springer (2009)
12. Kwak, G.H., et al.: Automatic mandibular canal detection using a deep convolutional neural network. Sci. Rep. **10**(1), 5711 (2020)
13. Lahoud, P., et al.: Development and validation of a novel artificial intelligence driven tool for accurate mandibular canal segmentation on cbct. J. Dent. **116**, 103891 (2022)
14. Lin, X., et al.: Accurate mandibular canal segmentation of dental cbct using a two-stage 3d-unet based segmentation framework. BMC Oral Health **23**(1), 551 (2023)
15. Oktay, O., et al.: Attention u-net: Learning where to look for the pancreas. arXiv preprint arXiv:1804.03999 (2018)
16. Ronneberger, O., Fischer, P., Brox, T.: U-Net: convolutional networks for biomedical image segmentation. In: Navab, N., Hornegger, J., Wells, W.M., Frangi, A.F. (eds.) Medical Image Computing and Computer-Assisted Intervention – MICCAI 2015: 18th International Conference, Munich, Germany, October 5-9, 2015, Proceedings, Part III, pp. 234–241. Springer International Publishing, Cham (2015). https://doi.org/10.1007/978-3-319-24574-4_28
17. Selvaraju, R.R., Cogswell, M., Das, A., Vedantam, R., Parikh, D., Batra, D.: Gradcam: Visual explanations from deep networks via gradient-based localization. In: Proceedings of the IEEE International Conference on Computer Vision, pp. 618–626 (2017)
18. Van der Velden, B.H., Kuijf, H.J., Gilhuijs, K.G., Viergever, M.A.: Explainable artificial intelligence (xai) in deep learning-based medical image analysis. Med. Image Anal. **79**, 102470 (2022)

Open Access This chapter is licensed under the terms of the Creative Commons Attribution 4.0 International License (http://creativecommons.org/licenses/by/4.0/), which permits use, sharing, adaptation, distribution and reproduction in any medium or format, as long as you give appropriate credit to the original author(s) and the source, provide a link to the Creative Commons license and indicate if changes were made.

The images or other third party material in this chapter are included in the chapter's Creative Commons license, unless indicated otherwise in a credit line to the material. If material is not included in the chapter's Creative Commons license and your intended use is not permitted by statutory regulation or exceeds the permitted use, you will need to obtain permission directly from the copyright holder.

Identification of Chimeric RNAs: A Novel Machine Learning Perspective

Paola Bonizzoni[1], Clelia De Felice[2], Yuri Pirola[1], Raffaella Rizzi[1], Rocco Zaccagnino[2(✉)], and Rosalba Zizza[2]

[1] Dip. di Informatica, Sistemistica e Comunicazione, University of Milano-Bicocca, Milan, Italy
[2] Dip. di Informatica, University of Salerno, Salerno, Italy
rzaccagnino@unisa.it

Abstract. *Chimeric RNAs* are transcripts generated by gene fusion and intergenic splicing events, thus comprising nucleotide sequences from different genes. Recent studies have shown that some chimeric RNAs can play a role in cancer development, and so can be used as diagnostics biomarkers when specifically expressed in cancerous cells and tissues. Most gene fusion prediction tools rely on an initial alignment step. However, alignments might be biased, especially for chimeric reads, creating many false positives. Therefore, developing alignment-free prediction methods of fusion genes would be helpful and may provide new insights into the genomic breakage phenomenon in the cell.

In this direction, machine learning could pave the way for new solutions, due to their success in predicting genomic regulatory elements and alternative junction events from the genomic context. To date, however, these techniques have had a marginal supporting role, and, furthermore, manually-curated data sets, that are crucial for model training, are often expensive, unreliable or simply unavailable.

Here we propose a novel ML-based method that learn to recognize the hidden patterns that allow us to identify chimeric RNAs deriving from oncogenic gene fusions. Preliminary comparison with another state-of-the-art method shows promising results.

Keywords: Chimeric RNA · RNA sequencing · Deep learning

1 Introduction

The increasing use of RNA-seq in transcriptomic analysis revealed additional complexity of the transcriptome, such as *chimeric RNAs* beyond gene fusion products. This complexity brings challenges and opportunities to understand the development and progression of human diseases, and thus to discover novel biomarkers and therapeutic targets. The disease process can be triggered by changes in genes, and differences in gene function alone or in combination.

The original version of the chapter has been revised. A correction to this chapter can be found at https://doi.org/10.1007/978-3-031-82768-6_24

Cancer is an example of an acquired genetic disease [18]. To date, numerous chimeric RNAs have been found in various cancers, becoming potential diagnostic biomarkers and promising therapeutic targets. However, recent studies also revealed that chimeric RNA itself is not a cancer or disease-specific phenomenon. Many chimeric RNAs are also uncovered in normal physiology, challenging some traditional views regarding cancer genetics [8,17]. Therefore, understanding the role of chimeric RNAs in cancer and rare diseases, and above all being able to detect them is an extremely challenging problem. With the term "chimeric RNA" we refer to a fusion transcript composed of exons, or fragments of exons from different genes at the RNA level. For the fusion events happening at the DNA level, the term "gene fusion" is usually used.

Most bioinformatics prediction tools rely on an initial alignment step. Discordant reads mapping to two different genes are then identified, and a series of filtering and/or realignment steps are applied to identify candidate chimeric RNAs. However, this approach is restricted by diverse combinations of limiting factors which create many false positives. Most of all, even though, if there is a robust, reproducible, and unbiased method, we cannot identify the fusion genes that were lowly expressed. Therefore, developing reference-free and alignment-free prediction methods of fusion genes would be helpful and may provide new insights into the genomic breakage phenomenon in the cell. To address these limitations, various methods based on *Machine Learning* (ML) have been proposed, in which, however, the marginal role of ML does not allow us to overcome limiting factors such as different conditions of sequencing depth, read length, and filtering criteria. Motivated by recent success of *Deep Learning* (DL) approaches in predicting the genomic regulatory elements and alternative splice events from the genomic context, in [12] the authors hypothesized that the exon junctional breakpoints of known fusion genes identified from the split reads of RNA-seq data can be used to construct a DL model for predicting the breakage tendency.

In this paper, we propose a novel DL-based model that learns to recognize the hidden patterns that allow us to identify chimeric RNAs deriving from oncogenic gene fusions. This consists of a double-classifier framework which first classifies the sequence of the k-mers of a read, and then infers the chimeric information by giving as input the list of k-mer classes to a *transformer*-based classifier. Preliminary experiments carried out on simulated data have shown how the proposed method is able in some cases to outperform state-of-the-art methods.

The source code and files produced are available at https://github.com/FLaTNNBio/gene-fusion-kmer.

2 Background

2.1 Gene Fusion Detection

Since 2009, the use of RNA sequencing has caused an explosion in the discovery of chimeric RNAs. Today, the detection of chimeric RNAs is an important application of RNA-seq, and a number of prediction methods have been developed to identify chimeric RNAs from RNA-seq data [16], such as `FusionSeq` [15]. In the last few years, several additional fusion detection tools have been released, such as FusionMetaCaller [13], JAFFA [4], IDPfusion [22], TRUP [7], FusionCatcher [14] and PRADA [20]. One of the main issues of these tools is that, in addition to missing true fusion events, they can also produce false positives [3].

Not many attempts have been made in trying to exploit the potential of ML to address the limitations of mapping-based gene fusion prediction. In [11], for example, the authors proposed a gene fusion detection algorithm, named scFusion, for single-cell RNA sequencing (scRNA-seq) data. scRNA-seq (opposed to *bulk* RNA-seq) enables researchers to study cell heterogeneity and to quantify cell type-specific gene expression in mixed cell populations, facilitating the identification of biomarkers to predict the clinical prognosis and therapeutic targets of cancers. Although many fusion detection tools have been developed for bulk RNA-seq data, fusion detection using scRNA-seq data is still challenging due to the heavy amplification step that may generate artificial chimeric reads, leading to false positive fusion candidates. To overcome this limitation, scFusion takes as input reads mapped by STAR [6]. It first follows a standard procedure for fusion detection in bulk RNA-seq to identify and cluster unique split-mapped reads and discordant reads mapped to different genes, and then applies a statistical model and a DL-based model to filter the potential false positives.

However, ML techniques have so far played a marginal role and its enormous possibilities have not yet been fully explored. Furthermore, manually-curated training data sets are often expensive, unreliable or simply unavailable. Even when reliable data sets are available, they may impose a ceiling on the performance.

2.2 Machine-Learning Background

Natural language processing (NLP) is a multidisciplinary field that deals with the interaction between human language and computers. The main goal of NLP is to enable computers to understand, interpret and generate human language in a meaningful way. This field has seen notable progress in recent years thanks to the advancement of DL, which allows the development of models that learn directly from the information contained in natural language and that can recognize complex patterns, make predictions and generate coherent and significant texts.

Some of the fundamental concepts of NLP regard the use of *transformers*, a Neural Network (NN) architecture that has revolutionized the field of NLP by introducing new attention-based modeling paradigms [21]. The fundamental concept of transformers is *multi-head attention*, which allows the network to assign different weights to different parts of the sequential input based on their relevance to the current context, thus allowing it to consider long-term dependencies and capture relationships complex semantics within the sequence. The inputs of a transformer are represented by vectors called *embeddings*, which capture the semantic information of the words within the sequence. During the transformation process, embeddings are passed through a series of attention layers and feed-forward NNs to model the relationships between elements of the sequence.

An important development in transformers for NLP is the *Bidirectional Encoder Representations from Transformers* (BERT) [5], a transformer-based language model able to learn bidirectional linguistic representations through an unsupervised training process on a large and diverse corpus of text. During the training process, BERT performs the so-called *masked language modeling* (MLM), in which it randomly masks some input words or tokens and tries to predict them based on the surrounding context. This MLM process allows BERT to learn deep, contextualized representations of words, which incorporate global text information. A distinctive feature of BERT is its ability to capture complex semantic relationships and perform natural language understanding tasks without needing to be adapted to a specific task or dataset.

3 ML-Based Gene Fusion Detection

In this work, we consider gene fusion as the chromosome rearrangement by which two genes are joined together into a single gene (*fusion gene*). As such, a *chimeric* transcript is given by the concatenation of two parts, each one originated from one of the genes joined together. We focus on detecting chimeric reads in a set of paired-end RNA-Seq reads.

3.1 Overview

Here we provide an overview of the method we propose to address the problem of classifying sequences as chimeric or non-chimeric. Specifically, we focus on: *(i)* how to represent the reads in order to exploit the potential of BERT and *(ii)* how to define a DL-based model capable of effectively exploiting this representation to address the chimeric read detection problem.

From k-mers to Sentences. The *k-mers*, i.e., all the substrings of length k which can be extracted from a DNA or RNA sequence, allow the local characteristics of the sequences to be considered while lessening the impact of sequencing errors.

In this work we represent a read using the list of its k-mers. This representation allows the model to learn the local characteristics of reads and perform accurate classification. The solution proposed is based on the definition of a model capable of analyzing and classifying lists of k-mers. More precisely, given a read of length n and a value k, we extract the list of its $n - k + 1$ k-mers. We split such list in consecutive segments of n_words k-mers. Then, the k-mers in a segment are joined together in a *sentence* by using a space character as a separator, thus producing a set of *sentences*. Figure 1 shows an example of sentences generated from an input read, using $k = 4$ and $n\_words = 3$ (that is, 3 k-mers per sentence).

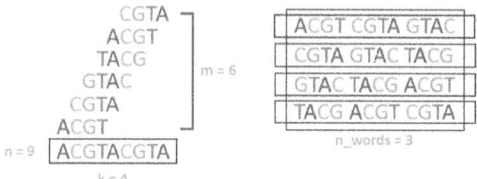

Fig. 1. Sentences generation on the read `ACGTACGTA`, with $k = 4$ and $n\_words = 3$.

The DL-Based Model. The sentence-based representation previously described above is in turn exploited by a DL-based model for the detection of chimeric reads, built as an ensemble of two sub-models: *Gene classifier* and *Fusion classifier*. The goal of *Gene classifier* is to classify a *sentence* into the gene from which it is generated. It is trained using all the sentences derived from non-chimeric reads extracted from the transcripts of a reference set of genes (see Fig. 2).

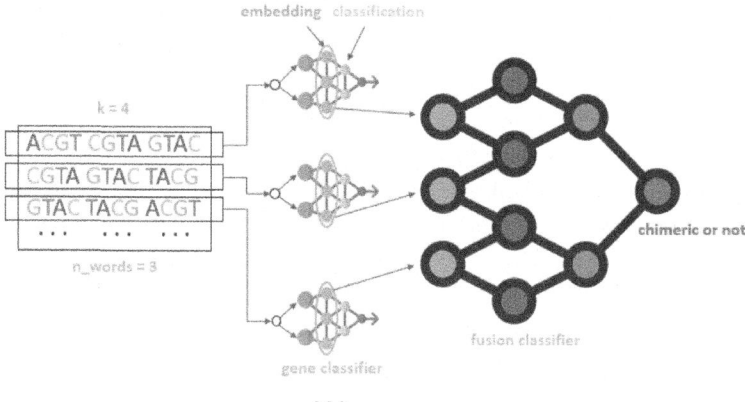

Fig. 2. Architecture of the proposed DL model: *Gene classifier* + *Fusion classifier*

To train *Fusion classifier*, a set of chimeric and non-chimeric reads is generated from the same reference set of genes used for training *Gene classifier*. Then, for each read all the sentences of length `n_words` are generated and then provided as input to *Gene classifier*, previously trained. *Gene classifier* includes an *embedding layer*, as well as several classification layers. The outputs of the embedding layer for all the generated sentences are grouped into a single *embedding matrix*, which constitutes the input for *Fusion classifier*. Then, *Fusion classifier* uses such embedding matrices to distinguish between reads that arise from the fusion of two genes and reads that originate from a single gene.

We note that both classifiers are based on NNs, as a result of preliminary experiments involving traditional ML models, in which the optimal results were obtained through NNs which we will describe below. Furthermore, the architectures of these models are the result of numerous preliminary experiments, in which we observed how using the embeddings generated by the *Gene classifier*, rather than the results of its classification, directly as input to the *Fusion classifier*, produces superior results. This superiority is attributed to the fact that these embeddings distinctly capture the representation constructed by the model, rather than relying on an approximation of the final classification.

3.2 The Methodology

ML [19] plays a pivotal role in bioinformatics, revolutionizing the analysis of biological data. Its ability to sift through vast datasets enables the identification of patterns, aiding in genomics, proteomics, and drug discovery. By automating data interpretation, ML accelerates research, enhances precision, and unveils hidden relationships within complex biological systems. This synergy empowers scientists to make informed decisions, driving advancements in personalized medicine and our understanding of intricate biological processes.

Gene Classifier.

Pre-processing. Data pre-processing phase consists of the following steps:

1. *Reads generation:* for each gene in a reference set of n selected genes, the set of transcripts is downloaded from *Ensembl*[1]. (release 109); then, using Genome Tools' shredder command, all the non-overlapping substrings of length `len_read` of the transcripts are generated, producing a set of reads in a `FASTA` file for each considered gene;
2. *k-mers generation:* all the reads obtained in the previous step are inserted into a single `CSV` file. For each read, the gene of origin to which the read belongs is reported, then, for each read all the k-mers are extracted (the number of k-mers generated for each read is `n_kmers = len_read - k + 1`) and the final result is a `CSV` file in which each line represents a read and contains its k-mers followed by a *label*, i.e., an integer value between 0 and $n - 1$ indicating the reference gene associated with the read.
3. *Set generation:* the `CSV` file is randomly divided into *training* (80% of the lines), *validation* (10% of the lines), and *test* (10% of the lines) set.
4. *Sentences generation:* for each set built by the previous step, all possible sentences of length `n_words` are generated; recall that a sentence is defined as a string composed of `n_words` consecutive k-mers, separated by a space.

Input Data Tokenization. Before training the model, the *tokenization* is required, which converts sentences into numerical vectors to be used as input for the model. For this purpose, the `DNATokenizer` method has been defined. It implements a vocabulary that maps all possible strings over the alphabet `{A,C,G,T,N}` of

---
[1] https://www.ensembl.org/index.html

length len_kmer into numerical values. The vocabulary therefore allows each string to be associated with a unique numeric *token*. In addition to alphabet strings, the vocabulary also contains definitions of five *special tokens*: *(i) masking* token (MASK), used for the *masking technique* during model training, *(ii) unknown* token (UNK), used for strings that are not present in the vocabulary, *(iii) padding* token (PAD), used to align sentences to a common length, *(iv) start of sequence* token (CLS), and *(v) end of sequence* token (SEP).

These special tokens provide the model with additional information about the context of the reads, allowing for adequate processing during the training.

The Model. Once the input data has been tokenized, the model is trained. We remark that our methodology enables an abstraction of the actual model used for gene classification. This means that different specific implementations for *Gene classifier* can be used, with respect to the data described above. Among the different solutions explored, the one proven to be particularly effective is DNABert [10], a knowledge transfer-based model that leverages the BERT architecture for processing DNA sequences. It consists of four main layers:

– *embedding layer*: responsible for converting the sentence-based representation of reads, i.e., sequence of tokens, into a dense vector representation.
– *encoder layer*: based on a transformer-based architecture, is responsible for capturing the local and global relationships between consecutive k-mers.
– *pooler layer*: responsible for aggregating the information extracted from the encoding layer into a compact and high-quality representation.
– *classification layer*: responsible for the final classification of the reads, by using dense NNs or traditional classification algorithms.

To monitor prevent overfitting, the *early stopping* technique is used, i.e., to terminate model training if no improvements in performance are observed on the validation set for a period of k epochs. In our case, best performance have been observed by setting $k = 10$. Once training is finished, the best configuration is evaluated on the test set, consisting of reads never encountered during training, so representing an unbiased evaluation of the model's generalization ability.

Fusion Classifier. This section provides details about pre-processing and training steps for *Fusion classifier*. We remark that the pre-processing is driven by an additional parameter called n_fusion, which specifies the number of fused transcripts that must be generated for each gene during the *simulation phase*. As we will see, simulating fused transcripts is crucial for the pre-processing as it allows to build synthetic data that represent reads of fused transcripts. The value of the parameter should be chosen depending on the number of genes that are considered so that the data distribution in the training dataset resembles that of real datasets. This helps the model to achieve its goal of distinguishing between chimeric and non-chimeric reads.

Pre-processing. Data pre-processing phase consists of the following steps:

1. *Fused transcripts generation:* `Fusim` [2] is used to simulates gene fusion events and to generate the fused transcript sequences; these transcripts are essentially the concatenation of transcripts coming from two different genes.
2. *Reads generation:* `ART Illumina` [9] is used to simulate reads from the fused transcripts using Illumina technology; as we will see, this helps improve the robustness and accuracy of the gene classification model.
3. *Read labeling:* each read is labeled as *chimeric* or *non-chimeric*, by observing if it overlaps (or not) the *breakpoint* of the transcript from which it has been generated, i.e., the point where the transcript of the first gene ends and the second gene transcript begins.
4. *k-mers generation:* each read is saved in single line of a `CSV` file, containing: *(i)* all the k-mers of the read, *(ii)* the identifiers of the two genes involved in the fusion, *(iii)* the *breakpoint* of the sequence, i.e., where the gene fusion occurs exactly, and *(iv)* the *label*: a binary value indicating whether the read is *chimeric* (0) or *non-chimeric* (1).
5. *Set generation:* the `CSV` file is randomly separated into *training* (80% of the lines), *validation* (10% of the lines) and *testing* (10% of the lines) set.

The Model. *Fusion classifier* exploits the embedding outputs of *Gene classifier* for the classification of reads. More specifically, for each input read: *(i)* all possible sentences of length `n_words` are generated, *(ii)* each sentence is given as input to *Gene classifier*, after tokenizing it using the technique described above, *(iii)* the output of the embedding layer of *Gene classifier* for each sentence is compacted into a single matrix which is given as input to *Fusion classifier*.

During our experiments, several architectures for *Fusion classifier* were explored. Among these, the most effective architecture turned out to be a simple *fully connected network* consisting of the following layers: *(i)* one *projection* layer which allows the model to explore and capture the correlations between the different sentences; *(ii)* a sequence of `n_hidden_layers` fully connected layers, each of size `hidden_size`, and each one followed by a *dropout* layer, which aims to reduce overfitting in the model; *(iii)* one *linear* layer which has a single output neuron to perform classification of the entire embedding matrix as a chimeric or non-chimeric; the output neuron returns a value between 0 and 1, which can be interpreted as the probability of the input read to be a chimeric read.

As in the case of *Gene Classifier*, during the training of *Fusion Classifier*, the early stopping criterion was adopted to monitor the progress of the training.

4 The Experiments

In this section, the results obtained from experiments on *Gene classifier* and *Fusion classifier* are presented, analyzed, and finally compared with the results obtained by `FusionCatcher` on simulated data.

4.1 Gene Classifier

In this section, first the data used for training and evaluating *Gene Classifier* are described, and then, the performance achieved are discussed.

Data Analysis. Seventeen different *reference genes* have been selected to first train *Gene classifier*, and then to simulate fused transcripts from which we generated the chimeric and non-chimeric reads used to train and evaluate *Fusion Classifier*. The reference genes are: RUNX1, ETV6, RIPOR1, CTCF, KMT2A, PAX5, EZR, PTEN, PMEL, TAL1, DUX4, CRLF2, MEF2D, BCL9, TCF3, ZNF384, and PBX1. These genes have been chosen based on their relevance and also to cover a variety of biological contexts. Furthermore, some of them are involved in known genetic fusions. The experiments were conducted using several values for the parameters described in Sect. 3.2. Here we only report the ones achieving best results: len_read = 150, len_kmer = 6, and n_words = 20.

After the pre-processing a total of 486,250 sentences were generated for the training. Each sentence consists of 20 consecutive k-mers, and is given as input to the model during the training. in the Supplementary materials [23] shows the data distribution for the training set. Figure S1 In the context of genomic data, where the creation of new data via data augmentation may not be possible or desirable, non-uniform distribution of data, as in our case, can pose a challenge. To address this problem and mitigate overfitting on the most supported classes, a *loss function* weighting strategy has been adopted. This weighting assigns a higher weight in the loss function into instances of underrepresented classes, in order to penalize the model if it tries to favor the most populated classes and to promote better generalization across all classes of the classification problem.

Figure S2 in the Supplementary materials [23] shows the distribution of the data in the test set, that follows the same trend observed in the training set. The match in the data distribution between the training set and the test set is an important element in evaluating the effectiveness of the model. If the distribution of the test data is similar to that of the training data, this suggests that the model has been exposed to a variety of examples during training and has the ability to generalize to new data from the same distribution.

Test Results. During the test phase, the model achieved the following performance: *accuracy* 88.76, *precision* 88.89, *recall* 88.76, and *F1 score* 88.73.

Figure 3 shows the confusion matrix which provides a visual representation of the results of *Gene Classifier*. In this matrix of size 17×17, the rows represent the *actual* gene class while the columns represent the *predicted* gene class. The C_{ij} value indicates the number of samples belonging to class i that were classified as class j. The confusion matrix allows you to evaluate the performance of the model in a multiclass classification, providing an overall overview of the performance for each class. In our case, it is evident that the highest values are concentrated along the main diagonal, indicating that achieved good classification ability.

In addition, Fig. S3 in the Supplementary materials [23] shows the ROC and AUC curves. The ROC curve is a graphical representation of the discrimination

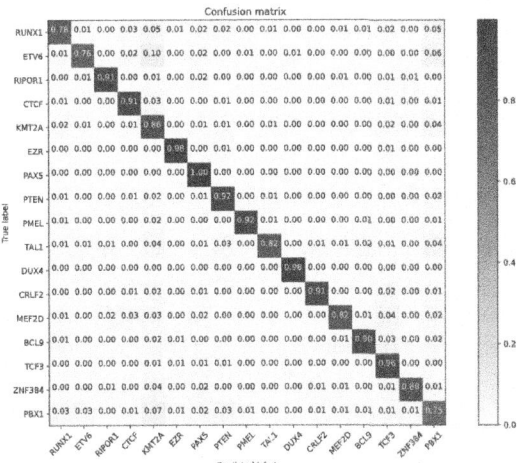

Fig. 3. Confusion matrix on test data for *Gene classifier*.

ability of a model as the classification threshold varies. Graphically, the ROC curve is drawn by plotting the true positive rate (TPR) on the y-axis and the false positive rate (FPR) on the x-axis. TPR represents the sensitivity of the model, i.e. the ability to correctly identify positive examples, while FPR represents the proportion of false positives compared to negative examples. In our case, we can observe that the model achieved a very good classification performance. The ROC curve tends the upper left corner, indicating that the model has good discrimination ability between classes. Furthermore, the AUC has a high value, close to 1, which confirms that the model has a very accurate classification ability.

4.2 Fusion Classifier

Data Analysis. To generate the chimeric and non-chimeric reads, n_fusion= 30 chimeric transcripts were generated. Then, by using ART Illumina, the chimeric or non-chimeric reads of length 150 from these transcripts were simulated. Table S1 in the Supplementary materials [23] shows the distribution of samples in the two classes in the training set. As we can see, the distribution of data is not uniform, with the number of chimeric reads representing only the 7% of the set. To address this challenge, also in this case we have adopted the loss function weighting approach. As shown in Table S1 in the Supplementary materials [23] the distribution of data int the test set follows the same distribution of the training set. This indicates that the test data is representative and can provide an accurate assessment of the model's performance.

Test Results. During the test phase, the model achieved the following performance: *accuracy* 79.09, *precision* 92.94, *recall* 76.27, and *F1 score* 81.95.

Figure 4 shows the confusion matrix, which provides a visual overview of the model predictions for chimeric and non-chimeric sequences. As we can see, the model achieves satisfactory results in both classes, with a significant number of correct predictions. The main diagonal of the matrix has high values, indicating a good ability of the model to discriminate between the two classes. Figure S4 in the Supplementary materials [23] shows the ROC curve of *Fusion Classifier*.

4.3 Comparison with FusionCatcher

To compare the proposed model with `FusionCatcher`, we have used `Fusim` to build a specific set of simulated reads from the set of 17 reference genes used to train the proposed models. Such a set consists of 5848 non-chimeric reads and 431 chimeric reads corresponding to 141 fusion gene events. As a result, `FusionCatcher` was able to detect 13 fusions out of the 141 total in the set, while the proposed model detected 128 (deducted by 345 reads detected as chimeric).

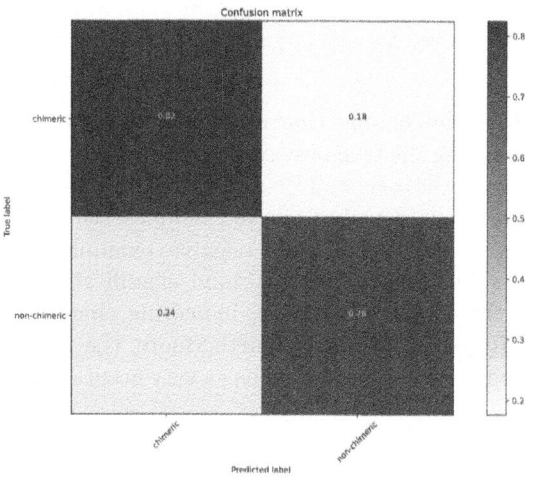

Fig. 4. Confusion matrix on test data for *Fusion classifier*.

We remark that the proposed model has also produced false positives, i.e., detected 31 not actually chimeric reads, corresponding to 8 false fusion events.

5 Conclusions

In this paper, we propose a novel DL-based model based on NLP and Transformers to identify chimeric RNAs deriving from oncogenic gene fusions. The encoding of RNA-seq sequences in our approach is based on k-mers. In a previous work we also explored a different encoding based on Lyndon words for RNA-seq data [1]. Experimental results show the high potential of the approach. However, various improvements can be applied. First, specific data structures for storing relevant k-mers used in the training can largely improve the performance. Experiments on real data will be necessary to tune the method, and to test the applicability of the tool for diagnostic purposes. Still, we need to better evaluate our model under different conditions of sequencing depth, read length, and filtering criteria.

References

1. Bonizzoni, P., et al.: Numeric lyndon-based feature embedding of sequencing reads for machine learning approaches. Inf. Sci. **607**, 458–476 (2022)
2. Bruno, A., et al.: FUSIM: a software tool for simulating fusion transcripts. BMC Bioinform. **14**(1), 1–5 (2013)
3. Carrara, M., et al.: State of art fusion-finder algorithms are suitable to detect transcription-induced chimeras in normal tissues? BMC Bioinform. **14**, 1–11 (2013)
4. Davidson, N., Majewski, I., Oshlack, A.: JAFFA: High sensitivity transcriptome-focused fusion gene detection. Genome Med. **7**(1), 1–12 (2015)
5. Devlin, J., Chang, M.W., Lee, K., Toutanova, K.: Bert: Pre-training of deep bidirectional transformers for language understanding. arXiv:1810.04805 (2018)
6. Dobin, A., et al.: STAR: ultrafast universal RNA-seq aligner. Bioinformatics **29**(1), 15–21 (2013)
7. Fernandez-Cuesta, L., et al.: Identification of novel fusion genes in lung cancer using breakpoint assembly of transcriptome sequencing data. Genome Biol. **16**, 1–11 (2015)
8. Finta, C., Zaphiropoulos, P.G.: Intergenic mRNA molecules resulting from trans-splicing. J. Biol. Chem. **277**(8), 5882–5890 (2002)
9. Huang, W., Li, L., Myers, J., Marth, G.: ART: a next-generation sequencing read simulator. Bioinformatics **28**(4), 593–594 (2012)
10. Ji, Y., Zhou, Z., Liu, H., Davuluri, R.: DNABERT: pre-trained bidirectional encoder representations from transformers model for DNA-language in genome. Bioinformatics **37**(15), 2112–2120 (2021)
11. Jin, Z., et al.: Single-cell gene fusion detection by scFusion. Nat. Commun. **13**(1), 1084 (2022)
12. Kim, P., et al.: FusionAI: predicting fusion breakpoint from DNA sequence with deep learning. iScience **24**(10) (2021)
13. Liu, S., et al.: Comprehensive evaluation of fusion transcript detection algorithms and a meta-caller to combine top performing methods in paired-end RNA-seq data. Nucleic Acids Res. **44**(5), e47 (2016)
14. Nicorici, D., et al.: FusionCatcher–a tool for finding somatic fusion genes in paired-end RNA-sequencing data. bioRxiv p. 011650 (2014)
15. Sboner, A., et al.: FusionSeq: a modular framework for finding gene fusions by analyzing paired-end RNA-sequencing data. Genome Biol. **11**, R104 (2010)
16. Singh, S., Li, H.: Comparative study of bioinformatic tools for the identification of chimeric RNAs from RNA sequencing. RNA Biol. **18**(sup1), 254–267 (2021)
17. Singh, S., et al.: The landscape of chimeric RNAs in non-diseased tissues and cells. Nucleic Acids Res. **48**(4), 1764–1778 (2020)
18. Strynatka, K., Gurrola-Gal, M., Berman, J., McMaster, C.: How surrogate and chemical genetics in model organisms can suggest therapies for human genetic diseases. Genetics **208**(3), 833–851 (2018)

19. Tan, P.N., Steinbach, M., Kumar, V.: Introduction to data mining. Pearson Education (2016)
20. Torres-García, W., et al.: PRADA: pipeline for RNA sequencing data analysis. Bioinformatics **30**(15), 2224–2226 (2014)
21. Vaswani, A., et al.: Attention is all you need. In: Advances in Neural Information Processing Systems, vol. 30 (2017)
22. Weirather, J., et al.: Characterization of fusion genes and the significantly expressed fusion isoforms in breast cancer by hybrid sequencing. Nucleic Acids Res. **43**(18), e116 (2015)
23. Identification of Chimeric RNAs: a novel machine learning perspective - Supplementary material. https://github.com/FLaTNNBio/gene-fusion-kmer

PartialFibers: An Efficient Method for Predicting Drug-Drug Interactions

Aysegul Bumin[(✉)], Kejun Huang, and Tamer Kahveci

University of Florida, Gainesville, FL, USA
{aysegul.bumin,kejun.huang,tkahveci}@ufl.edu

Abstract. Drug resistance is one of the fundamental challenges in modern medicine. Using combinations of drugs is an effective solution to counter drug resistance as is harder to develop resistance to multiple drugs simultaneously. Finding the correct dosage for each drug in the combination remains to be a challenging task. Testing all possible drug-drug combinations on various cell lines for different dosages in wet-lab experiments is infeasible since there are many combinations of drugs as well as their dosages yet the drugs and the cell lines are limited in availability and each wet-lab test is costly and time-consuming. Efficient and accurate in silico prediction methods are surely needed. Here we present a novel computational method, *PartialFibers*, to address this challenge. Unlike existing prediction methods *PartialFibers* takes advantage of the distribution of the missing drug-drug interactions and effectively predicts the dosage of a drug in the combination. Our results on real datasets demonstrate that *PartialFibers* is more flexible, scalable, and achieves higher accuracy in less time than the state of the art algorithms.

Keywords: drug-drug combination · drug resistance · tensor completion · fiber-based approach

1 Introduction

One of the fundamental challenges in modern medicine is that over time, many diseases develop resistance to drugs. For infectious diseases, this often happens as the microbial species evolve, and as a result, drug resistant species replace those that are not resistant [3]. For genetic disorders, such as cancer, drug resistance arises naturally as each cancer is a result of unique mutations for individuals, while drugs target the condition of an average patient belonging to a population of cancer patients of varying mutation types [25]. New mutations yield cancer cells, which may not fit into the average cancer profile, leading to drug resistant cancer cells [12].

The effectiveness of anti-cancer drugs is often measured by their effects on the growth and death of cancer cells relative to control samples. Two standard metrics for measuring drug potency and efficacy are the half-maximal inhibitory concentration (IC50) and the fraction of viable cells at the highest drug concentration (Emax) [11]. One way to combat drug resistance is to use combinations of drugs [2,13,17]. Doing so aims to increase the efficacy of the drugs by taking advantage of drug-drug interactions, and thus yields an effect on the targeted disorder more than the effect of the individual drugs if they were used independently. This improvement in drug efficacy happens for several reasons. First, the drug combination reduces the chance of developing resistance, for when cancer cells are exposed to multiple drugs, it is more difficult for them to develop resistance to all of them at the same time. This is because, a synergic drug combination accesses context specific multi-target mechanisms, and makes it harder for a cancer cell to develop resistance to multiple mechanisms of action [14]. Secondly, it reduces the risk of toxicity side effects as synergistic combination often reduces the dosage needed for each drug [14,16,22].

There are several databases which maintain the interactions among drugs. Among them, $GDSC^2$ hosts the efficacy of approximately 2000 clinically relevant drug combinations. These combinations encompass 125 characterized breast, colorectal, and pancreatic cancer cell lines [13]. The synergy between drugs is rare and highly context-dependent. However, when the targeted drugs are combined, they are more likely to be synergistic [13]. At the heart of the drug-drug interaction identification problem lies the fact that trying all possible drug-drug combinations on various cell lines and finding the right dosage of the drug-drug combination sufficient to eliminate a targeted percentage of given cancer cell lines is not feasible. This is because there are many drug-drug combinations, let alone the amounts of cell lines to be tested for each drug. This challenge necessitates developing computational strategies for estimating the right drug-drug combinations to be effective on the targeted cell lines. Many prediction methods for single drug response are developed with efforts of National Cancer Institute (NCI) and AstraZeneca-Sanger Drug Combination Prediction Dialog for Reverse Engineering Assessments and Methods (DREAM) Challenge [7]. Some of them are based on matrix factorization [26], recommender system based on cell line similarities [23], drug/cell line network [27] or manifold learning [1]. As we include drug-drug combinations, the problem size increases, bringing further challenges. Existing methods for predicting drug-drug interactions consider this data as a set of independent samples [9,28]. The data available for drug-drug interactions however varies both in terms of quantity and distribution across different drugs as well as cell lines. This inherent nature of variability of drug-drug interaction data introduces substantial disadvantages to existing methods in terms of their accuracy and efficiency.

Our contributions. In this paper, we consider the problem of predicting drug-drug interactions. We develop a novel algorithm named *PartialFibers* to efficiently and accurately predict the response of a given cell line to a given pair of drugs. More specifically, assume that we are given an incomplete three dimensional dataset, where the first dimension corresponds to a drug (called *library drug*), the second dimension also corresponds to a drug called *anchor drug* (i.e., a drug with a specific dosage,) and the third dimension corresponds to cell lines. The value at each entry of this tensor denotes IC50 value of the library drug for that cell line when that library drug is combined with the corresponding anchor drug. *We aim to estimate the missing values in this dataset accurately and efficiently.* Our method takes advantage of the topological properties of the three dimensional drug-drug interaction data. Fiber structure (i.e., the vector consisting of all values in the given 3D matrix, tensor, when the coordinates of two of the dimensions are fixed) is a common topological pattern in the missing data [5,19,24]. Drug-drug interaction data, however, does not have a uniform fiber structure as some values may be known and others may be missing in each fiber. We call such patterns *partial fibers*. Our method allows us to learn from multiple dimensions of the given data. Given 3D drug-drug interaction data, at each iteration, our algorithm randomly picks one of the three dimensions and learns from the partial fibers along that dimension. Our method provides a simultaneous update for all the samples along the same partial fiber. Our results demonstrate that *PartialFibers* efficiently and accurately predicts the drug responses for the cell lines and it shows superior performance as compared to state-of-the-art methods in terms of both accuracy and running time performance. It benefits from the intrinsic structure of the data for leading to an efficient per-iteration update performing a faster and more accurate imputation of drug-drug interactions.

2 Methods

In this section, we describe our *PartialFibers* method for efficient and accurate imputation of the missing drug-drug interactions data. In Sect. 2.1, we introduce our notations and the preliminary concepts of tensor completion needed to explain our algorithm. In Sect. 2.2, we describe our *PartialFibers* algorithm in detail.

2.1 Notation and Preliminaries

Here, we introduce the notation that we use for describing the core concepts required to understand our algorithm.

Baseline Tensor Completion. The three-dimensional tensor completion problem aims to estimate the missing values in a given tensor, such that a nonempty strict subset of the entries of this tensor is known apriori. The objective here is to find three-factor matrices such that their inner product produces a tensor of the same size, and the values at the tensor locations corresponding to the apriori known values are as close as possible to the given known values. Once such a tensor is produced, this technique estimates the values generated in the positions corresponding to the missing values as the generated values at those positions of the new tensor.

Let us denote a three-dimensional tensor with I rows, J columns and K depth with $\boldsymbol{X} \in \mathbb{R}^{I \times J \times K}$, and the value of the entry in this tensor at index (i, j, k) with $x_{i,j,k}$. We denote the rank of this tensor as R ($R \leq \min\{I, J, K\}$ is a positive integer), and the three factor matrices as $A \in \mathbb{R}^{I \times R}$, $B \in \mathbb{R}^{J \times R}$, $C \in \mathbb{R}^{K \times R}$. We denote the ith row of the factor matrix A with a_i, the jth row of the factor matrix B with b_j and the kth row of the factor matrix C with c_k. Similarly, we represent the value at index (i,r) of the factor matrix A with a_{ir}, the value at index (j,r) of the factor matrix B with b_{jr} and the value at index (k,r) of the factor matrix C with c_{kr}. A nonempty subset of the entries in X are known. We denote the set of indices (i,j,k) for which the values of $x_{i,j,k}$ are available with Ω. Using the notation above, the traditional CANDECOMP/PARAFAC (CP) formulation of tensor completion [6,10] is

$$\underset{a_i, b_j, c_k}{\text{minimize}} \sum_{(i,j,k) \in \Omega} \|x_{ijk} - \sum_r a_{ir} b_{jr} c_{kr}\|^2. \qquad (1)$$

In (1), we present the optimization problem to find the factor matrices A, B, C that are going to be used to complete the missing entries in X. Classic CP tensor completion algorithm iterates over the available tensor entries (Ω) one by one ($\forall (i,j,k) \in \Omega$) and updates the corresponding entries of the three factor matrices A, B, C [4]. Once the objective (1) converges, the complete data matrix can be constructed using the learned factor matrices.

Tensor Completion with Fibers. The distribution of the missing entries over the data tensor X is not always uniform across the three dimensions. This provides opportunities to benefit from the structure of the missing data. A common structure for the missing data is *fiber*. A fiber is the vector of all index values (i, j, k) when the values of any two of these three variables are fixed. For instance, given the values of i and j, fiber is the vector of all the values along the kth dimension [15]. Thus, a fiber in a three dimensional tensor may consist of a varying index in any of the three dimensions. Recent research shows that the optimization function (1) can be computed more efficiently when the missing values of the tensor are along the same dimension (i.e., they make up a fiber) as multiple entries of the factor matrices can be computed in one iteration [5].

In this paper, we formulate the drug-drug interaction prediction problem, as the tensor completion problem. The fundamental challenge here is the tensor corresponding to this data does not follow the fiber structure we explain above. To solve this problem, we define a multi-dimensional notation which has more expressive power than the classic tensor notation. We use the "*" symbol to represent the dimension at which the fiber extends. For instance, we denote the fiber which has fixed values j and k along the last two dimensions and varies in the first dimension with $x_{*jk} = [x_{1,j,k}, x_{2,j,k}, \ldots, x_{I,j,k}]^T$. Similarly, we denote the fiber which has fixed values i and k along the first and third dimensions and varies in the second dimension with $x_{i*k} = [x_{i,1,k}, x_{i,2,k}, \ldots, x_{i,J,k}]^T$. Finally, we denote the fiber which has fixed values i and j along the first and second dimensions and varies in the third dimension with $x_{ij*} = [x_{i,j,1}, x_{i,j,2}, \ldots, x_{i,j,K}]^T$. Assume that, regardless of the direction, for all the fibers, within a fiber, either all entries are missing or all entries are available. Let us represent the Hadamard (element-wise) product with $\otimes$. Let us represent the set of all available fibers along the first dimension with Ψ_1. Similarly, let us represent the set of all available fibers along the second and third dimensions with Ψ_2 and Ψ_3, respectively.

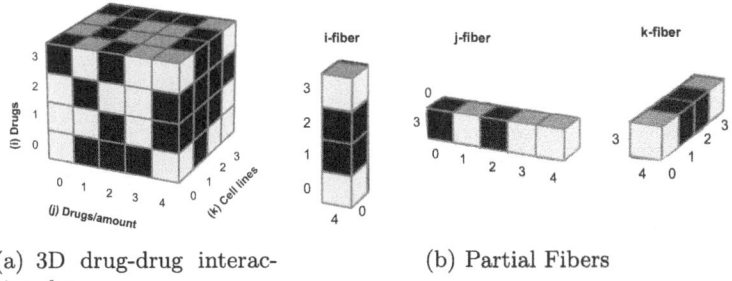

(a) 3D drug-drug interaction data

(b) Partial Fibers

Fig. 1. (a) Three dimensional tensor representation and the partial fibers. Rows, columns, and depth represent the library drugs, anchor drugs with amounts, and cell lines respectively. Entries with black color represent the available values, and the entries with gray color represent the missing values. (b) Examples of partial fibers, (Color figure online)i-fiber, j-fiber and k-fiber. The i-fiber is a fiber where the j and k th dimensions are fixed and fiber is built using the dimension i. The same logic applies to the j-fiber and k-fiber

Using this notation, depending on the direction of the fiber, we reiterate the original fiber update formulation and rewrite (1) as follows. If the fiber is along the first dimension, we write

$$\underset{A,b_j,c_k}{\text{minimize}} \sum_{(*,j,k)\in\Psi_1} \|x_{*jk} - A(b_j \otimes c_k)\|^2. \tag{2}$$

Similarly, we rewrite the optimization function based on the second dimension and the third dimension to update the factor matrices, B and C. We update the factor matrices iteratively until the factor matrices represent the actual data with a small loss, i.e., until convergence.

Updating the factor matrices in fiber provides faster convergence than the classic CP tensor completion algorithm because the entire fiber can be updated simultaneously. However, this strategy works only if all the values in a given fiber are missing. Thus, it cannot be applied to solve the drug-drug interaction prediction problem considered in this paper, as this data does not have this characteristic.

2.2 Tensor Completion with Partial Fibers

We first introduce the drug-drug interaction dataset and highlight the properties of the dataset. Then, we discuss our *PartialFibers* algorithm designed for drug-drug interaction dataset.

Algorithm 1. *PartialFibers*, SGD for tensor completion using partial fibers

Input: X (the 3D incomplete data)
1: initialize A, B, and C
2: **repeat**
3:   Randomly select a dimension i, j, or k. (Assume that the dimension is i)
4:   Randomly draw an i-fiber, $(*, j, k)$ from $\Psi_{*,j,k}$
5:   Initialize D
6:   Calculate the gradients
7:   $\nabla_A f_{*jk} = -(x_{*jk} - DA(b_j \otimes c_k))(b_j \otimes c_k)^\top$
8:   $\nabla_{b_j} f_{*jk} = -DA^\top(x_{*jk} - DA(b_j \otimes c_k)) \otimes c_k$
9:   $\nabla_{c_k} f_{*jk} = -DA^\top(x_{*jk} - DA(b_j \otimes c_k)) \otimes b_j$
10: $\begin{cases} DA \leftarrow DA + \gamma(x_{*jk} - DA(b_j \otimes c_k))(b_j \otimes c_k)^\top, \\ b_j \leftarrow b_j + \gamma DA^\top(x_{*jk} - DA(b_j \otimes c_k)) \otimes c_k, \\ c_k \leftarrow c_k + \gamma DA^\top(x_{*jk} - DA(b_j \otimes c_k)) \otimes b_j. \end{cases}$
11: **until** convergence

We use the drug-drug interaction data in the GDSC[2] dataset [13] for pairs of drugs. We refer to the first drug in a pair of drugs in this dataset as *library drug*. Unlike the library drug, this dataset associates the second drug with a specific dosage of that drug. We refer to the second drug at that dosage as *anchor drug*. For the given dosage of the anchor drug, this dataset provides the IC50 value of the library drug when the two drugs are applied on a given cell line. We represent this data using a three-dimensional tensor with I rows, J columns and K depth, $X \in \mathbb{R}^{I \times J \times K}$. Row i, represents the ith library drug. Column j, represents the jth anchor drug along with its dosage. Depth k, represents the kth cell line. In summary, each entry $x_{i,j,k}$, denotes the dosage of the ith library drug required to eliminate 50% of the cell line k, when it is combined with the pre-defined dosage of anchor drug j. Figure 1 illustrates a representation of the drug-drug interaction data. Black and gray entries represent available and missing data, respectively.

We categorize fibers in two groups: *complete fiber* and *partial fiber*. We say that a fiber is *complete* if all the entries of that fiber are available or they are all missing. Otherwise, we call it a *partial fiber*. Orthogonal to this grouping, we further categorize fibers into 3 groups based on their directions; *i*-fibers, *j*-fibers, and *k*-fibers. *I*-fibers represent all library drugs for a fixed anchor drug/amount, j, and for a fixed cell line. For instance, in Fig. 1b, the *i*-fiber represents all the library drugs for the anchor drug/amount in index 4, and the cell line in index 0. Similarly, the *j*-fiber represents all the anchor drugs/amounts for the library drug in index 3, and the cell line in index 0. The *k*-fiber represents all the cell lines for the library drug in index 3, and the anchor drug/amount in index 4. In Fig. 1b, all three fibers are partial fibers since some values are available and some are missing in each. As the figure demonstrates, each fiber in the 3D representation may have a different number of missing entries. Also, the missing and present entries can be in any position as well as the (positive) number of positions.

We solve the optimization problem (1) using partial fiber updates. Algorithm 1 presents our *PartialFibers* method. The algorithm takes the incomplete tensor X as input. We start by randomly initializing the factor matrices A, B and C (line 1). At each iteration, we randomly select a dimension among the three dimensions with equal probability (line 3). Upon selecting the dimension, we randomly choose one fiber among all the fibers in the selected direction with a uniform probability of being selected (line 4). In order to keep Algorithm 1 short, we write the subsequent steps assuming that we select the first dimension. However, it is important to note that any of the three dimensions can be chosen at each iteration. The subsequent steps still work the same by using the correct dimensions as indices in the formulation. Selecting the first dimension implies that the current fiber will be selected among all available i-fibers. Thus, a fiber along the first dimension for a randomly selected j, k is the vector $[x_{1jk}, x_{2jk}, \ldots, x_{Ijk}]^T$. Next, we construct an $I \times I$ diagonal matrix D as follows. For all indices $i \in 1, 2, \ldots, I$, if the value x_{ijk} in the given tensor is known (i.e., $(i, j, k) \in \Omega$), then we set the diagonal entry $D[i, i] = 1$. We set all the remaining entries in D to 0 (line 5 of Algorithm 1). The purpose of this diagonal matrix is to mask the unavailable entries within a fiber and to focus on only the available entries. We represent the missing data with the underscore "_". As an example, let us assume that $I = 4$, and the first and third indices along dimension i are missing. We represent the (j,k)th fiber with $x_{*jk} = [\_, x_{2,j,k}, \_, x_{4,j,k}]$. The corresponding diagonal matrix $D \in R^{4 \times 4}$ for this example is, $D = \begin{bmatrix} 0 & 0 & 0 & 0 \\ 0 & 1 & 0 & 0 \\ 0 & 0 & 0 & 0 \\ 0 & 0 & 0 & 1 \end{bmatrix}$. The next step is to perform parameter update, i.e. we need to update the indices of each factor matrix by using the information we learned from the available data. We use traditional stochastic gradient descent (SGD) as a baseline stochastic optimization algorithm [20]. Traditional SGD requires using one sample at a time. We redesign the traditional SGD to perform an efficient update by simultaneously learning from all the selected fiber entries. We design our SGD algorithm following from the observation that samples come in the form of x_{*jk} fibers, which involves the entire matrix A and the variables b_j, c_k. We represent the objective function for the (j, k)th fiber with f_{*jk} and rewrite the optimization function as; $\text{minimize}_{A, b_j, c_k} \sum_{(*,j,k) \in \Psi_{*,j,k}} f_{*jk} = \|x_{*jk} - DA(b_j \otimes c_k)\|^2$. We represent the gradient of the objective function f_{*jk} with respect to A, b_j, and c_k with $\nabla_A f_{*jk}$, $\nabla_{b_j} f_{*jk}$ and $\nabla_{c_k} f_{*jk}$ respectively. The stochastic gradients are as in the lines 7, 8, 9 of Algorithm 1.

We represent the step size in SGD with γ. Then, the SGD algorithm takes the form as in line 10 of Algorithm 1. We repeat iterations (lines 2 to 11) until convergence, i.e. until the factor matrices are representative of the available data. The same convergence guarantees for the SGD apply for the partial fibers too, hence after convergence, we construct the complete data matrix using the learned factor matrices. Notice that it is also possible to adapt (2.2) to other stochastic gradient descent algorithms, such as Adam [18], Adagrad [8], and SPPA [4] through similar algebraic manipulations.

Advantages. There are three main advantages of partial fibers over the complete fiber setting:

- **Data flexibility.** If a specific entry in the tensor (i, j, k) is missing, the complete fiber approach requires at least one of the fibers x_{*jk}, x_{i*k} or x_{ij*} to be complete. This is an extremely restrictive constraint which is often not observed in drug-drug interaction datasets. *PartialFibers* provides a flexible condition which works for any distribution of missing data. We allow the data to have a different number of available entries, the entries can be in any order along the same dimension, we benefit by grouping the missing items together and perform the partial fiber update efficiently.
- **Algorithmic Flexibility.** Complete fiber approach requires all fibers to be along the same dimension. *PartialFibers*, however, does not have this requirement. It selects the fiber dimension randomly with the same probability to benefit from every dimension equally.
- **Running Time.** *PartialFibers* provides further improvement in running time as it performs a parameter update using all the samples within the same fiber. It allows these samples to be non-sequential and of any size. The more number of available items would mean more simultaneous updates, which results in further time improvement. In the worst case, there will be only one such training sample which is equivalent to considering each sample one by one. (Fully stochastic, non-fiber state).

3 Experimental Evaluation

Dataset. We use the $GDSC^2$ dataset which contains anchor drugs, library drugs, and their impact on a set of cell lines along with their IC50 values [13]. The dataset contains 160 anchor drugs, 65 library drugs, and 125 cell lines. Approximately 77% of the data is missing.

Experimental Setup. We separate the available drug-drug interaction data into two disjoint sets: training and testing, for a given percentage $p \in (0 : 100)$ as $p\%$ and $(100 - p)\%$ of the available drug-drug interaction data, respectively. In our experiments, we select the p to be 85, and this designates 15% of the available data to be used for testing and the remaining for training. However, we do not use all of the remaining data for training. In order to test the potential of our method for different scenarios of data availability, we further sample from the 85% of the available data to cover $50\%, 60\%, 70\%, 80\%$ of the entire available data. For instance, for 50%, we only use a randomly selected subset of only 60% of the data (although more data is available, and hide the rest of the data from *PartialFibers*). To ensure the consistency in our experiments, as we grow the training dataset with more samples, we made sure that every anchor drug, library drug, and cell line pair in 50% is contained within 60%, every sample in 60% is contained within 70%, and every sample in 70% is contained within 80%.

Methods Compared. We compare the prediction performance of *PartialFibers* with two state-of-the-art methods. These are non-fiber, well-known tensor completion algorithm [6,10] and Medians of Means (MoM) [21] approaches.

Other Details. We implement all the algorithms in Python and perform the experiments on a Mac machine with 8-Core Intel Core i9 processors running at 2.3 GHz, and 16GBs RAM.

3.1 Advantage of Partial Fiber Structure in Loss

We start by evaluating how well *PartialFibers* predicts the missing values. We apply our *PartialFibers* algorithm, non-fiber and MoM algorithms separately to compute the missing values in drug-drug interactions dataset. Then, we calculate the normalized mean squared prediction error over all test samples with different percentages of data availability. Figure 2 presents the normalized mean squared loss values in the presence of different data availabilities. We observe that regardless of how much data is available, *PartialFibers* performs better than non-fiber [10] and MoM [21] algorithms. *PartialFibers* reaches to a smaller loss value than both methods in less than 400 seconds. Increasing the available data from 50% to 70%, we see that the normalized MSE loss value decreases the most for *PartialFibers* among the three methods. This implies that as partial fibers get denser (i.e., has more data points) our algorithm converges faster. We observe that the three compared algorithms benefit from the increasing data availabilities in different amounts. The converged loss value for *PartialFibers*, Non Fiber, and MoM [21] decreases by 24.2%, 15.80% and 1.86% respectively as we increase the available data percentage from 50% to 70%. We observe the smallest change in MoM. This implies that the newly added data do not have too much of an impact on the overall prediction error of MoM. We conjecture that this is because MoM selectively uses only subsets of the new data depending on the tensor entry to be predicted.

We hypothesize that the more data we have, the more simultaneous updates *PartialFibers* can provide. This implies that we expect that *PartialFibers* will converge faster with more data available. Figure 3 shows the convergence of our method when 50% and 80% of data is available. We observe that for both data availabilities, *PartialFibers* successfully achieves a small normalized mean squared loss value. This is because, as we provide more data, the number of indices that are updated simultaneously is expected to increase. This allows us to get a smaller loss value in less time. Our results support our hypothesis: Our method for 80% data availability is always below that for 50% in Fig. 3. This means that we can reach the same loss value in less time by using more data.

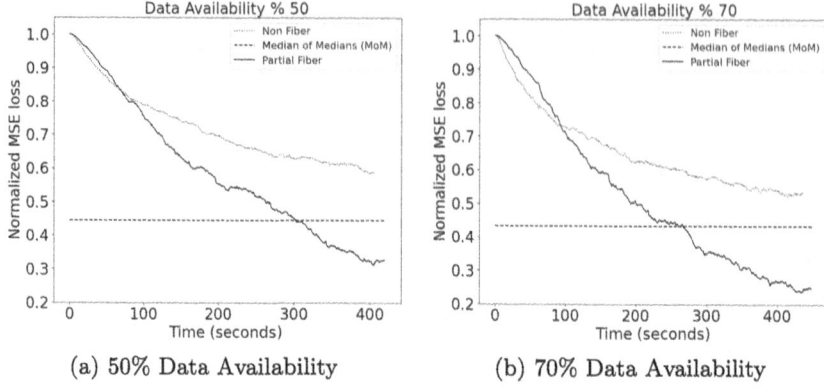

(a) 50% Data Availability   (b) 70% Data Availability

Fig. 2. Change of loss values over time (seconds) for different algorithms using different percentages of data availabilities. Figure 2a, and Fig. 2b represent the normalized mean squared loss values for the Partial Fiber, the Non Fiber and MoM Algorithms using 50%, and 70% of the available data respectively.

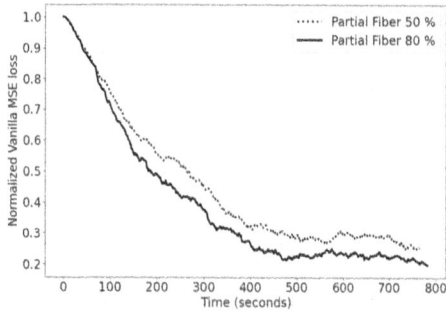

Fig. 3. The normalized mean squared error, loss values over time (seconds) for *PartialFibers* with different data availabilities. The solid line represents the *PartialFibers* with 80% data availability, and the dotted line represents the *PartialFibers* with 50% data availability.

3.2 Drugs Which Have the Most/Least Accurate Predictions

Here, we present our results that focus on the drugs which have the most/least accurate predictions. We calculate the individual loss values for each of the anchor drugs and library drugs, both with partial fiber setting and non-fiber setting. We sort the drugs based on their loss values and share the name of the drugs that have the highest loss in each setting. We share the detailed results in Fig. 4. We observe that more than half (∼ 20) of the anchor drugs, and more than half of the library drugs (∼ 40) reach a very small loss value, in both fiber and non-fiber settings. This implies that both of the methods are efficiently predicting the drug response for most of the drugs. Our results demonstrate that loss values are smaller for the *PartialFibers* for both anchor and library drugs. When we

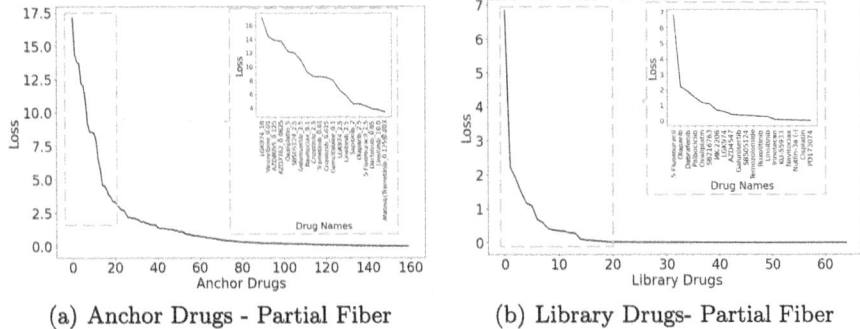

(a) Anchor Drugs - Partial Fiber   (b) Library Drugs - Partial Fiber

Fig. 4. The loss values for each anchor drug and each library drug after convergence and the sorted loss values with a focus on the top 20 drugs with the highest loss values. Figure 4a and Fig. 4b represent the loss values for anchor drugs and library drugs after completion with *PartialFibers*.

focus on 20 drugs which have the highest loss, for anchor drugs we see that 12 out of 20 drugs are common for both *PartialFibers* and non-fiber approaches, and if we narrow down and look at the top 10 drugs we see that 7 of them are same for both of the algorithms. The order might be different but for the same set of drugs we observe poor prediction performance, and we see that the loss is always less with *PartialFibers*. The common anchor drugs are as follows; Crizotinib, Trametinib, Oxaliplatin, SB505124, LGK974, AZD8055, Vinorelbine, AZD7762, 5-Fluorouracil, Navitoclax, Olaparib, Sapitinib. Each drug in this set targets different type of cancer. It is important to note the weaknesses before we can make a trustworthy data completion.

4 Conclusion

In this paper, we presented a novel algorithm called *PartialFibers* for predicting the drug dosage in drug-drug combination setting. Given partial information on the dosage of drug-drug combinations, *PartialFibers* effectively predicts the IC50 values for the missing entries by specifically benefiting from the structure of the data without further assumptions. We showed that *PartialFibers* is a time efficient, flexible, scalable computational method which can provide accurate predictions.

References

1. Ahmadi Moughari, F., Eslahchi, C.: ADRML: anticancer drug response prediction using manifold learning. Sci. Reports **10**(1) (2020). https://doi.org/10.1038/s41598-020-71257-7
2. Al-Lazikani, B., Banerji, U., Workman, P.: Combinatorial drug therapy for cancer in the post-genomic era. Nature Biotechnol. **30**(7), 679–692 (2012). https://doi.org/10.1038/nbt.2284

3. Ayukekbong, J.A., Ntemgwa, M., Atabe, A.N.: The threat of antimicrobial resistance in developing countries: causes and control strategies. Antimicrob. Resistance Infect. Contr. **6**(1),(2017). https://doi.org/10.1186/s13756-017-0208-x
4. Bumin, A., Huang, K.: Efficient implementation of stochastic proximal point algorithm for matrix and tensor completion. In: European Signal Processing Conference. IEEE (2021)
5. Bumin, A., Ritz, A., Slonim, D., Kahveci, T., Huang, K.: FiT: fiber-based tensor completion for drug repurposing. In: ACM International Conference on Bioinformatics (2022)
6. Carroll, J.D., Chang, J.-J.: Analysis of individual differences in multidimensional scaling Via an N-way generalization of "Eckart-Young" decomposition. Psychometrika **35**(3), 283–319 (1970). https://doi.org/10.1007/BF02310791
7. Costello, J.C., et al.: A community effort to assess and improve drug sensitivity prediction algorithms. Nature Biotechnol. **32**(12), 1202–1212 (2014)
8. Duchi, J., Hazan, E., Singer, Y.: Adaptive subgradient methods for online learning and stochastic optimization. J. Mach. Learn. Res. **12**(7) (2011)
9. Fan, K., Cheng, L., Li, L.: Artificial intelligence and machine learning methods in predicting anti-cancer drug combination effects. Brief. Bioinform. **22**(6), bbab271 (2021)
10. Harshman, R.A.: Foundations of the PARAFAC procedure: Models and conditions for an "explanatory" multi-modal factor analysis. UCLA Work. Papers Phonet. **16**, 84 (1970)
11. Havel, H., et al.: Nanomedicines: from bench to bedside and beyond. AAPS J. **18**, 1373–1378 (2016)
12. Housman, G., et al.: Drug resistance in cancer: an overview. Cancers **6**(3), 1769–1792 (2014)
13. Jaaks, P., et al.: Effective drug combinations in breast, colon and pancreatic cancer cells. Nature **603**(7899), 166–173 (2022)
14. Keith, C.T., Borisy, A.A., Stockwell, B.R.: Multicomponent therapeutics for networked systems. Nature Rev. Drug Discov. **4**(1), 71–78 (2005)
15. Kolda, T.G., Bader, B.W.: Tensor decompositions and applications. SIAM Rev. **51**(3), 544–500 (2009)
16. Lehár, J., et al.: Synergistic drug combinations tend to improve therapeutically relevant selectivity. Nature Biotechnol. **27**(7), 659–666 (2009)
17. Lopez, J.S., Banerji, U.: Combine and conquer: challenges for targeted therapy combinations in early phase trials. Nature Rev. Clin. Oncol. **14**(1), 57–66 (2017)
18. Madani Tonekaboni, S.A., Soltan Ghoraie, L., Manem, V.S.K., Haibe-Kains, B.: Predictive approaches for drug combination discovery in cancer. Brief. Bioinform. **19**(2), 263–276 (2018)
19. Orekhov, V.Y., Ibraghimov, I., Billeter, M.: Optimizing resolution in multidimensional nmr by three-way decomposition. J. Biomol. NMR **27**(2), 165–173 (2003)
20. Ruder, S.: An overview of gradient descent optimization algorithms. arXiv preprint arXiv:1609.04747 (2016)
21. Sapashnik, D., et al.: Cell-specific imputation of drug connectivity mapping with incomplete data. Plos one **18**(2), e0278289 (2023)
22. Sharom, J.R., Bellows, D.S., Tyers, M.: From large networks to small molecules. Current Opinion Chem. Biol. **8**(1), 81–90 (2004)
23. Suphavilai, C., Bertrand, D., Nagarajan, N.: Predicting cancer drug response using a recommender system. Bioinformatics **34**(22), 3907–3914 (2018)
24. Tomasi, G., Bro, R.: Parafac and missing values. Chemometrics and Intelligent Laboratory Systems **75**(2), 163–180 (2005)

25. Vasan, N., Baselga, J., Hyman, D.M.: A view on drug resistance in cancer. Nature **575**(7782), 299–309 (2019)
26. Wang, L., Li, X., Zhang, L., Gao, Q.: Improved anticancer drug response prediction in cell lines using matrix factorization with similarity regularization. BMC Cancer **17**(1) (2017). https://doi.org/10.1186/s12885-017-3500-5
27. Wei, D., Liu, C., Zheng, X., Li, Y.: Comprehensive anticancer drug response prediction based on a simple cell line-drug complex network model. BMC Bioinformatics **20**(1), 1–15 (2019). https://doi.org/10.1186/s12859-019-2608-9
28. Wu, L., et al.: Machine learning methods, databases and tools for drug combination prediction. Brief. Bioinform. **23**(1), bbab355 (2022)

Optimizing Deep Learning for Biomedical Imaging

Ayush Chaturvedi[1(✉)], Guohua Cao[2], and Wu-chun Feng[1]

[1] Virginia Polytechnic Institute and State University, Blacksburg, VA, USA
ayushchatur@vt.edu, wfeng@vt.edu
[2] School of Biomedical Engineering, ShanghaiTech University, Shanghai, China
caogh@shanghaitech.edu.cn

Abstract. With the significant increase in the use of deep learning (DL) for biomedical imaging, the corresponding DL models have become increasingly complex and computationally intensive to achieve high accuracy. This work presents both architecture-aware optimizations and sparsity optimizations to efficiently utilize underlying parallel hardware resources and reduce the computational demand of DL models while maintaining their accuracy. We demonstrate the efficacy of our optimization techniques on an existing DL model in the biomedical domain, i.e., DDNet, short for Densenet and Deconvolution Network, that is designed to enhance the quality of CT images. Overall, our optimization techniques in concert reduce the total training time by 1.94× while maintaining accuracy.

Keywords: AI · deep learning · image de-noising · chest CT · COVID-19

1 Introduction

Computed tomography (CT) has been pivotal in detecting abnormalities within the human body. With recent advances in artificial intelligence (AI) and the availability of open-source CT image datasets, deep learning (DL) models are actively being trained to help detect abnormalities in CT images [10,15,21]. The accuracy of these DL models depends heavily on the quality of CT images in the datasets, which, in turn, correlate to amount of radiation dosage from a CT scan. While a standard-dosage CT scan can generate high-quality CT images, it increases the attributable risk of death from cancer by up to 0.1% [2]. Thus, medical institutions worldwide use a low-dosage CT (LDCT) scan, resulting in low-quality CT images. In turn, scientists rely on DL models to improve the quality of CT images generated from low-dosage CT scans [9,14,23]. However, these LDCT images further exacerbate the computational needs of DL, as training these DL models requires substantial data and state-of-the-art computing resources. Thus, optimization techniques are needed to train these DL models more efficiently with fewer computational resources. To this end, we apply and demonstrate our optimization techniques on our DL image enhancement model called DDNet [27], short for Densenet and Deconvolution Network.

Figure 1 shows the auto-encoder-decoder architecture of DDNet, consisting of a convolution network and deconvolution network, both connected via skip connections. The convolution network features four denseblocks [11], each containing five densely connected convolution layers for efficient feature extraction. Along with the dense blocks, the network consists of 37 convolution layers in total. The images used to train the network model consist of high-quality (HQ) and low-quality (LQ) chest CT images of size 512 × 512 in 32-bit grayscale. For the loss computation, the network uses a complex loss function that combines the mean square error (MSE) and multi-scale structural similarity index metric (MS-SSIM) [25].

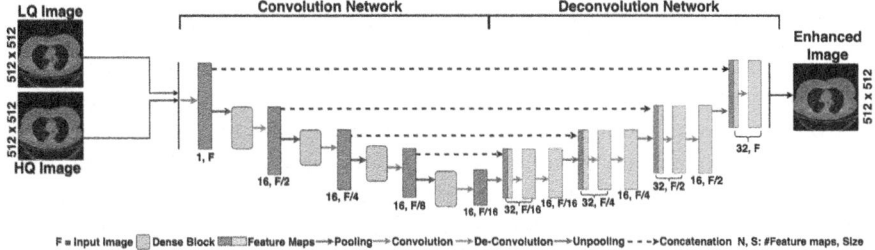

Fig. 1. Architecture of DDNet.

While DDNet improves the quality of chest CT scans and, in turn, results in better COVID-19 detection [8], its architecture requires very large GPU memory and extensive training time. Although both the human brain and a DL model consist of millions of interconnected neurons that serve as fundamental processing units, the human brain, unlike traditional DL models, exhibits a *sparse structure* [7,12], i.e., not all the neurons in the brain are always interconnected. Hence, we present algorithmic strategies and, in turn, optimizations that incorporate *sparsity* into DL models to reduce their computational demands during training while maintaining accuracy.[1]

In summary, our work improves current state-of-the-art DL-based CT imaging via the following contributions:

- Novel realization of sparse algorithms to reduce complexity and training time of deep-learning (DL) models by accounting for their neural architecture.
- An optimized data loader that mitigates the data-movement latency associated with training DL models on small datasets in the PyTorch framework.
- A mixed-precision algorithm for convolution neural networks that leverages a specialized data format and 'tensor cores' in modern NVIDIA GPUs.

---

[1] These sparse techniques should *not* be confused with "sparse reconstruction techniques" in biomedical imaging [1]. Thus, for lucidity, we define sparsity in DL models as referring to *the magnitude of zero entries in their programmatic representation*.

2 Sparse Optimizations

Like the human brain, a deep learning (DL) model consists of many layers with millions or billions of neurons. These layers are represented as a combination of parameters (i.e., weights and biases) stored in huge multi-dimensional matrices, also called *tensors*. Having larger layers with more parameters improves accuracy, but operations on huge tensors require significant computational resources. Thus, to reduce the number of effective parameters, sparse techniques modify specific entries in a tensor to zero in a process called *pruning*, thus creating sparsity in tensors and the overall DL model. In this work, we leverage three kinds of sparse techniques for our DDNet model: random unstructured, structured, and magnitude-based sparse optimizations.

2.1 Unstructured and Structured Sparsity

The sparse taxonomy is defined by the structure imposed on the tensors while pruning the values from a DL model. *Structured* sparsity addresses the dimensions of the tensors associated with weights, bias, and filter values, whereas *unstructured* sparsity only formulates the criteria for selecting values from these tensors in a DL model. We use both techniques and apply them to DDNet.

In DDNet, the convolution and deconvolution layers collectively form a significant part of the overall computation in the forward pass, i.e., matrix-multiply-add (MMA) operations on multi-dimensional tensors. Moreover, skip connections across the network and shortcut connections within the dense blocks require intermediate results to be held in memory, thus increasing memory requirements for DDNet. To reduce compute and memory overhead, we introduce sparsity by pruning tensors engaged in skip connections, convolution, and deconvolution.

Random Unstructured: Randomly pruning parameters of the DL layers falls under the class of unstructured sparse techniques [6,17]. Figure 2(a) contrasts convolution (red triangle) over a chest CT image without (above) and with (below) randomly pruned tensors. As depicted, random entries in the tensors of the convolution filter and weights are set to zero, thus ceasing to contribute to the model and reducing the number of effective parameters in DDNet.

Structured: Fig. 2(b) incorporates structured sparsity into DDNet by pruning entire dimensions in the weights tensors and blocks in the convolution filters. We only prune those dimensions that are *not* associated with skip connections in the weights tensors. In filters, blocks on the upper left corner and lower right corner are pruned. The resulting tensors are modified to a dense representation so that only non-zero entries are brought to the memory. Moreover, pruning blocks or channels in these tensors reduce their effective dimensions and the total number of convolution operations. As a result, the convolution layers are now calculated faster because they require fewer MMA operations and less memory.

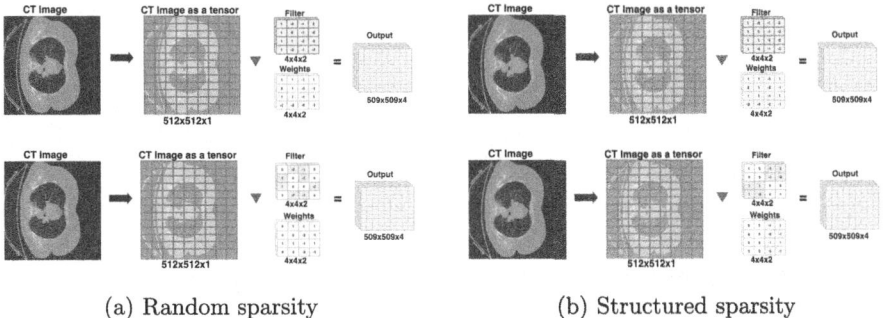

(a) Random sparsity   (b) Structured sparsity

Fig. 2. Convolution operation with sparse filter and weights using (a) random sparse and (b) structured sparse techniques, respectively.

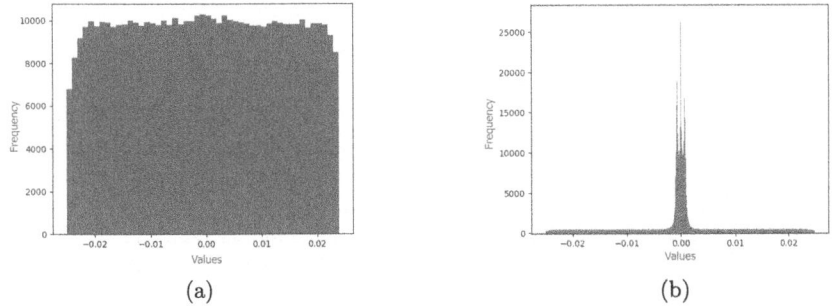

Fig. 3. Normal distribution of parameters for (a) dense DDNet and (b) DDNet with 50% sparsity.

Magnitude-Based Sparsity: To intelligently prune from a DL model, we employ another sparsification technique that imposes 'criteria' to select parameters to be pruned. For example, 'Top-K' uses the absolute values of the model parameters as a proxy for their importance. The underlying assumption is that parameters with the smallest magnitude will contribute the least to the DL model. To select which parameters to prune, we analyze the normal distribution of all parameter values associated with every layer in DDNet in Fig. 3(a). By sorting these values based on their magnitude and then pruning K% of the values from the lower half of the sorted distribution, we get a distribution of parameter values, as shown in Fig. 3(b), where K is set to 50.

While sparsity provides significant performance benefits, it comes at the expense of information loss due to the removal of data via pruning. Consequently, the accuracy of the DDNet model suffers, as shown in Table 1. As remediation, the model needs to be *retrained*, which we articulate in §2.2 with our proposed hybrid training schedule.

Table 1. Accuracy with sparse and dense DDNet. Hyperparameters: batch size = 1; learning rate = 1e-4; decay rate = 0.95; training epochs = 50.

Model	% Sparsity	Sparsity Type	MS-SSIM	Training Time
DDNet (dense)	0	None	97.22 ± 1.49	235 min
DDNet (sparse)	50	Random	42.39 ± 2.10	175 min

2.2 Hybrid Training Schedule

While sparse optimizations reduce the required number of parameters, there are certain parameters that encode patterns that are critical for the model's accuracy. Removing such parameters leads to a decrease in the model's ability to accurately predict the target variable (see MS-SSIM column in Table 1).

Figure 4(a) compares the total training and validation loss for the dense DDNet model (black and green line) and 50% sparse DDNet model (red and blue line). The total training loss of the model with sparsity (blue line) starts at a lower value and keeps decreasing until 10 epochs, at which point it suddenly explodes to finally converge at a final value that is orders of magnitude worse than its dense DDNet counterpart. Thus, to recover the accuracy lost via pruning, the parameters that remain (after pruning) must be subject to a certain amount of re-training. To identify the point in time during training as to 'when' sparsity and re-training should be done, we visualized the gradients of the total loss values during the training and validation phase.

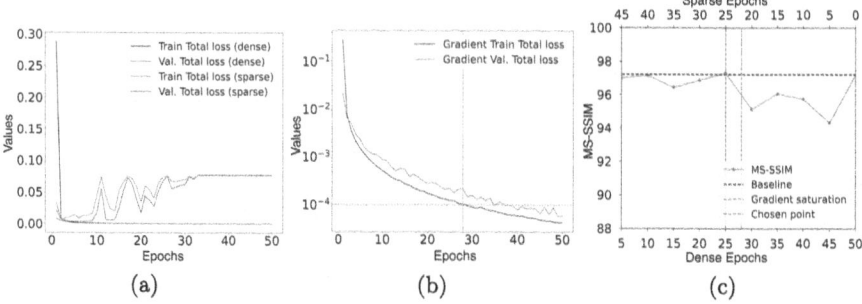

Fig. 4. Results of Hybrid Training: (a) Training Loss Comparison Between Dense and Sparse DDNet, (b) Progression of Gradients of the Total Loss During Training, (c) Accuracy with Different Hybrid Schedules: Dense Epochs Followed by Sparse Epochs.

Figure 4(b) shows a logarithmic plot for the progression of the gradients of the total (training and validation) loss values for a dense DDNet model. The figure shows that the change in the gradients spans multiple orders of magnitude across the first 28 epochs of training (red dotted lines). Thereafter, the loss values continue to diminish but remain within the same order of magnitude, indicating that the overall loss value is approaching a global minima.

Removing parameters until the gradients saturate will hurt the model's accuracy. Thus, we propose a 'hybrid training schedule' wherein dense training is followed by pruning (i.e., sparsity) and then re-training, thus preserving accuracy while improving performance. Using such a hybrid schedule, Fig. 4(c) shows the variation in accuracy (i.e., MS-SSIM) with different combinations of dense (X-axis below) and sparse epochs (X-axis above) that total 50 epochs. The horizontal black dashed line represents the accuracy corresponding to complete dense training (baseline). The vertical green dashed line highlights the point having an ideal balance of dense and sparse epochs that results in the same accuracy.

3 Architecture-Aware Optimizations

To complement the sparsity optimizations, we propose three architecture-aware optimizations to fully utilize the underlying Nvidia Ampere GPU hardware.

3.1 DoLL: Efficient Data-Loader for Small Datasets

Distributed data parallelism (DDP) scales the training of DL models that utilize large datasets. In PyTorch, DDP capabilities are supported by a Distributed Data Loader (DDL) library that prefetches data to the GPUs via multi-threaded worker processes in the background. However, with small datasets, DDL is inefficient for two reasons: (1) worker threads can sit idle after working on their corresponding chunk of the dataset (or mini-batch) and consume resources for the rest of the training period and (2) operations on the respective mini-batches, e.g., index distribution, batch sample preparation, and transformations, occur on the CPU while training occurs on the GPU. To remediate these issues, we design an efficient data-loader for small datasets, i.e., DoLL.

Architecture of DoLL Fig. 5 shows the DoLL architecture, which uses the large GPU memory by staging the entire dataset on it, in parallel, for each replica process. Unlike DDL, DoLL does *not* initialize communication queues and inter-process communication (IPC) for index distribution or data preparation; instead, it leverages the NVIDIA Collective Communications Library (NCCL) so data is directly communicated between GPUs within each node over the NvLink interconnect, thus bypassing the CPU and PCIe bus. While staging the entire dataset on GPU memory takes significantly longer than moving a small mini-batch sample, the overall cost is significantly less than moving small mini-batches from CPU to GPU repeatedly during each training epoch. As a result, *more* operations can now be performed on the *faster* GPU with DoLL.

3.2 Mixed Precision and Tensor Cores

Mixed-precision training leverages a combination of higher- and lower-precision storage formats to accelerate computation in the training process by utilizing higher clock speeds for arithmetic operations in lower precision.

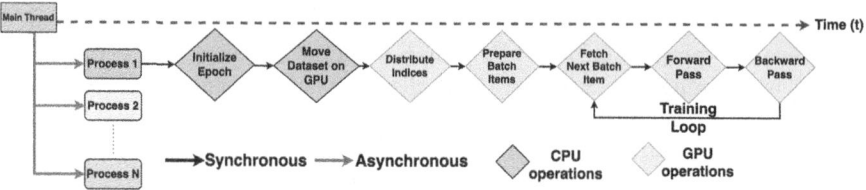

Fig. 5. Architecture of the new data loader, i.e., DoLL.

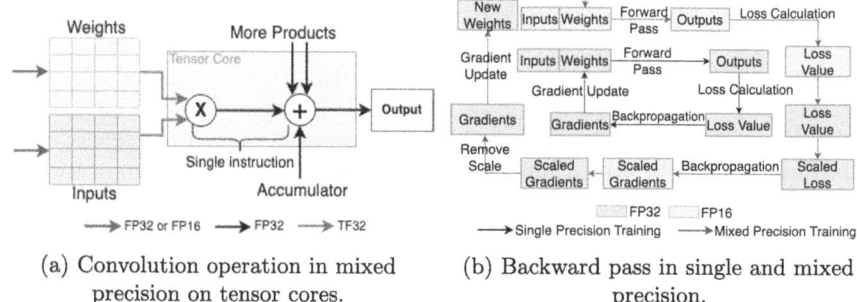

(a) Convolution operation in mixed precision on tensor cores.

(b) Backward pass in single and mixed precision.

Fig. 6. Mixed precision: (a) convolution in the forward pass and (b) backward pass.

In the forward pass, tensor cores perform matrix-multiply-add (MMA) operations in TF32 format [20] for convolution, as shown in Fig. 6(a), and deconvolution.

Performing the entire MMA operation in mixed precision delivers high performance for the following two reasons: (1) tensor cores perform fused-matrix-multiply-add (FMMA) operations in a single clock cycle, delivering more throughput than a CUDA core, which takes two clock cycles to complete an FMMA, and (2) using a specialized format, TF32, the FMMA operations require less memory because of the reduced number of mantissa bits.

For the backward pass, loss calculations and gradient updates are computed in half-precision with values scaled with a scalar value. Simultaneously, the original values are stored and respectively updated in higher single-precision (with the same scale factor) to preserve accuracy. Figure 6(b) contrasts the training workflow between single precision (black arrows) and mixed precision (blue arrows).

3.3 Graph Capture Optimization

Deep learning (DL) frameworks often use a loose coupling of low-level hardware-specific binaries (written in CUDA, HIP, C, and C++) for performance-critical operations with user-friendly APIs for programming productivity. This results in reduced performance and increased overhead, e.g., CUDA kernel launches.

In PyTorch, the CUDA kernel launch overhead is the latency experienced due to the repetitive launch of the same CUDA kernels (on Nvidia GPUs) to compute layers in the DL model during the training iterations. To address this,

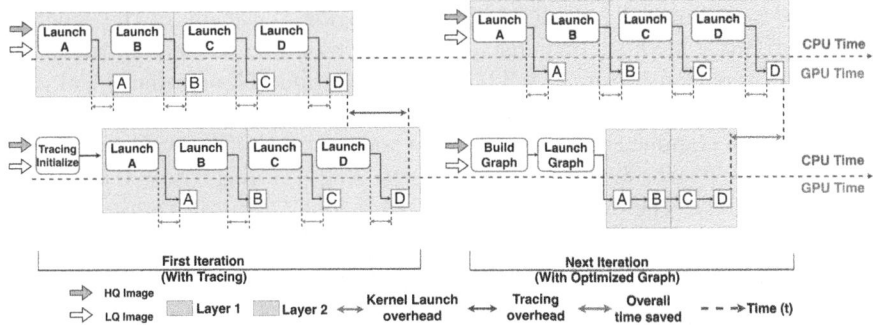

Fig. 7. An example demonstrating graph capture optimization (GCO).

we leverage the CUDA Graphs API with PyTorch. Figure 7 illustrates the optimization through an example of a DL model with two layers undergoing two training epochs with (below) and without (above) our graph capture optimization (GCO). The two layers, Layer 1 and Layer 2, launch two GPU kernels, each containing (A, B) and (C, D), respectively, via the CUDA backend API; in reality, these layers may launch multiple GPU kernels. The GCO initializes an alternative CUDA stream to capture the runtime information of the DL workflow; this stage is called 'tracing.' The new CUDA stream records information about the control and data path in the main CUDA stream (where training happens), such as the execution order of the GPU kernels (A, B, C, and D), their input parameters, their sizes, and data types. As a consequence, tracing consumes additional GPU memory that is needed for secondary buffers to match the size, data type, and dimensions of the input and output tensors of each GPU kernel. As a result, the entire iteration is slower due to the 'tracing overhead' (blue arrow).

Once the tracing finishes, the alternative stream holds a serialized version of a static CUDA graph, which has the same data path and execution order of GPU kernels as the main CUDA stream. Then, at the start of the next epoch, PyTorch's just-in-time (JIT) compiler uses the captured information and instantiates a static CUDA graph containing the GPU kernels in the main CUDA stream. The alternate stream and memory buffers are discarded, and a static CUDA graph is launched on the GPU scheduler. The launched CUDA graph is replayed for the rest of the training duration, mitigating the need to repeatedly launch and destroy GPU kernels at each epoch. Thus, the subsequent epochs run faster, amortizing the latency of the tracing in the initial epochs.

4 Results and Analysis

In this section, we evaluate the *accuracy* of our optimizations with respect to MS-SSIM (multi-scale structural similarity index) [25] and *performance* with respect to total time to train DDNet with a target MS-SSIM value of 97.22 ± 1.49. We construct a dataset of chest CT scans from four public biomedical data sources: Mayo Clinic [18], BIMCV Medical Imaging Databank of the Valencia Region, MIDRC: Medical Imaging and Data Resource Center [19] and Lung Image Database Consortium (LIDC) Image Collection. These radiological data sources contain 3D chest CT scans composed of 2D image slices, each of size 512× 512 pixels. The training setup for the dense and sparse DDNet hyperparameters is as follows: 50 epochs for training, batch size = 32, learning rate = 0.0001, and decay rate = 0.95. The original (dense) DDNet took 79 min to converge to the target MS-SSIM value, which, in turn, we use as our reference training time.

4.1 Sparse Optimizations

Table 2 shows the speedup and accuracy drop using different sparse techniques. With a hybrid training schedule of 25 dense epochs and 25 sparse epochs, all the sparsity optimizations provide a similar speedup of 1.14× with no loss in accuracy because all three sparse optimizations target only the same convolution and deconvolution layers in the DDNet model. Moreover, all optimizations prune 50% of entries in the weights and bias tensors associated with these layers to benefit from the 2:4 sparsity support in the 2nd generation of tensor cores on the Nvidia Ampere GPU [20]. With the same 50% sparsity in all three optimization techniques, an equal number of effective parameters translates to an equal number of FMMA operations in all sparse techniques. As a result, the final sparse models show the same degree of optimization and training time reduction.

Table 2. Speedup vs. accuracy with different sparse optimizations.

Sparsity Type	Speedup	MS-SSIM	%Accuracy Drop
Structured	1.14×	97.75 ± 1.62	0
Random Unstructured	1.14×	97.30 ± 2.17	0
Top-K	1.14×	96.98 ± 2.02	0.02

4.2 Architecture-Aware Optimizations

We apply and evaluate our architecture-aware optimizations individually for both sparse and dense models and then in concert. First, mixed precision delivers a speedup of 1.49× for the dense model and 1.34× for the sparse models (see Fig. 8(a)). Additionally, with mixed precision, the memory consumption of the two models reduces from 62 GB to 27 GB because using lower precision (TF32 in the forward pass and half-precision in the backward pass) requires fewer bytes.

Second, DoLL provides a constant speedup of 1.21× (see Fig. 8(a)) for the dense and sparse models since the amount of training data is the same for both models. Figure 9(a) compares the profiles for data movement using two data

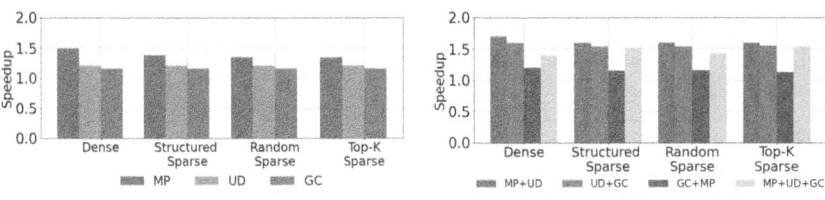

(a) Individual optimizations.   (b) All optimizations in concert.

Fig. 8. Speedup with combinations of optimizations. MP: Mixed precision, UD: Using DoLL, GCO: Graph capture optimization.

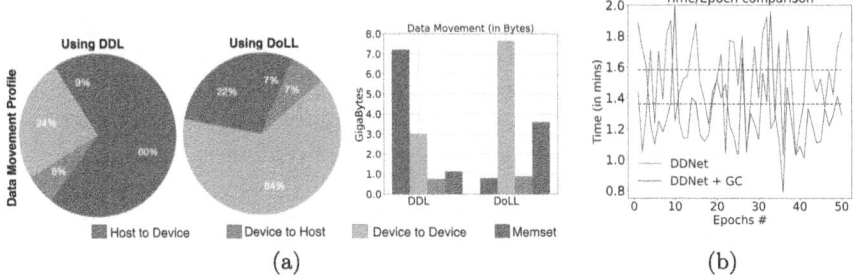

Fig. 9. Improvement with architecture-aware optimizations. (a) Data movement profile of DDNet by percentage and total bytes, (b) per epoch time comparison with GCO.

loaders: PyTorch's DDL and our DoLL. With DDL, 60% of all data movement is due to the repetitive CPU-to-GPU, i.e., host-to-device (H2D), transfers of incremental portions of the dataset. DoLL reduces this H2D transfer overhead from 60% down to 7% by moving the entire dataset to the GPU before training starts, thus minimizing H2D transfers. In terms of the amount of data moved between the CPU and GPU. With DoLL, the H2D transfer is less than 1 GB, which is in contrast to the 7 GB with DDL. However, the DL model now consumes more GPU memory, as expected and shown in Fig. 9(a), where the data movement for device-to-device increases from 24% to 64% and memset from 9% to 22%.

Third, graph capture optimization (GCO) delivers a 1.16× speedup. Figure 9(b) shows a per-epoch time comparison of the baseline DDNet model (blue line) to one with GCO (maroon line). Due to the tracing overhead, the initial five epochs are slower, but the overall mean per epoch time is faster.

When all three optimizations are combined (see Fig. 8(b)), we observe an overall speedup of 1.7× for the baseline model and 1.6× each, for the three sparse models. A concerted overall speedup for all the optimizations is not realized because GCO does not work well with mixed precision. When GCO is combined with mixed precision, the data types and the size of the input tensors in GPU kernels vary at each iteration due to the mixed precision. As a result, the graph optimization skips such GPU kernels from the CUDA graph, and the CUDA kernel launch latency for such GPU kernels is thus not mitigated.

5 Related Work

This section presents related work from (1) deep learning-based CT image enhancement and (2) sparsity in deep learning (DL). Li et al. [16] present an extensive survey of DL-based image post-processing techniques for improving CT image quality. DNN architectures, such as auto-encoders [22], deep convolution neural networks (CNNs) [4], generative adversarial networks (GANs) [26], and transformers [28] have shown potential in improving CT image quality. However, these architectures generally ignore the computational cost of training and the hardware support needed in the CT equipment to deploy such DNNs.

To address the above, *sparse optimizations* reduce the complexity of DL models, including CNNs [3], GANs [24] and transformer-based neural networks [5]. Hoefler et al. present an extensive survey on sparsity in deep learning [13].

6 Conclusion

With the increasing use of deep learning (DL) in the biomedical domain, researchers need to focus on both the accuracy *and* performance of DL models. While an accurate model is essential, the high computational cost of training such a model limits its accessibility. Therefore, this work presents a combination of architecture-aware and sparse optimizations for DL models in the biomedical domain. Our architecture-aware optimizations deliver a speedup of up to 1.7× for the baseline (dense) DDNet model without losing any accuracy. By introducing a hybrid dense+sparse training schedule to the aforementioned architecture-aware optimizations, we achieve an additional 1.14× speedup, resulting in an aggregate speedup of 1.94× while maintaining accuracy.

References

1. Bian, J., et al.: Evaluation of sparse-view reconstruction from flat-panel-detector cone-beam ct. Phys. Med. Biol. **55**(22), 6575–6599 (2010)
2. Brenner, D.J., Hall, E.J.: Computed tomography - an increasing source of radiation exposure. N. Engl. J. Med. **357**(22), 2277–2284 (2007)
3. Changpinyo, S., Sandler, M., Zhmoginov, A.: The power of sparsity in convolutional neural networks (2017)
4. Chen, H., et al.: A low-dose ct via convolutional neural network. Biomed. Opt. Express **8**(2), 679 (2017)
5. Child, R., Gray, S., Radford, A., Sutskever, I.: Generating long sequences with sparse transformers (Apr 2019)
6. Frankle, J., Dziugaite, G.K., Roy, D.M., Carbin, M.: Pruning neural networks at initialization: Why are we missing the mark? In: 9th International Conference on Learning Representations (May 2021)

7. Friston, K.: Hierarchical models in the brain. PLoS Comp. Biology **4**(11) (2008)
8. Goel, G., Gondhalekar, A., Qi, J., Zhang, Z., Cao, G., Feng, W.: Computecovid19+: Accelerating covid-19 diagnosis and monitoring via high-performance deep learning on ct images. In: Proc. of the 50th International Conference on Parallel Processing. Association for Computing Machinery (2021)
9. Gong, W., et al.: Deep learning-based low-dose ct for adaptive radiotherapy of abdominal and pelvic tumors. Front. Onco.l **12** (2022)
10. Gupta, R.K., Bharti, S., Kunhare, N., Sahu, Y., Pathik, N.: Brain tumor detection and classification using cycle generative adversarial networks. Interdiscip. Sci.: Comput. Life Sci. **14**(2), 485–502 (2022)
11. Hasan, N., Bao, Y., Shawon, A., Huang, Y.: DenseNet convolutional neural networks application for predicting COVID-19 using CT image. SN Comput. Sci. **2**(5) (2021). https://doi.org/10.1007/s42979-021-00782-7
12. Herculano-Houzel, S., Mota, B., Wong, P., Kaas, J.H.: Connectivity-driven white matter scaling and folding in primate cerebral cortex. Proc. Nat'l Acad. Sci. **107**(44), 19008–19013 (2010)
13. Hoefler, T., Alistarh, D., Ben-Nun, T., Dryden, N., Peste, A.: Sparsity in deep learning: pruning and growth for efficient inference and training in neural networks. J. Mach. Learn. Res. **22**(1) (Jan 2021)
14. Jiang, B.: Deep learning reconstruction shows better lung nodule detection for ultra-low-dose chest ct. Radiology **303**, 202–212 (2022)
15. Lei, Y., et al.: Breast tumor segmentation in 3d automatic breast ultrasound using mask scoring r-cnn. Med. Phys. **48**(1), 204–214 (2020)
16. Li, D., Ma, L., Li, J., Qi, S., Yao, Y., Teng, Y.: A comprehensive survey on deep learning techniques in ct image quality improvement. Med. Biol. Eng. Compu. **60**(10), 2757–2770 (2022)
17. Liu, S., et al.: The unreasonable effectiveness of random pruning: Return of the most naive baseline for sparse training. arXiv preprint arXiv:2202.02643 (2022)
18. McCollough, C. et al: Data from Low Dose CT Image and Projection Data (2021)
19. Medical Imaging & Data Resource Ctr.: (2021-04-09). https://www.midrc.org/
20. Mishra, A., et al.: Accelerating sparse deep neural networks (2021)
21. Nasrullah, N., Sang, J., Alam, M.S., Mateen, M., Cai, B., Hu, H.: Automated lung nodule detection and classification using deep learning combined with multiple strategies. Sensors **19**(17), 3722 (2019)
22. Ronneberger, O., Fischer, P., Brox, T.: U-net: Convolutional networks for biomedical image segmentation (May 2015)
23. Sanaat, A., Shiri, I., Arabi, H., Mainta, I., Nkoulou, R., Zaidi, H.: Deep learning-assisted ultra-fast/low-dose whole-body pet/ct imaging. Eur. J. Nucl. Med. Mol. Imaging **48**(8), 2405–2415 (2021)
24. Wang, Y., Wu, J., Hovakimyan, N., Sun, R.: Double dynamic sparse training for gans (Feb 2023)
25. Wang, Z., Simoncelli, E., Bovik, A.: Multiscale structural similarity for image quality assessment. In: The 37th Asilomar Conference on Signals, Systems; Computer (2003)
26. Wolterink, J.M., Leiner, T., Viergever, M.A., Išgum, I.: Generative adversarial networks for noise reduction in low-dose ct. IEEE Tran. on Med. Imag. **36**(12), 2536–2545 (2017)

27. Zhang, Z., Liang, X., Dong, X., Xie, Y., Cao, G.: A sparse-view CT reconstruction method based on combination of densenet and deconvolution. IEEE Tran. on Med. Imag. **37**(6), 1407–1417 (2018)
28. Zhao, J., Hou, X., Pan, M., Zhang, H.: Attention-based generative adversarial network in medical imaging: a narrative review. Comput. Biol. Med. **149**, 105948 (2022)

Exploring a Solution Curve in the Phase Plane for Extreme Firing Rates in the Izhikevich Model

Chu-Yu Cheng[(✉)] and Chung-Chin Lu

Department of Electrical Engineering, National Tsing Hua University, Hsinchu City 30013, Taiwan
s109061803@m109.nthu.edu.tw, cclu@ee.nthu.edu.tw

Abstract. The Izhikevich neuron model is a widely adopted computational neuron model that comprises a set of quadratic differential equations involving two variables. Consequently, obtaining a closed-form solution is unattainable, making it challenging to perform further rate-coding analysis. In this study, we establish a balanced background noise Izhikevich neuron model with periodic signal input. Treating the system of differential equations as a velocity vector field, we are able to compute the Hamiltonian energy function for this model. The interspike-interval firing rate function is then derived with the aid of the Hamiltonian function. Using the firing rate function, we propose a solution curve on the novel γ-γ' phase plane for better understanding the timing when extreme values of the firing rate function occur. Additionally, we address a phase advance phenomenon that occurs between the sinusoidal current injection and the interspike-interval firing rate curve, attempting to provide a qualitative explanation for this phenomenon.

Keywords: Izhikevich model · Balanced background noise · Sinusoidal current · Velocity vector field · Hamiltonian energy function · Interspike-interval (ISI) firing rate · Extreme value · γ-γ' phase plane · Phase advance

1 Introduction

Computational neuron models are widely applied in various disciplines, including the contemporary frontiers of machine learning and artificial intelligence (AI), where neural models play vital roles in the development of spiking neural networks (SNN) and convolutional neural networks (CNN) [3,12]. Among numerous types of neuron models, one category that stands out is the spiking neuron models, as discussed by Gerstner [10]. Spiking neuron models often rely on mathematical equations to elucidate the temporal dynamics of membrane potentials on neuronal cell membranes. There is a diversity of neuron models, ranging from current-based ones like simplified phenomenological integrate-and-fire (IAF) models, leaky IAF (LIF) models [1], and Connor-Stevens models [5] to conductance-based models such as Hodgkin-Huxley models [14], and their customized versions like Morris-Lecar models [20] and FitzHugh-Nagumo models [9,21]. The latter category, while more electrophysiologically accurate and biologically realistic, involves higher time and space complexity, posing challenges for simulations [6].

The Izhikevich models, as presented in the work by Izhikevich in 2003 [15], provide a well-balanced solution, combining dynamism and ease of simulation. These neuron models incorporate an ample number of parameters, enabling a broad range of distinct firing patterns. The Izhikevich model has garnered significant adoption across various scientific domains [13,16]. Nevertheless, the presence of at least one quadratic differential equation with multiple variables within the model introduces a notable challenge. Specifically, the lack of an analytical solution poses a substantial obstacle when attempting to investigate the phase shift phenomenon between input periodic signals and the interspike-interval (ISI) firing rate. This analysis holds a central role in facilitating further explorations related to the encoding of information associated with firing rates within a neuron model.

The Hamiltonian energy function revolutionized the reformulation of equations of motion, offering advantages like energy preservation and enhanced integration methods. As time passed, the Hamiltonian framework transcended classical mechanics, reaching into diverse fields including quantum mechanics [17], fluid dynamics [22], and control theory [8]. In recent years, the use of Hamiltonian energy functions has expanded into computational neuroscience [11], providing an alternative approach to understanding neural network dynamics. This Hamiltonian energy function empowers us to scrutinize phase shifts between input periodic signals and interspike-interval (ISI) firing rates within an Izhikevich neuron model.

The remainder of this paper is structured as follows: In Sect. 2, we introduce a customized version of the Izhikevich neuron model, which incorporates sinusoidal current injection in combination with balanced background noises. We then proceed to formulate the corresponding Hamiltonian energy function. In Sect. 3, we derive the ISI firing rate function for the balanced Izhikevich model using the Hamiltonian operator. Additionally, we present the γ-γ' phase plane indicating when the ISI firing rate of a neuron reaches its extreme values, including a discussion of the phase advance phenomenon. We summarize our findings and draw relevant conclusions in Sect. 4.

2 Methods

2.1 The Balanced Izhikevich Model

Izhikevich's Quadratic Neuron Model. The core component of the Izhikevich neuron model consists of a system of ordinary differential equations governed by two sets of parameters: c_1, c_2, c_3, c_4, c_5, and a, b, c, d. The author has introduced a universal configuration for the first set of parameters, namely,

$c_1 = 0.04, c_2 = 5, c_3 = 140, c_4 = 1, c_5 = 1$, tailored to represent a prototypical cortical neuron [15]. The membrane potential V and an additional recovery variable U are described by the following system of differential equations:

$$\frac{dV(t)}{dt} = c_1 V^2(t) + c_2 V(t) + c_3 - c_4 U(t) + c_5 I_{\text{ext}}(t)$$

$$\frac{dU(t)}{dt} = a(bV(t) - U(t)), \qquad (1)$$

where I_{ext} represents the external current injection. It is important to note that, in the original work by the author, the variables V and U, as well as the parameters a, b, c, d, are all dimensionless for the sake of convenience. When the membrane potential reaches the firing threshold of $V_{\text{th}} = 30$ mV, the model initiates a spike and resets according to the following conditions:

$$V(t) \leftarrow c \quad \text{and} \quad U(t) \leftarrow U(t) + d.$$

It is evident that the parameter c corresponds to the reset potential V_{reset}, while the parameter a functions as the reciprocal of the time constant of U. Through the adjustment of these parameters, the Izhikevich model can capture various neuronal behaviors. The author has proposed eight classic neuron types, including Regular Spiking (RS), Intrinsically Bursting (IB), Chattering (CH), Fast Spiking (FS), Thalamo-Cortical (TC) with high and low initial membrane potential V_0, Resonator (RZ), and Low-Threshold Spiking (LTS). Figure 1 shows the phase portrait analysis of all types of Izhikevich models. The parameter values for these eight original Izhikevich neuron model types are detailed in Table 1.

Table 1. Parameters for 8 types of Izhikevich's neuron model.

c_1	c_2	c_3
0.04	5	140
a	b	c
0.02 / 0.1(FS, RZ)	0.2 / 0.25(TC, LTS)/ 0.26(RZ)	-65 / -55(IB) / -50(CH)
c_4	c_5	
1	1	
d	V_0	
2 / 4(IB) / 8(RS) / 0.05(TC)	-74 / -63 or -87(TC)	

Our Balanced Izhikevich Model. In our previous research [4], both excitatory and inhibitory background noises played significant roles that influenced the ISI firing rate [2,25]. Consequently, it is essential to account for these factors when calculating the agility measure of a neuron model [4]. To accommodate

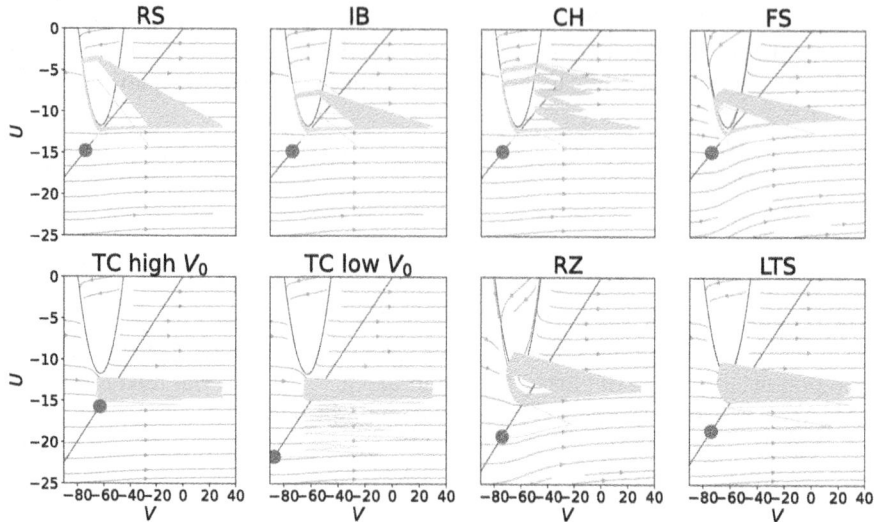

Fig. 1. Phase portrait of eight types of Izhikevich models. The phase portrait displays eight distinct types of Izhikevich models. The parameters of the current injection are: $I_0 = 3.0$ nA, $I_1 = 1.5$ nA, $f = 10$ Hz. In each subplot, the horizontal axis represents the membrane potential V, while the vertical axis represents the recovery variable U. The blue quadratic curve indicates the V nullcline, where $\frac{dV}{dt} = 0$, and the blue straight line represents the U nullcline, where $\frac{dU}{dt} = 0$. The simulation's starting point is denoted by a red dot, and the simulation duration spans 500 ms.

these background noises, two additional terms were incorporated into our customized version of the Izhikevich model. This background noise configuration is based on the research conducted by Troyer and Miller [24]. The model we employed is described by the following set of equations:

$$\frac{dV(t)}{dt} = c_1 V^2(t) + c_2 V(t) + c_3 - c_4 U(t) + c_5 I_{\text{ext}}(t) + g_{\text{ex}}(E_{\text{ex}} - V(t)) + g_{\text{in}}(E_{\text{in}} - V(t))$$

$$\frac{dU(t)}{dt} = a(bV(t) - U(t)) \qquad \frac{dg_{\text{ex}}}{dt} = -\frac{g_{\text{ex}}}{\tau_{\text{ex}}} \qquad \frac{dg_{\text{in}}}{dt} = -\frac{g_{\text{in}}}{\tau_{\text{in}}}. \qquad (2)$$

The updating functions for the background noises are given by:

$$g_{\text{ex}} \leftarrow g_{\text{ex}} + n_{\text{ex}} d_{\text{ex}} \qquad \text{and} \qquad g_{\text{in}} \leftarrow g_{\text{in}} + n_{\text{in}} d_{\text{in}}$$

where g_{ex} and g_{in} represent the excitatory and inhibitory synaptic conductance, while E_{ex} and E_{in} denote the respective reversal potentials. Furthermore, n_{ex} and n_{in} are the quantities of postsynaptic potentials received within each time step dt, and d_{ex} and d_{in} are the corresponding synaptic input weightings. The synaptic conductance of the background noises exponentially decreases over time with time constants τ_{ex} and τ_{in}, respectively. The values of these parameters in our model can be found in Table 2, and for illustrative purposes, we provide sample simulation results in Fig. 2. The external periodic current source in our model is $I_{\text{ext}}(t) = I_0 + I_1 \cos \omega t$, where $\omega = 2\pi f$ and f varies at different frequencies, ranging from 1 Hz to 100 Hz.

Table 2. Parameters for excitatory versus inhibitory background noises in our balanced Izhikevich model.

Excitatory		Inhibitory
800	N	200
7	λ (Hz)	15
0.01	d	0.05
5	τ (ms)	10
0	E (mV)	-70

2.2 Hamiltonian Energy Function for Balanced Izhikevich Model

The approach we use to derive the Hamiltonian energy function is based on the methodology described by Yang et al. [26]. The system of differential equations governing a neuron's dynamics can be analogously conceptualized as the motion of a fluid in a velocity vector field denoted as $\mathbf{f}(*)$, and this vector field can be decomposed into two distinct components: a vortex subvector field $\mathbf{f}_v(*)$ and a gradient subvector field $\mathbf{f}_g(*)$, i.e., $\mathbf{f}(*) = \mathbf{f}_v(*) + \mathbf{f}_g(*)$. Viewed from the perspective of the vector field, the system can be expressed as:

$$\left[\frac{dV}{dt} \; \frac{dU}{dt} \; \frac{dg_{\text{ex}}}{dt} \; \frac{dg_{\text{in}}}{dt} \right]^T = \mathbf{f}_v(V, U, g_{\text{ex}}, g_{\text{in}}) + \mathbf{f}_g(V, U, g_{\text{ex}}, g_{\text{in}}), \tag{3}$$

where

$$\mathbf{f}_v(V, U, g_{\text{ex}}, g_{\text{in}}) \& = \left[c_3 - c_4 U + c_5 I_{\text{ext}} + g_{\text{ex}} E_{\text{ex}} + g_{\text{in}} E_{\text{in}} \; abV \; 0 \; 0 \right]^T \tag{4}$$

$$\mathbf{f}_g(V, U, g_{\text{ex}}, g_{\text{in}}) \& = \left[c_1 V^2 + c_2 V - g_{\text{ex}} V - g_{\text{in}} V \; -aU \; -\frac{g_{\text{ex}}}{\tau_{\text{ex}}} \; -\frac{g_{\text{in}}}{\tau_{\text{in}}} \right]^T. \tag{5}$$

The Hamiltonian energy function $\mathcal{H}$, which represents the total energy of this system, must satisfy the following conditions:

$$\nabla \mathcal{H}^T \mathbf{f}_v(V, U, g_{\text{ex}}, g_{\text{in}}) = 0 \quad \text{and} \quad \nabla \mathcal{H}^T \mathbf{f}_g(V, U, g_{\text{ex}}, g_{\text{in}}) = \frac{d\mathcal{H}}{dt}. \tag{6}$$

Therefore, we obtain the general solution for $\mathcal{H}$ which is also the reliable solution satisfying (6):

$$\mathcal{H} = [c_3 - c_4 U + c_5 I_{\text{ext}} + g_{\text{ex}} E_{\text{ex}} + g_{\text{in}} E_{\text{in}}]^2 + abc_4 V^2. \tag{7}$$

The behavior of the Hamiltonian energy $\mathcal{H}$ for four type of Izhikevich model under sinusoidal current injection is depicted in Fig. 3. In contrast to the original system of differential equations, the Hamiltonian energy function presents a more general mathematical framework, rendering it better suited for subsequent derivations and analyses of the firing rate formula in our research.

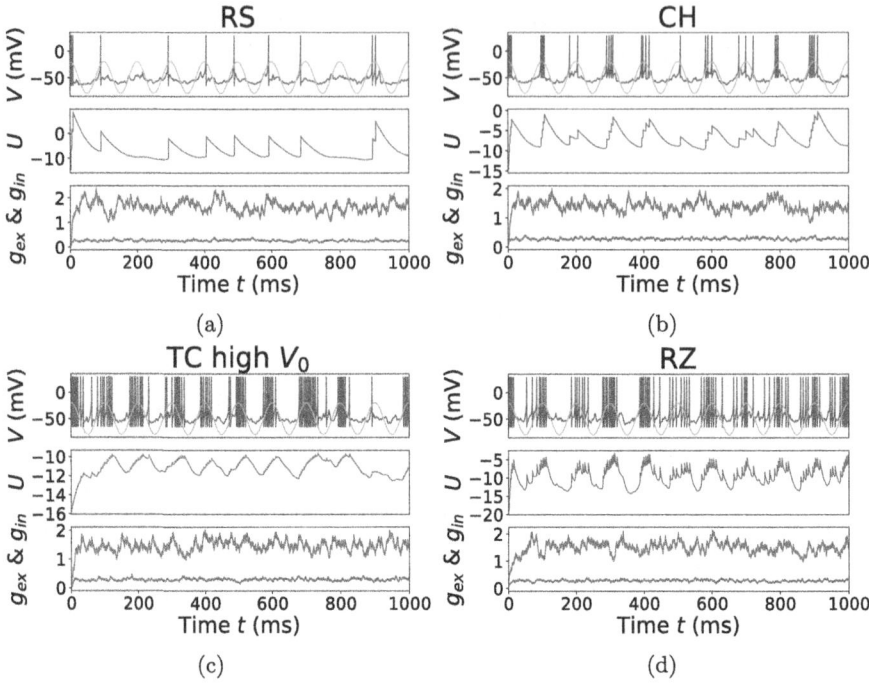

Fig. 2. Comparison of four balanced Izhikevich models. The parameters of current injection and background noises are: $I_0 = 3.0$ nA, $I_1 = 1.5$ nA, $f = 10$ Hz, $N_{\text{ex}} = 800$, and $N_{\text{in}} = 200$. In panels (a)-(d), the top panel for each neuron type displays the membrane potential V with a blue curve, while the orange curve represents the sinusoidal current injection I_{ext}. The second panel depicts the recovery variable U. In the bottom panel, the background noise conductance g_{ex} (in green) and g_{in} (in red) are plotted. The four different types of balanced Izhikevich models are labeled as follows: (a) RS: Regular Spiking, (b) CH: Chattering, (c) TC high V_0: Thalamo-Cortical with a high starting membrane potential, and (d) RZ: Resonator.

3 Results and Discussion

3.1 ISI Firing Rate Function for Balanced Izhikevich Model

As established in the previous section, the Hamiltonian energy function of our balanced Izhikevich model is of the form:

$$\mathcal{H}(t) = [c_3 - c_4 U(t) + c_5(I_0 + I_1 \cos \omega t) + g_{\text{ex}} E_{\text{ex}} + g_{\text{in}} E_{\text{in}}]^2 + abc_4 V(t)^2. \quad (8)$$

Now, when we focus on the specific time point t within one period in the steady state that the membrane potential reaches the action potential V_{th}, triggering a spike, and then resets to V_{reset}, in this case, the Hamiltonian energy function can be calculated as follows:

$$\mathcal{H}(t) = [c_3 - c_4 U(t) + c_5(I_0 + I_1 \cos \omega t) + g_{\text{ex}} E_{\text{ex}} + g_{\text{in}} E_{\text{in}}]^2 + abc_4 V_{\text{reset}}^2. \quad (9)$$

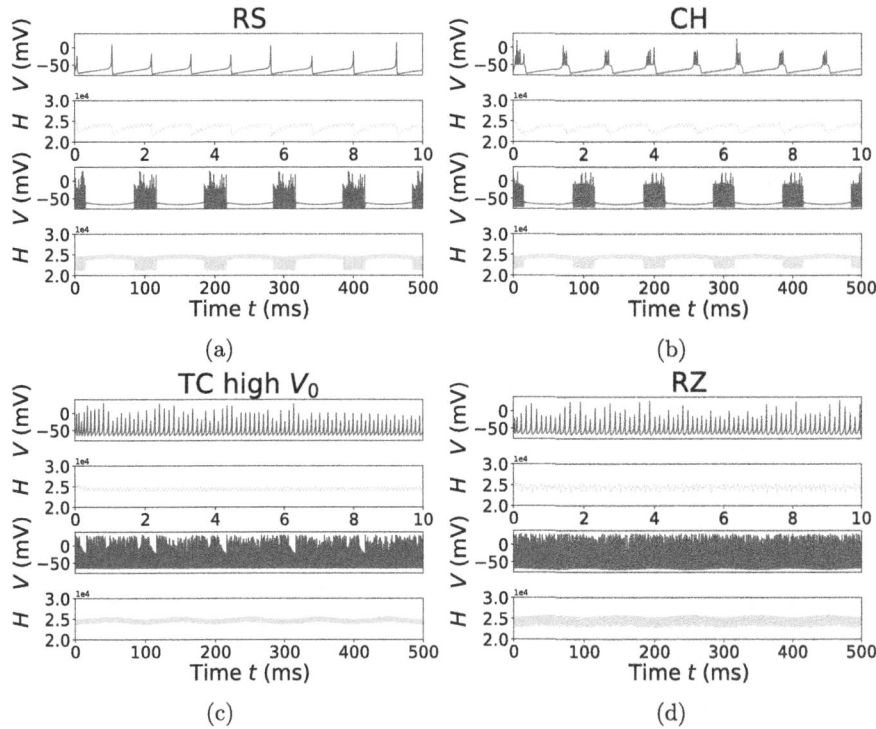

Fig. 3. Hamiltonian energy of four types of Izhikevich models. The parameters of current injection are: $I_0 = 3.0$ nA, $I_1 = 1.5$ nA, $f = 10$ Hz. In panels (a)-(d), the membrane potential V is displayed in the top and third panels, while the corresponding Hamiltonian energy $\mathcal{H}$ is presented in the second and bottom panels. The top two panels in each subplot are zoom-in views for more spiking details at a time scale of 10 ms, while the bottom two panels cover 500 ms. The four different types of Izhikevich models are labeled as follows: (a) RS: Regular Spiking, (b) CH: Chattering, (c) TC high V_0: Thalamo-Cortical with a high starting membrane potential, and (d) RZ: Resonator.

Let Δt be the time interval between two adjacent spikes, then we have

$$\mathcal{H}(t+\Delta t) = [c_3 - c_4 U(t+\Delta t) + c_5(I_0 + I_1 \cos(\omega t + \omega \Delta t)) + g_{\text{ex}} E_{\text{ex}} + g_{\text{in}} E_{\text{in}}]^2 + abc_4 V(t+\Delta t)^2. \quad (10)$$

Since $V(t) = V(t+\Delta t) = V_{\text{reset}}$, and the difference between these two equations should be zero, assuming the time interval Δt is sufficiently long for all other effects to decay, i.e., $U(t+\Delta t) \approx U(t)$, it implies the non-trivial solution

$$\cos \omega(t+\Delta t) + \cos \omega t = \frac{-2c_3 + 2c_4 U(t) - 2c_5 I_0 - 2g_{\text{ex}} E_{\text{ex}} - 2g_{\text{in}} E_{\text{in}}}{c_5 I_1}. \quad (11)$$

Therefore we have

$$\Delta t = \frac{1}{\omega}\cos^{-1}\left(\frac{-2c_3 + 2c_4 U(t) - 2c_5 I_0 - 2g_{\text{ex}}E_{\text{ex}} - 2g_{\text{in}}E_{\text{in}}}{c_5 I_1} - \cos\omega t\right) - t. \tag{12}$$

Note that the last term t in (12) is actually restricted within a single period, which can be substituted with $(t \mod \frac{1}{f})$ for a more generalized version, and Δt can also be denoted as t_{isi}, and the steady state interspike-interval firing rate function $r_{\mathcal{H}s}(t)$ derived from $\mathcal{H}$ can be obtained by taking the reciprocal of the t_{isi}, i.e.,

$$r_{\mathcal{H}s}(t) = \frac{1}{t_{\text{isi}}} = \frac{\omega}{\cos^{-1}\left(\alpha U(t) + \beta - \cos\omega t\right) - t} \tag{13}$$

where

$$\alpha = \frac{2c_4}{c_5 I_1} \quad \text{and} \quad \beta = \frac{-2c_3 - 2c_5 I_0 - 2g_{\text{ex}}E_{\text{ex}} - 2g_{\text{in}}E_{\text{in}}}{c_5 I_1}.$$

3.2 The γ-γ' Phase Plane and the Solution Curve

Taking derivative of $r_{\mathcal{H}s}(t)$ with respect to t, we have

$$\frac{d}{dt}r_{\mathcal{H}s}(t) = \frac{\omega\left[\alpha\frac{dU}{dt} + \omega\sin(\omega t) + \sqrt{1 - (\alpha U(t) + \beta - \cos\omega t)^2}\right]}{[\cos^{-1}\left(\alpha U(t) + \beta - \cos\omega t\right) - t]^2 \sqrt{1 - (\alpha U(t) + \beta - \cos\omega t)^2}}. \tag{14}$$

Observing that the $r_{\mathcal{H}s}$ reaches its maximum or minimum when the brackets in the numerator equals zero. Let $\gamma(t) = \alpha U(t) - \cos\omega t + \beta$, then $\gamma'(t) = \alpha\frac{dU(t)}{dt} + \omega\sin\omega t$ and this term becomes:

$$\gamma'(t_{\text{ev}}) + \sqrt{1 - \gamma(t_{\text{ev}})^2} = 0, \tag{15}$$

where t_{ev} represents a discrete sequence of time points at the moment when ISI firing rate function reaches its extreme values. Equation (15) further implies the following form:

$$\gamma^2 + \gamma'^2 = 1, \text{ with } -1 \leq \gamma \leq 1, \quad \gamma' \leq 0 \tag{16}$$

The solution curve of (16) on the phase plane of γ' with respect to γ is actually the lower half of the unit circle centered at the coordinates $(0,0)$ with radius 1 (see Fig. 4). This indicates that every time the trajectory representing the neuron's real-time dynamics intersects with this curve, the extreme value of the ISI firing rate of the neuron occurs at that exact moment.

3.3 The Phase Advance Phenomenon

Our prior computational simulations using the modified balanced Izhikevich model have acquired some intriguing findings: for certain neuronal types and

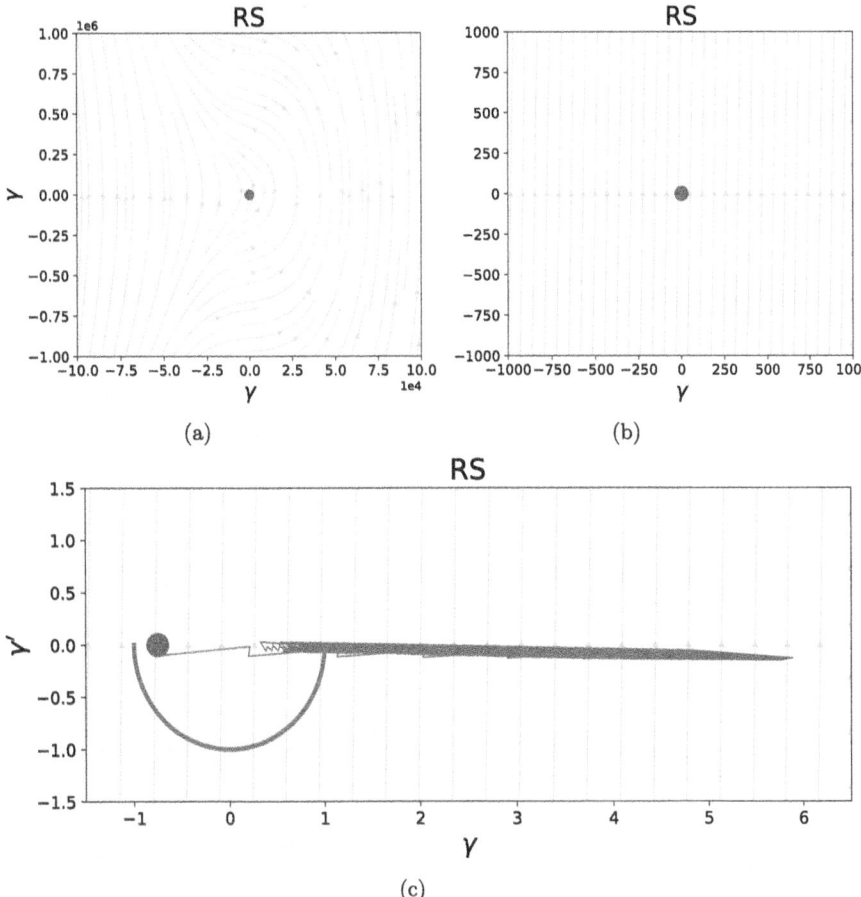

Fig. 4. The γ-γ' phase plane of an Izhikevich RS type model. (a) The whole picture of the γ-γ' phase plane, and (b) a zoom-in version of (a). (c) The further zoom-in view of (b), where the red lower half circle is the solution curve for the extreme value of ISI firing rate function, and the green dot is the simulation starting point.

specific combinations of sinusoidal current amplitudes I_0, I_1, and frequency f, we have observed an unexpected phase advance phenomenon in the instantaneous firing rate of the neuron model. This behavior is demonstrated in Fig. 5. Note that the phase advance phenomenon is not typically observed in current-based neuron models like IAF and LIF models, and has been regarded as a violation of causality. However, recent research has suggested that the phase advance phenomenon might be not only possible but also prevalent in specific neural network contexts [7,18,19,23].

We need to emphasize that the studies mentioned here predominantly address the phase advance phenomenon within the scope of entire neural networks or localized neural networks. They do not assert the phase advance capability at

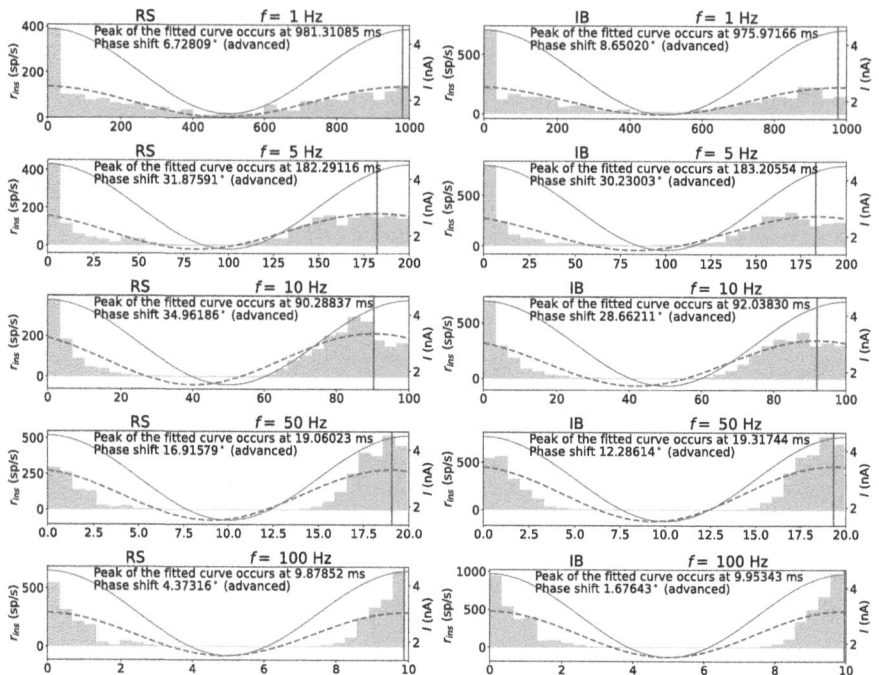

Fig. 5. Phase advance phenomenon. The parameters of current injection are: $I_0 = 3.0$ nA, $I_1 = 1.5$ nA, $f = 10$ Hz. In each panel, the horizontal axis represents the time interval (ms) of a single period. The left-hand blue vertical axis indicates the instantaneous firing rate, represented by the blue histogram in the background, which is the simulation result. Each spike belonging to a certain time bin within one period is map in the corresponding bar. The right-hand red vertical axis represents the current intensity for the red solid curve, which represents the sinusoidal current input. The green dashed curve represents the fitted firing rate curve, and the vertical green solid straight line indicates the position of the peak of the fitted firing rate curve. The left column of panels shows one of the eight different types of balanced Izhikevich models: Regular Spiking (RS), and the right column shows another type: Intrinsically Bursting (IB).

the level of individual neurons. Nevertheless, in our findings, a single balanced Izhikevich model clearly exhibits this phenomenon. The phase advance phenomenon, as discussed here, is not a universal trait applicable to arbitrary input signals; it is specifically associated with continuous, identical periodic input signals. This prompts us to explore several plausible explanations that do not defy causality. One such proposition posits that the so-called predictive capability of neurons concerning incoming signals may be due to the repetitive and single-patterned nature of the periodic input. Because of the sustained influence of the preceding high-current input signal phase within the same period, which has not yet faded, the threshold for the subsequent response remains lowered, and hence even before the arrival of the next period's high-current signal phase, the neuron has already accumulated sufficient energy during an earlier, low-current phase and fired multiple spikes in response.

4 Conclusion

Considering that the original Izhikevich model is defined by a system of quadratic differential equations involving two variables, it is widely recognized that finding explicit solutions for this system is a formidable task. The methods employed in our previous research endeavors, unfortunately, cannot be adapted to address this specific challenge. Nevertheless, by directing our attention toward the Hamiltonian energy function $\mathcal{H}$, we have successfully derived the steady state interspike-interval (ISI) firing rate function (13) for the balanced Izhikevich model. Furthermore, using this ISI firing rate function as a foundation, we have gone on to discover the solution curve (16) on the γ-γ' phase plane (Fig. 4), which offer an intuitive way to show when the ISI firing rate function of the Izhikevich model attains its maximum or minimum values.

References

1. Abbott, L.F.: Lapicque's introduction of the integrate-and-fire model neuron (1907). Brain Res. Bull. **50**(5–6), 303–304 (1999)
2. Burkitt, A.N., Meffin, H., Grayden, D.B.: Study of neuronal gain in a conductance-based leaky integrate-and-fire neuron model with balanced excitatory and inhibitory synaptic input. Biol. Cybern. **89**(2), 119–125 (2003)
3. Cao, Y., Chen, Y., Khosla, D.: Spiking deep convolutional neural networks for energy-efficient object recognition. Int. J. Comput. Vision **113**, 54–66 (2015)
4. Cheng, C.Y., Lu, C.C.: The agility of a neuron: Phase shift between sinusoidal current input and firing rate curve. J. Comput. Biol. **28**(2), 220–234 (2021)
5. Connor, J., Stevens, C.: Voltage clamp studies of a transient outward membrane current in gastropod neural somata. J. Physiol. **213**(1), 21 (1971)
6. Dayan, P., Abbott, L.F., Abbott, L.: Theoretical Neuroscience: Computational and Mathematical Modeling of Neural Systems. MIT press Cambridge, MA (2001)
7. DePiero, V.J., Borghuis, B.G.: Phase advancing is a common property of multiple neuron classes in the mouse retina. Eneuro **9**(5) (2022)
8. Feichtinger, G., Hartl, R.F.: On the use of Hamiltonian and maximized Hamiltonian in nondifferentiable control theory. J. Optim. Theory Appl. **46**, 493–504 (1985)
9. FitzHugh, R.: Impulses and physiological states in theoretical models of nerve membrane. Biophys. J . **1**(6), 445–466 (1961)
10. Gerstner, W., Kistler, W.M.: Spiking neuron models: Single neurons, populations, plasticity. Cambridge university press (2002)
11. Greydanus, S., Dzamba, M., Yosinski, J.: Hamiltonian neural networks. Advances in neural information processing systems **32** (2019)
12. Hazan, H., et al.: BindsNET: a machine learning-oriented spiking neural networks library in Python. Front. Neuroinform. **12**, 89 (2018)
13. Heidarpur, M., Ahmadi, A., Ahmadi, M., Azghadi, M.R.: CORDIC-SNN: On-FPGA STDP learning with Izhikevich neurons. IEEE Trans. Circuits Syst. I Regul. Pap. **66**(7), 2651–2661 (2019)

14. Hodgkin, A.L., Huxley, A.F.: A quantitative description of membrane current and its application to conduction and excitation in nerve. J. Physiol. **117**(4), 500 (1952)
15. Izhikevich, E.M.: Simple model of spiking neurons. IEEE Trans. Neural Networks **14**(6), 1569–1572 (2003)
16. Izhikevich, E.M.: Solving the distal reward problem through linkage of STDP and dopamine signaling. Cereb. Cortex **17**(10), 2443–2452 (2007)
17. James, D., Jerke, J.: Effective Hamiltonian theory and its applications in quantum information. Can. J. Phys. **85**(6), 625–632 (2007)
18. Lundstrom, B.N., Fairhall, A.L., Maravall, M.: Multiple timescale encoding of slowly varying whisker stimulus envelope in cortical and thalamic neurons in vivo. J. Neurosci. **30**(14), 5071–5077 (2010)
19. McLelland, D., Paulsen, O.: Neuronal oscillations and the rate-to-phase transform: mechanism, model and mutual information. J. Physiol. **587**(4), 769–785 (2009)
20. Morris, C., Lecar, H.: Voltage oscillations in the barnacle giant muscle fiber. Biophys. J . **35**(1), 193–213 (1981)
21. Nagumo, J., Arimoto, S., Yoshizawa, S.: An active pulse transmission line simulating nerve axon. Proc. IRE **50**(10), 2061–2070 (1962)
22. Salmon, R.: Hamiltonian fluid mechanics. Annu. Rev. Fluid Mech. **20**(1), 225–256 (1988)
23. Smith, G.D., Cox, C.L., Sherman, S.M., Rinzel, J.: Fourier analysis of sinusoidally driven thalamocortical relay neurons and a minimal integrate-and-fire-or-burst model. J. Neurophysiol. **83**(1), 588–610 (2000)
24. Troyer, T.W., Miller, K.D.: Physiological gain leads to high ISI variability in a simple model of a cortical regular spiking cell. Neural Comput. **9**(5), 971–983 (1997)
25. Van Vreeswijk, C., Sompolinsky, H.: Chaos in neuronal networks with balanced excitatory and inhibitory activity. Science **274**(5293), 1724–1726 (1996)
26. Yang, Y., Ma, J., Xu, Y., Jia, Y.: Energy dependence on discharge mode of Izhikevich neuron driven by external stimulus under electromagnetic induction. Cogn. Neurodyn. **15**, 265–277 (2021)

Cancer and Tissue Prediction Using Mutational Signatures in Highly Mutated Cancers

Julia Cordes and Jaime Davila(✉)

Department of Mathematics, Statistics and Computer Science, St Olaf College, Northfield, MN 55057, USA
davila3@stolaf.edu

Abstract. Around three to five percent of all cancers have unknown primary origin and identifying their tissue type is crucial for clinical purposes, especially for highly mutated cancers which can benefit from immunotherapy.

A mutational signature describes a distinct pattern of mutations caused by a specific mutagenic process and is usually associated with a specific tissue type. For example, tobacco exposure causes a high number of C to A mutations which are frequent in lung cancer, while UV light induces a high amount of CC to TT mutations, which occur in melanomas. The previous observation motivates the goal for our study, which is to use mutational signatures contributions to predict the cancer and tissue type of highly mutated tumor samples.

We use the Mutational Signatures v.3.3 cohort from the Catalogue of Somatic Mutations in Cancer (COSMIC) and consider only nine highly mutated cancer types resulting in a set of 1,477 samples. We remove artifactual signatures and consider frequently occurring signatures, resulting in a core set of twenty signatures which we used as features for our models.

We tested regression and tree-based models to predict cancer and tissue type. Random forests produced superior results predicting cancer type with an accuracy, specificity and sensitivity of 83.4%, 97.9%, and 76.4%, and predicting tissue type with an accuracy, specificity and sensitivity of 89.5%, 98.0%, and 84.8%. Our approach is limited in cancers that share similar mutational signatures, e.g. our lowest accuracy (76.7%) occurs in defective mismatch repair cases from endometrial, stomach, and colorectal cancers.

Keywords: Mutational Signatures · Random Forests · Cancer Tissue Prediction · Highly Mutated Cancers

1 Introduction

A mutational profile represents mutation frequencies that arise in a particular dinucleotide context. A catalogue of 63 well-characterized mutational signatures corresponding to different mutational processes is available from the Catalogue Of Somatic Mutations in Cancer (COSMIC) [1, 2]. The COSMIC mutational signatures were initially established using Non-negative Matrix Factorization (NMF), an approach that approximates mutational profiles as a linear combination of a set of mutational signatures [3]. It is well known that particular signatures occur more frequently in particular tissue types. For example, COSMIC mutational signature SBS4, which is caused by tobacco smoking and is characterized by an excess of C > T mutation, occurs frequently in lung cancer [4].

The tumor mutation burden (TMB) refers to the density of mutations in a cancer sample. High TMB cases have received increased attention since they are associated with better response rate to immunotherapy treatment [5]. Also, of importance, the increased number of mutations in high TMB cases results in a better estimation of the abundance of known mutational signatures, even when using clinical genomic panels which cover a small fraction of the genome.

About three to five percent of neoplasms are cancers of unknown primary (CUP) which are usually identified by the presence of metastatic disease with no identified primary tumor. Predicting the tissue of origin of CUPs is of clinical importance since precise treatment depends on the identification of the primary tumor site of CUPs [6].

Existing methods for cancer type prediction rely on gene expression quantification using transcriptome microarrays or RNA-sequencing (RNA-seq), however the use of gene expression approaches is limited in the clinical setting [7, 8]. Importantly, existing clinical cancer genomic tests provide mutational signature information, supporting their utilization in cases of CUP [9].

As a first step towards the goal of using mutational signatures to predict cancer and tissue type in clinical samples with high TMB, we leverage the COSMIC dataset. Using the available mutational signature contributions from high TMB cases in COSMIC, we set up to establish the feasibility of predicting cancer and tissue types using machine learning approaches.

2 Cohort Selection

We utilize the mutational profiles from COSMIC v 3.3, which contains the estimated contribution for 63 mutational signatures across 9,337 cancer samples. We exclude the contribution of 18 mutational signatures that are likely artifactual, obtaining a reliable set of 45 mutational signatures. We tally the contribution of these 45 mutational signatures and calculate the Tumor Mutational Burden (TMB) for each sample. Samples with a TMB higher than 10 mutations/mega base are considered highly mutated (Fig. 1). About 19% (1,763 out of 9,337) of the samples in our cohort are highly mutated.

Since we are interested in predicting cancer and tissue type for highly mutated samples, we decided to consider only cancer types where over 20% of the samples were highly mutated (Fig. 2). Our resulting cohort has 9 out of the 43 total cancer types, and their types and sample sizes are listed in Table 1. This cohort contains about 15% (1,477 out 9,337) of the samples from the original COSMIC cohort.

3 Feature Selection

The original dataset contains the contribution of 45 mutational signatures. However not all these mutational signatures are active in highly mutated samples. In order to consider a relevant and reliable set of signatures for the highly mutated cases, we consider only

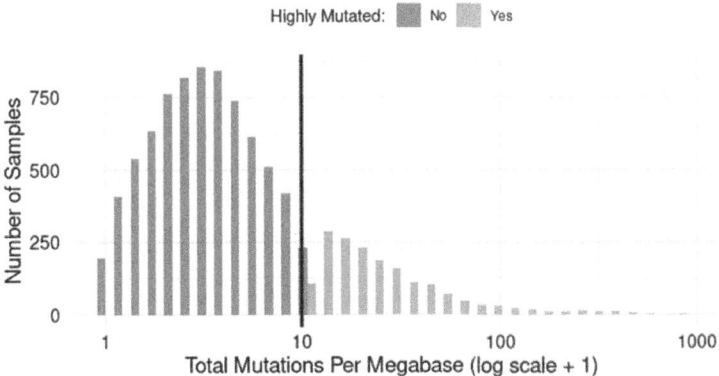

Fig. 1. Distribution of Tumor Mutational Burden (TMB) using log scale across COSMIC Mutational Signatures v.3.3 cohort. Samples with a high TMB (>10 mut/Mb) are shown in yellow and samples with low mutation burden are shown in blue

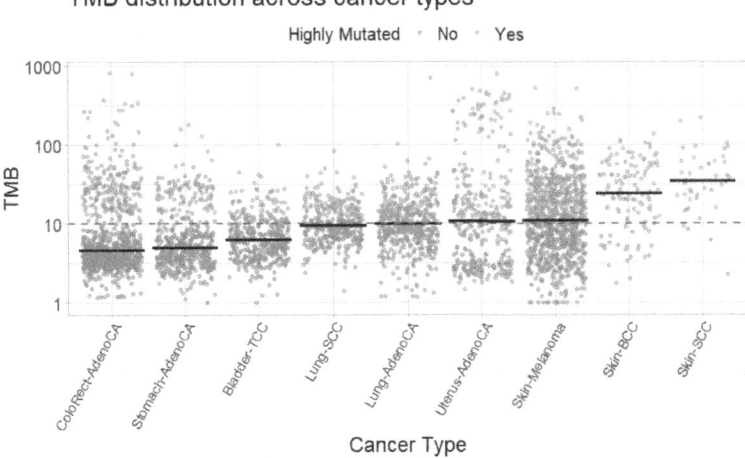

Fig. 2. Distribution of Tumor Mutation Burden across nine highly mutated cancer types. A cancer type was considered highly mutated if over 20% of its samples were high TMB

mutational signatures that were present in at least 10 samples and that when present, contributed to at least 20% of the mutational profiles on average, in at least one of the nine selected cancer types (Fig. 3).

The resulting set of twenty signatures can be grouped according to the putative process that originates them and the results are summarized in Table 2. Our resulting set of twenty signatures can be divided into roughly nine processes which include Mismatch Repair Deficiency, Tobacco Smoking, UV Light Exposure, and APOBEC among others (Table 2).

Table 1. Included cancer types, abbreviations, and sample sizes

Cancer Type	Acronym	Sample Size
Bladder Urothelial Carcinoma	TCC	83
Lung Adenocarcinoma	LUAD	200
Lung Squamous Cell Carcinoma	LUSC	133
Colon Adenocarcinoma	COAD	215
Stomach Adenocarcinoma	STAD	114
Skin Basal Cell Carcinoma	BCC	93
Skin Cutaneous Melanoma	SKCM	430
Cutaneous Squamous Cell Carcinoma	cSCC	38

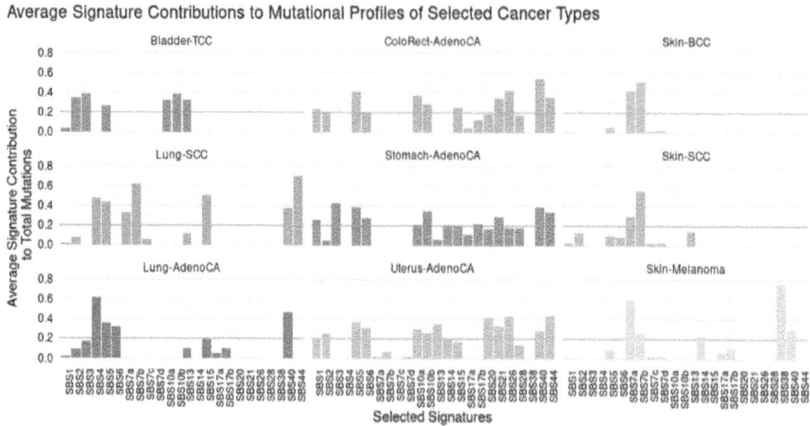

Fig. 3. Average contribution of selected signatures across nine cancer types. Only signatures that were present in at least 10 samples and that had an average contribution of at least 20% in a cancer type were selected. The grey line indicates the 20% threshold

4 Cancer Type Prediction

Initially, we were interested in predicting cancer type based on the contributions from each of the 20 selected mutational signatures. To do this we tested four different machine learning approaches: multinomial logistic regression, LASSO regression, random forests, and boosting. More details on our approach follows.

We use R version 4.3.0 and the tidymodels framework version 1.1.1 [10]. We use the multinomial logistic regression implementation from nnet version 7.3-18 [11]. Our LASSO regression model uses the glmnet 4.1-8 library, and we optimized the penalty parameter using 10-fold cross validation [12]. We use the random forest implementation from ranger 0.15.1, and optimized the parameters *num_trees* and *mtry* using 10-fold cross validation [13]. Our boosting models use the xgboost version 1.7.5.1 and we optimized the learning rate using 10-fold cross validation [14]. We train our models using 80% of

Table 2. Included Signatures

Proposed Process	Signatures
Mismatch Repair Deficiency	SBS6, SBS15, SBS21, SBS26, SBS44
Defective Recombination Repair	SBS3
Tobacco Smoking	SBS4
UV Light Exposure	SBS7a, SBS7b, SBS38
Aging	SBS1, SBS5
APOBEC	SBS2, SBS13
POLE	SBS10a, SBS10b, SBS14
POLD	SBS20
Unknown	SBS17b, SBS40

our data and calculate their performance metrics including accuracy, kappa, sensitivity and specificity, using a testing dataset with 20% of our data.

4.1 Cancer Type Prediction Using Absolute Contributions from Mutational Signatures

On our first approach we predict cancer type based on the absolute number of mutations contributed by each of the 20 selected mutational signatures. The performance metrics of our approach are summarized in Table 3.

Table 3. Performance metrics for cancer type prediction using absolute number of mutations from mutational signatures for cross-validation and testing datasets

Model	Accuracy	Kappa	Sensitivity	Specificity
Cross-validation				
LASSO	72.8%	67.3%	67.2%	96.5%
Random Forest	81.7%	77.9%	74.3%	97.6%
Boosting	81.1%	77.4%	75.4%	97.6%
Testing				
Multinomial Logistic	69.9%	63.5%	56.8%	96.0%
LASSO	67.6%	71.2%	68.9%	96.9%
Random Forest	80.7%	76.9%	70.1%	97.5%
Boosting	79.4%	75.3%	71.2%	97.3%

Using our testing dataset, our models ranged in accuracy from 67.6% using our LASSO approach to around 80% using random forests and boosting. All of our models have specificity higher than 96%, but their sensitivity ranges from 56.8% (multinomial logistic regression) to slightly over 70% for random forests and boosting.

Our tree-based models (random forest and boosting) have similar performance characteristics, however we settle on using random forests given their slightly better accuracy and kappa value. We visualize the misclassification pattern from our random forest using the confusion matrix (Fig. 4). The highest number of misclassifications occurs in confusing two types of lung cancer (LUAD and LUSC). This is likely caused by both cancers sharing a high presence of SBS4 (Fig. 3), which is caused by tobacco smoking [4].

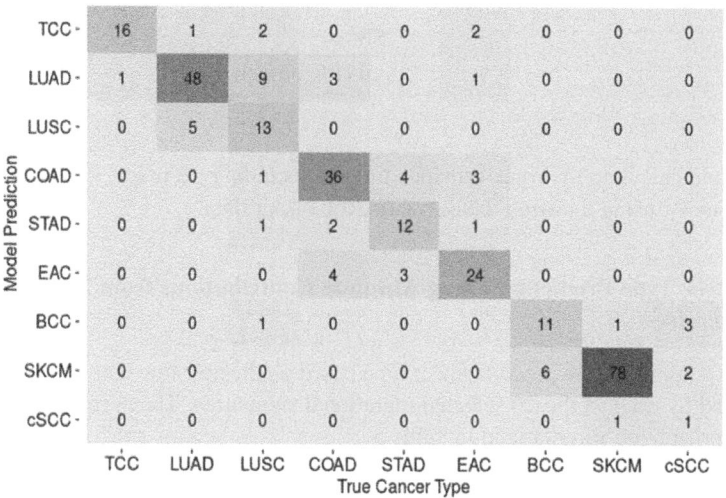

Fig. 4. Confusion matrix for random forest cancer type classification using absolute number of mutations per mutational signature type

4.2 Cancer Type Prediction Using Percentual Mutational Signature Contribution

Encouraged by our first set of results, we build a new set of models, this time using the percentual contributions for each of the selected 20 mutational signatures. The performance metrics of our approach are summarized in Table 4.

When compared with our initial approach we obtain higher performance metrics across all the tested models. Our accuracy ranges from about 75% in our LASSO and multinomial regression to about 83% in our tree-based models. All of our approaches have high specificity values around 97%. Our sensitivity ranges from about 67% in our LASSO regression model to over 76% in our random forest model. Since our random forest obtained better results across the board we decided to visualize the corresponding confusion (Fig. 5).

As in Fig. 4, our highest number of misclassifications occurs between two different lung cancer types (LUAD and LUSC), however we obtain only six errors when compared with the nine of our first approach. The shape of our confusion matrix suggests

Table 4. Performance metric for cancer type prediction using percentual signature contributions

Model	Accuracy	Kappa	Sensitivity	Specificity
Cross-validation				
LASSO	74.6%	69.5%	68.3%	96.8%
Random Forest	81.2%	77.4%	76.5%	97.6%
Boosting	80.6%	76.8%	76.8%	97.5%
Testing				
Multinomial Logistic	75.7%	70.9%	68.9%	96.9%
LASSO	75.3%	70.5%	67.2%	96.8%
Random Forest	83.4%	80.3%	76.4%	97.9%
Boosting	82.8%	79.5%	75.7%	97.8%

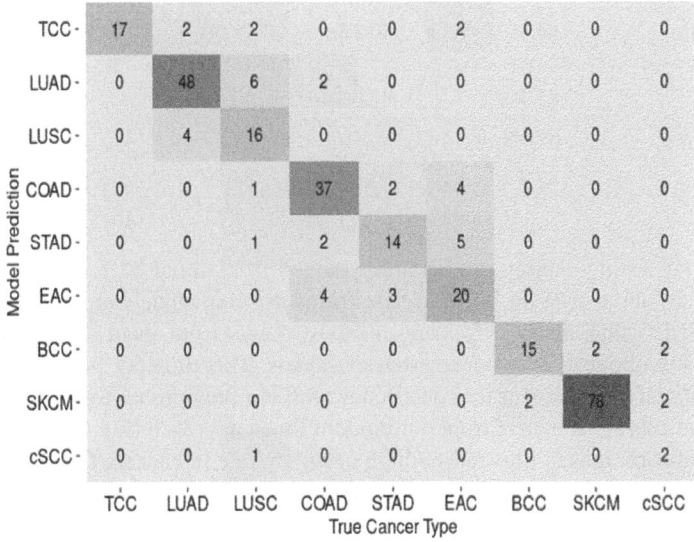

Fig. 5. Confusion matrix for random forest classification of cancer types using percentual signature contribution

that misclassification usually occurs across adjacent cancer types. It is worth noticing that cancers are sorted according to their active mutational signature types (Fig. 3). In most cases such mutational signatures coincide with tissue type. For example, UV light signature processes are active in BCC, SKCM, and cSCC (skin cancers), while tobacco smoking signatures are present in LUSC and LUAD (lung cancers) [15, 16]. This observation motivates our next question which is the prediction of the tissue type based on the mutational signature contributions.

5 Tissue Type Prediction

In this section we explore the question of predicting the tissue type based on the percentual contributions of the twenty selected signatures. We group our cancer types based on their tissue of origin: skin (BCC, SKCM, and cSCC), lung (LUAD and LUSC), bladder (TCC), colorectal (COAD), stomach (STAD), and uterus (EAC). The performance metrics of our approach are summarized in Table 5.

Table 5. Tissue type prediction performance characteristics

Model	Accuracy	Kappa	Sensitivity	Specificity
Cross-validation				
LASSO	84.2%	79.2%	77.0%	97.0%
Random Forest	88.9%	85.3%	84.3%	97.9%
Boosting	87.8%	83.9%	82.2%	97.7%
Testing				
Multinomial Logistic	84.8%	80.1%	78.5%	97.0%
LASSO	85.1%	81.0%	79.0%	97.2%
Random Forest	89.5%	86.3%	84.8%	98.0%
Boosting	89.9%	86.7%	84.9%	98.0%

Using our testing dataset, our accuracy ranged from about 85% in our regression approaches to about 89% in our tree-based models. Our models are highly specific, with specificity values above 97%. Our sensitivity ranges from about 78% in our regression models to about 85% in our tree-based models. This time our boosting approach came slightly on top, however for consistency with our previous sections we decided to visualize the confusion matrix from our random forest approach (Fig. 6).

The confusion matrix illustrates a high accuracy rate in cancers from the skin and bladder. The most misclassified tissue is uterus, which is confused with colorectal six times and with stomach five times. This can be explained as processes related to mismatch defect repair and POLE activation are common across gastric and endometrial cancers (Fig. 3), resulting in similar patterns of mutational signature use [17, 18].

6 Discussion

In the current work, we explore the problems of prediction of cancer type and tissue in highly mutated cases using machine learning models that use mutational signature contributions. Our most accurate approaches were tree-based models using percentual mutational signature contributions. In the cancer type prediction setting our random forest approach achieved accuracy, specificity, and sensitivity of 83.4%, 97.9%, and 76.4%. In the tissue prediction setting our random forest model achieved accuracy, specificity and sensitivity of 89.5%, 98.0%, and 84.8%.

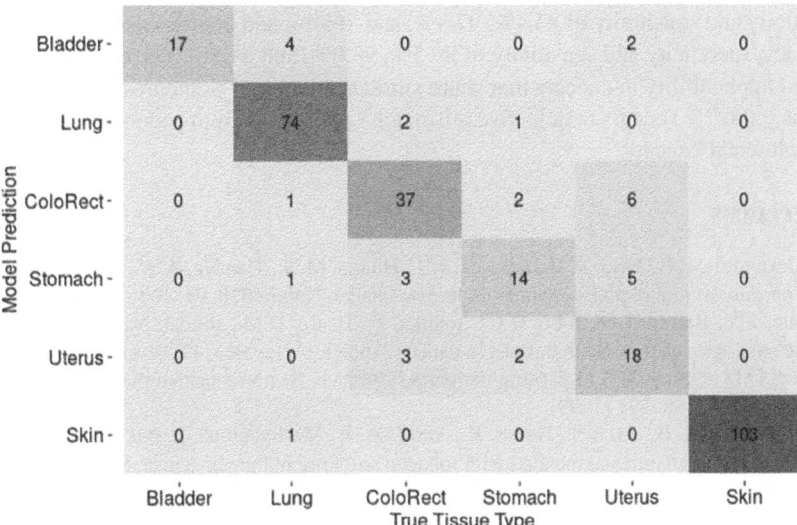

Fig. 6. Confusion matrix for random forest classification of tissue types using signature percentual contribution

Although our approaches were highly specific, our models had challenges when trying to tell apart cancer or tissue types that shared similar mutational signatures contributions. For example, in the cancer prediction setting, our model misclassified about 20% of the Lung Squamous Cell Carcinoma (LUSC) as Lung Adenocarcinomas (LUAD) due to both cases having a high contribution from signature SBS4, which is caused by tobacco smoking [15].

Similarly, in the tissue classification setting our approach misclassified about 30% of the cancers from the uterus, as cancers from either colorectal or stomach, since these cancers share common signatures associated with microsatellite instability/defective mismatch repair, and POLE hypermutation [17].

These limitations argue for the use of additional features which might include variables like age, sex, metastatic site, and mutational signature clonality among others, to create models with higher accuracy.

Another limitation of our approach is that it depends on the available decomposition of mutational profiles into a canonical set of signatures from COSMIC, as well as the use of thresholds for selecting highly mutated samples, highly mutated cancer types, and mutational signatures, as described in Sects. 2 and 3. This suggests further work where some of these limitations can be addressed by leveraging the variable selection properties of our random forest models using the original sample mutational profiles.

7 Conclusions

In this paper we tested regression and tree-based models to predict cancer and tissue type. Our most accurate approach involved random forests using percentual mutational signature contributions. Our approach is able to predict cancer type with an accuracy, specificity and sensitivity of 83.4%, 97.9%, and 76.4%, and predicts tissue type with an accuracy, specificity and sensitivity of 89.5%, 98.0%, and 84.8%. The current work has limited applicability in cancers that share similar mutational signatures, e.g. our lowest accuracy (76.7%) occurs in defective mismatch repair cases from endometrial, stomach, and colorectal cancers.

References

1. Alexandrov, L.B., Kim, J., Haradhvala, N.J., Huang, M.N., Tian Ng, A.W., Wu, Y., et al.: The repertoire of mutational signatures in human cancer. Nature **578**, 94–101 (2020)
2. Tate, J.G., Bamford, S., Jubb, H.C., Sondka, Z., Beare, D.M., Bindal, N., et al.: COSMIC: the catalogue of somatic mutations in cancer. Nucleic Acids Res. **47**, D941-d947 (2019)
3. Lee, D.D., Seung, H.S.: Learning the parts of objects by non-negative matrix factorization. Nature **401**, 788–791 (1999)
4. Alexandrov, L.B., Ju, Y.S., Haase, K., Van Loo, P., Martincorena, I., Nik-Zainal, S., et al.: Mutational signatures associated with tobacco smoking in human cancer. Science **354**, 618–622 (2016)
5. Goodman, A.M., Sokol, E.S., Frampton, G.M., Lippman, S.M., Kurzrock, R.: Microsatellite-stable tumors with high mutational burden benefit from immunotherapy. Cancer Immunol. Res. **7**, 1570–1573 (2019)
6. Qaseem, A., Usman, N., Jayaraj, J.S., Janapala, R.N., Kashif, T.: Cancer of unknown primary: a review on clinical guidelines in the development and targeted management of patients with the unknown primary site. Cureus. **11**, e5552 (2019)
7. He, B., Zhang, Y., Zhou, Z., Wang, B., Liang, Y., Lang, J., et al.: A neural network frame-work for predicting the tissue-of-origin of 15 common cancer types based on RNA-Seq data. Front Bioeng Biotechnol. **8**, 737 (2020)
8. Divate, M., Tyagi, A., Richard, D.J., Prasad, P.A., Gowda, H., Nagaraj, S.H.: Deep learning-based pan-cancer classification model reveals tissue-of-origin specific gene expression signatures. Cancers (2022)
9. Milbury, C.A., Creeden, J., Yip, W.-K., Smith, D.L., Pattani, V., Maxwell, K., et al.: Clinical and analytical validation of FoundationOne®CDx, a comprehensive genomic profiling assay for solid tumors. PLoS ONE **17**, e0264138 (2022)
10. Kuhn, M., Silge, J.: Tidy Modeling with R, 1st edn. O'Reilly Media (2022)
11. Venables, W.N., Ripley, B.D.: Modern Applied Statistics With S. Springer, New York (2002). https://doi.org/10.1007/978-0-387-21706-2
12. Friedman, J.H., Hastie, T., Tibshirani, R.: Regularization paths for generalized linear models via coordinate descent. J. Stat. Softw. **33**, 1–22 (2010)
13. Wright, M.N., Ziegler, A.: Ranger: a fast implementation of random forests for high dimensional data in C++ and R. J. Stat. Softw. **77**, 1–17 (2017)
14. Chen, T., Guestrin, C.: XGBoost. In: Proceedings of the 22nd ACM SIGKDD Inter-national Conference on Knowledge Discovery and Data Mining. ACM (2016)
15. Kucab, J.E., Zou, X., Morganella, S., Joel, M., Nanda, A.S., Nagy, E., et al.: A compendium of mutational signatures of environmental agents. Cell **177**, 821-836.e16 (2019)
16. Nik-Zainal, S., Kucab, J.E., Morganella, S., Glodzik, D., Alexandrov, L.B., Arlt, V.M., et al.: The genome as a record of environmental exposure. Mutagenesis **30**, 763–770 (2015)
17. DiGuardo, M.A., Davila, J.I., Jackson, R.A., Nair, A.A., Fadra, N., Minn, K.T., et al.: RNA-Seq reveals differences in expressed tumor mutation burden in colorectal and endometrial cancers with and without defective DNA-mismatch repair. J. Mol. Diagn. **23**, 555–564 (2021)
18. Meier, B., Volkova, N.V., Hong, Y., Schofield, P., Campbell, P.J., Gerstung, M., et al.: Mutational signatures of DNA mismatch repair deficiency in C. elegans and human cancers. Genome Res. **28**, 666–75 (2018)

On the Hardness of Wildcard Pattern Matching on de Bruijn Graphs

Arnab Ganguly[1], Daniel Gibney[2](✉), Arghya Kusum Das[3], and Sharma V. Thankachan[4]

[1] University of Wisconsin - Whitewater, Whitewater, WI 53190, USA
gangulya@uww.edu
[2] University of Texas at Dallas, Richardson, TX 75080, USA
daniel.gibney@utdallas.edu
[3] University of Alaska Fairbanks, Fairbanks, AK 99775, USA
akdas@alaska.edu
[4] North Carolina State University, Raleigh, NC 27695, USA
svalliy@ncsu.edu

Abstract. In the pattern matching on labeled graphs problem, given an edge labeled graph $G = (V, E)$ and a string P, one seeks to identify if there exists a walk in the graph whose concatenation of edge labels (approximately) matches P. This is an elementary subproblem for utilizing genome graphs to represent collections of genetic sequences where patterns arise as reads in the sequencing data. Unfortunately, for general graphs, it is known that an algorithm running in $O(|E||P|^{1-\varepsilon}+|E|^{1-\varepsilon}|P|)$ time for constant $\varepsilon > 0$ is not possible under the Strong Exponential Time Hypothesis (SETH). De Bruijn graphs provide a valuable exception, allowing for a path exactly matching a pattern to be found in $O(|E|+|P|)$ for constant-sized alphabets. This property has led de Bruijn graphs to be applied as indexes in the popular tool vg-toolkit. In this work, we consider the case where wildcards (that match with any edge label) are included in the pattern, and the graph is a de Bruijn graph. We demonstrate that adding these wildcards to the pattern is enough to again prove quadratic lower bounds conditioned on SETH for pattern matching on de Bruijn graphs, even when restricted to alphabets of size at most three and k-mer length $\Theta(\log |V|)$.

Keywords: Pattern Matching · Genome Graphs · de Bruijn Graphs

De Bruijn graphs are an important tool in Computational Biology. Given a set of strings of length k each, called k-mers, the de Bruijn graph of this set consists of a vertex for each k-mer and a directed edge from vertex v to v' if the k-mer for v is of the form αL and the vertex v' is of the form $L\beta$ for two symbols α and β. We consider the edge from v to v' as being labeled with β. An elementary problem for de Bruijn graphs, and more generally all edge-labeled graphs, is given the graph G and a string P, find a walk in G whose concatenation of edge labels matches or (approximately matches) P. For general graphs, hardness results conditioned on the hardness of the Orthogonal Vectors Problem (OVP) [6] (discussed more below), and likely more general assumptions in circuit complexity [8], state that determining the existence of a path exactly matching P cannot be done in time $O(|E|^{1-\varepsilon}|P| + |E||P|^{1-\varepsilon})$ for any constant $\varepsilon > 0$. These results hold even when restricted to DAGs of maximum total degree at most three and imply the hardness for the cases where some number of mismatches or edits are allowed to the pattern. It was also previously discovered that the Longest Common Subsequence problem can be reduced to pattern matching on labeled graphs with mismatches, which implies some hardness not captured by OVP, for example, hardness even for quantum computers [5].

For de Bruijn graphs, the exact matching problem can be solved in time $O(|E| + |P|)$ for constant-sized alphabets [7]. This linear time pattern matching capability has led to de Bruijn graphs being used as an indexing data structure for seed-and-extend techniques on larger genome graphs in the popular software tool vg-toolkit [11]. It was recently shown that these results do not extend to approximate pattern matching on de Bruijn graphs when a specified number of mismatches are allowed between the pattern and the edge labels of the graph, as a quadratic lower bound based on OVP was shown [9]. This hardness result stands in contrast to both linear time exact matching on de Bruijn graphs mentioned above and matching with mismatches on texts of length n (equivalent to the case where G is a path with n edges) where there exists a $O(n\sqrt{|P|})$ time algorithm [1].

In the wild card matching problem considered here, the pattern P can contain wildcard symbols # that can match any edge label in G, whereas non-wildcard symbols must match the aligned edge label. When G is equivalent to a string, i.e., G is a path with n edges, this problem can be solved in time $O(n \log |P|)$ [3,4], suggesting it might be an easier problem on de Bruijn graphs than pattern matching with mismatches. Furthermore, being able to perform such matches would also have immediate applicability to read mapping as it is often the case that real-world read sets contain unspecified bases indicated by 'N' where the nucleic acid symbol could be 'A', 'T', 'C', or 'G'. The result presented in this paper, summarized in Theorem 1, indicates that this is not the case and that a strongly subquadratic time algorithm is not possible assuming quadratic hardness of OVP.

Theorem 1. *Given a de Bruijn graph $G = (V, E)$ and string P containing wildcards, determining if there exists a walk in G whose concatenation of edge labels matches P cannot be done in time $O(|E|^{1-\varepsilon}|P| + |E||P|^{1-\varepsilon})$ for any constant $\varepsilon > 0$ under the assumption of OVP-hardness. Moreover, this holds for de Bruijn graphs that are acyclic, have edge alphabets of size 3 or greater, are constructed from k-mers with $k = \Theta(\log |V|)$, and pattern alphabets of size 2 or greater (not including the wildcard symbol).*

An algorithm with $O(|E||P|)$ time complexity is achieved by a straightforward modification of the alignment graph techniques by Amir et al. [2], or dynamic programming formulation by Navarro [10]. One simply weighs the edges between the *alignment graph* at each level corresponding to a wildcard with weight 0. We refer the reader to [2] for more details on the alignment graph. Although we prove the result for edge label alphabets of size three or greater, this is mainly to simplify the presentation, and we believe that the result holds for binary edge label alphabets as well. We conclude by discussing some alternative, theoretically interesting ways to parameterize the problem.

Reducing OVP to Wildcard Pattern Matching on de Bruijn Graphs

The Orthogonal Vectors Problem (OVP) is defined as follows: given two sets of d-dimensional binary vectors $\mathcal{A}$ and $\mathcal{B}$, determine if there exist vectors $a_i \in \mathcal{A}$ and $b_j \in \mathcal{B}$ such that a_i and b_j are orthogonal. Here, we consider $|\mathcal{A}| = |\mathcal{B}| = n$ and $d = \Omega(\log n)$. Williams showed that an algorithm for OVP running in time $O(d^{\Theta(1)} n^{2-\varepsilon})$ for any constant $\varepsilon > 0$ would violate the Strong Exponential Time Hypothesis (SETH) [12]. Alternatively, the possibly weaker assumption that OVP can not be solved in strongly subquadratic, $O(d^{\Theta(1)} n^{2-\varepsilon})$ for constant $\varepsilon > 0$, time can be used directly. OVP-hardness has been the basis for numerous fine-grained hardness results, the most closely related to this work being the exact pattern matching on general graphs result by Equi et al. [6]. The major difference in this work is that the graph we construct is a de Bruijn graph.

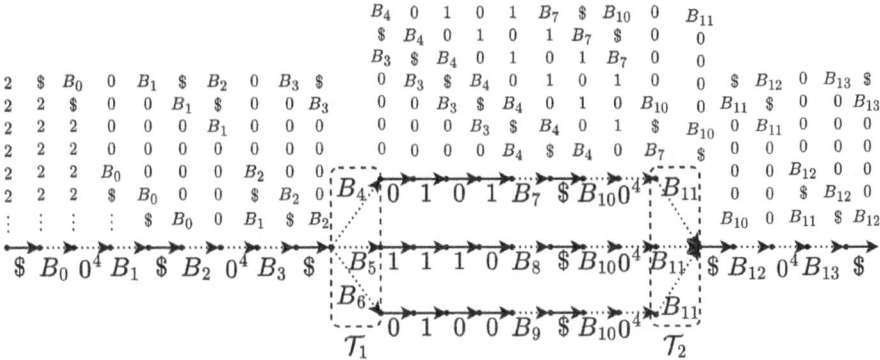

Fig. 1. The de Bruijn graph constructed from the set of binary vectors $\mathcal{A} = \{0101, 1110, 0100\}$. Dashed edges represent multiple character labeled edges. k-mer labels can be read from bottom to top. Only the k-mers for the top path following T_1 are shown. Here $d = 4$, $|B| = \lceil \log(6 \cdot 3 - 4) \rceil = 4$ and $k = d + 1 + 2|B| = 13$ If $\mathcal{B} = \{1001, 0111, 1011\}$, then $P = \$\#^{|B|}0\#\#0\#^{|B|}\$\#^{|B|}\#000\#^{|B|}\$\#^{|B|}0\#00\#^{|B|}\$$.

The Reduction. The OVP vector sets $\mathcal{A}$ and $\mathcal{B}$ are 0-indexed to facilitate binary encodings. The vector components are 1-indexed. Let B_i denote the binary encoding of i using $\lceil \log(6n-4) \rceil$ bits for $i \in [0\mathinner{.\,.} 6n-3]$. Let $|B| = \lceil \log(6n-4) \rceil$ and $k = d + 1 + 2|B|$. We create our graph $G = (V, E)$ as follows: we start with a path having edge labels $(\bigcirc_{i=0}^{n-2} \$ B_{2i} 0^d B_{2i+1})$, where $\bigcirc$ denotes concatenation.

We next construct the trie T_1 for the set of strings $\{B_i \mid i \in [2n-2..3n-3]\}$ and direct the edges from the root of T_1 to its leaves. The root-to-leaf paths of T_1 are labeled with the corresponding B_i for each leaf. We bring the root of T_1 into correspondence with the last vertex on the previously created path. For the leaf of T_1 corresponding to string $\$B_{2n-2+i}$ we attach a new path labeled with $a_i[1]a_i[2]\ldots a_i[d]B_{3n-2+i}\$B_{4n-2}0^d$ for $i \in [0..n-1]$. Next, we construct a trie T_2 for $\{(B_i)^R \mid i \in [3n-2..4n-3]\}$ with edges oriented towards the root. We bring the last vertex on the path labeled $a_i[1]a_i[2]\ldots a_i[d]B_{3n-2+i}\$B_{4n-2}0^d$ into correspondence with the leaf in T_2 corresponding to $(B_{3n-2+i})^R$ for $i \in [0..n-1]$. We then label every leaf-to-root path in T_2 with B_{4n-1}. Finally, we create the path with edge labels $(\bigcirc_{i=0}^{n-3} \$B_{4n+2i}0^d B_{4n+2i+1})\$$ and bring the first vertex of this path into correspondence with the root of T_2. See Fig. 1.

For the pattern P, we define $f(x) = 0$ if $x = 1$ and $f(x) = \#$ if $x = 0$. Then $P = \left(\bigcirc_{i=0}^{n-1} \left(\$\#^{|B|} \left(\bigcirc_{j=1}^{d} f(b_i[j])\right) \#^{|B|}\right)\right)\$$.

Correctness

Lemma 1. *The graph constructed as above is a de Bruijn graph for k-mer size $k = d + 1 + 2|B|$.*

Proof. **For all vertex $v \in V$, every path with k edges ending at v matches the same length k string:** By construction, the only vertices where there are multiple length k paths ending at them are ones in T_2 or on the path starting at the root of T_2 and at most k edges from its root. It suffices to show that this property holds for vertices in T_2 since all edge labels following vertices in T_2 match. Suppose for a vertex v in T_2 it has two paths ending at it matching $B_i[x..|B|]\$B_{4n-2}0^d B_{4n-1}[1..y]$ and $B_j[x..|B|]\$B_{4n-2}0^d B_{4n-1}[1..y]$. Then the depth of v relative to the root of T_2 is $|B| - y$. Suppose for the sake of contradiction that B_i and B_j have a rightmost differing character at position $z \in [x..|B|]$. Then the longest common suffix of $B_i[x..|B|]$ and $B_j[x..|B|]$ is of length $|B| - z$. By the construction of T_2, the depth of v in T_2 must be less or equal to the length of the longest common suffix of B_i and B_j, implying $|B| - y \leq |B| - z$, or $z \leq y$. However, we also have $k = d + 1 + 2|B| = (|B| - x + 1) + 1 + |B| + d + y$ implying $y = x - 1$. Combined, this implies $z \leq x - 1$, contradicting that $z \in [x..|B|]$.

For the vertices v where the longest path ending at v has fewer than k edges, we consider the vertex label for v as having a suffix consisting of the labels of the longest path ending at v followed by all 2's. For a given vertex v, it is now well defined to call the string matching any path with k edges ending at v its *vertex label*. We will next show that the set of vertex labels is equal to the set of k-mers whose de Bruijn graph is G.

Each vertex Label Occurs Only Once: If a vertex label contains no $\$$, it contains 2's, and the uniqueness of each vertex label containing 2's is immediate. For a vertex v with a label containing a $\$$ symbol, suppose that $\$$ symbol occurs at index $x \in [1..k]$ in its label. Then, the vertex label can only potentially occur again at a vertex v' where its $\$$ is also at index x. If $x \in [1 .. k - |B|]$, then the $\$$ in the label for v is followed by some encoding B_i. Momentarily excluding the case where $i = 4n - 2$ (B_{10} in Fig. 1), in the label for v' it is followed by some B_j where $i \neq j$ and hence the labels do not match.

- In the case where $x \in [k - |B| + 1 .. k]$ as long as v and v' are not in $\mathcal{T}_1$ the encoding $B_{i'}$ preceding the $\$$ symbol in the label for v and the encoding $B_{i''}$ preceding the $\$$ symbol in the label for v' are distinct by construction whenever $v \neq v'$. If both v and v' are in $\mathcal{T}_1$, then since $\mathcal{T}_1$ is a trie for the suffixes of their vertex labels, they are distinct vertices if and only if the suffixes of their labels differ.
- In the case where $x \in [1 .. k - |B|]$ and $i = 4n - 2$, suppose the vertex label for v is $B_{i'}[x..|B|]\$B_{4n-2}0^d B_{4n-1}[1..y]$ and the vertex label for v' is $B_{i''}[x..|B|]\$B_{4n-2}0^d B_{4n-1}[1..y]$ and the right most difference between $B_{i'}$ and $B_{i''}$ is at some $z \in [1..x-1]$ (otherwise the labels are distinct), then the leaves in $\mathcal{T}_2$ for the path with $B_{i'}$ and the path with $B_{i''}$ have a lowest common ancestor at depth $|B| - z \geq |B| - x + 1$ in $\mathcal{T}_2$ and the depth of v and v' in $\mathcal{T}_2$ is equal to $|B| - y = |B| - x + 1$. Since v and v' are both ancestors of leaves for $B_{i'}$ and $B_{i''}$ and the lowest common ancestor of the leaves for $B_{i'}$ and $B_{i''}$ has greater depth than v and v', we have v and v' must be the same vertex in $\mathcal{T}_2$.

There are No Missing Edges: We claim that no edges are missing from any vertex v with label αL to vertex v' with label $L\beta$. Suppose the $\$$ is at position $x \in [1..k]$ in the vertex label for v. If $x > 1$, the $\$$ symbol must be at position $x - 1$ in the vertex label for v', and if $x = 1$, at position k. In the case where $x > 1$ and $\alpha \neq \$$, there being no edge from v to v' in G implies that L contains at least one complete binary encoding that is not found in the length $k-1$ prefix of the vertex label for v'. When $x = 1$, then $\alpha = \$$ and there being no edge from v to v' in G implies that L contains two binary encodings distinct from those in the $k - 1$ length prefix of the vertex label for v'. □

Lemma 2. *There is an orthogonal pair of vectors $a_i \in \mathcal{A}$ and $b_j \in \mathcal{B}$ if and only if a wildcard match of P to G exists.*

Proof. First, suppose such an orthogonal pair $a_i \in \mathcal{A}$ and $b_j \in \mathcal{B}$ exists. We align the substring $\$\#^{|B|}(\bigcirc_{h=1}^{d} f(b_j[h]))\#^{|B|}$ of P starting at the $\$$ in G immediately preceding $\mathcal{T}_1$ with the path matching $\$B_{2n-2+i}(\bigcirc_{h=1}^{d} a_i[h])B_{3n-2+i}$. The remaining prefix and suffix of P are aligned to the only possible remaining paths, which have lengths at least as long as the remaining prefix and suffix of P, respectively. This alignment is possible since, due to the orthogonality of a_i and b_j, $a_i[h] = 1$ implies $b_j[h] = 0$. Then, from the definition of f, $f(b_j[h]) = \#$; so the corresponding symbol in P and edge label can be aligned. Also by the orthogonality of a_i and b_j, if $b_j[h] = 1$ then $a_i[h] = 0$ and $f(b_j[h]) = 0$ and the edge label for $a_i[h]$ is 0. If both components of $a_i[h] = b_j[h] = 0$, the wildcard that 0 is mapped to under f again allows an alignment.

Conversely, if there exists a alignment of P to G, then at least one substring of P of the form $\$\#^{|B|}(\bigcirc_{h=1}^{d} f(b_j[h]))\#^{|B|}$ is aligned starting at the $\$$ in G immediately preceding T_1. On the path matching $\$B_{2n-2+i}(\bigcirc_{h=1}^{d} a_i[h])B_{3n-2+i}$ that this substring of P is aligned to, each edge labeled with 1 due to $a_i[h] = 1$ must be aligned with a $\#$ symbol in P (since P contains no 1's). The $\#$ symbol corresponds to $b_j[h] = 0$. Hence, $a_i[h] = 1$ implies $b_j[h] = 0$ and a_i and b_j are orthogonal vectors. □

To conclude the reduction, we note that $|E|$, $|V|$ and $|P|$ are all $\Theta(n(d + \log n))$. Hence, an algorithm for wildcard matching on de Bruijn graphs running in $O(|E||P|^{1-\varepsilon} + |E|^{1-\varepsilon}|P|)$ for some constant $\varepsilon > 0$ would imply an algorithm for OVP running in $O(d^{2-\varepsilon} n^{2-\varepsilon} \log^{2-\varepsilon} n)$ time, contradicting OVP-hardness.

Discussion

The reduction presented in this work utilized $\tilde{\Theta}(|P|)$ wildcard characters in the pattern P. An interesting direction is algorithms for wildcard matching on de Bruijn graphs whose time complexity is parameterized by the number of wildcards in the pattern. Trivially, this can be accomplished in $O(w^\sigma(|P| + |E|))$ where w is the number of wildcards and σ is the size of the edge label alphabet. Can we do better and provide an algorithm with a time complexity that transitions smoothly from $O(|P| + |E|)$ for $w = 0$ to $O(|E||P|)$ as w becomes large?

Acknowledgement. S. Thankachan is partially supported by the U.S. National Science Foundation (NSF) award CCF-2316691.

References

1. Abrahamson, K.R.: Generalized string matching. SIAM J. Comput. **16**(6), 1039–1051 (1987)
2. Amir, A., Lewenstein, M., Lewenstein, N.: Pattern matching in hypertext. J. Algorithms **35**(1), 82–99 (2000)
3. Clifford, P., Clifford, R.: Simple deterministic wildcard matching. Inf. Process. Lett. **101**(2), 53–54 (2007)
4. Cole, R., Hariharan, R.: Verifying candidate matches in sparse and wildcard matching. In: Reif, J.H. (ed.) Proceedings on 34th Annual ACM Symposium on Theory of Computing, 19–21 May 2002, Montréal, Québec, Canada, pp. 592–601. ACM (2002)
5. Darbari, P., Gibney, D., Thankachan, S.V.: Quantum time complexity and algorithms for pattern matching on labeled graphs. In: Arroyuelo, D., Poblete, B. (eds.) String Processing and Information Retrieval - 29th International Symposium, SPIRE 2022, Concepción, Chile, 8–10 November 2022, Proceedings. Lecture Notes in Computer Science, vol. 13617, pp. 303–314. Springer (2022)
6. Equi, M., Mäkinen, V., Tomescu, A.I., Grossi, R.: On the complexity of string matching for graphs. ACM Trans. Algorithms **19**(3), 21:1–21:25 (2023)

7. Gagie, T., Manzini, G., Sirén, J.: Wheeler graphs: A framework for BWT-based data structures. Theor. Comput. Sci. **698**, 67–78 (2017)
8. Gibney, D., Hoppenworth, G., Thankachan, S.V.: Simple reductions from formula-sat to pattern matching on labeled graphs and subtree isomorphism. In: Le, H.V., King, V. (eds.) 4th Symposium on Simplicity in Algorithms, SOSA 2021, Virtual Conference, 11–12 January 2021, pp. 232–242. SIAM (2021)
9. Gibney, D., Thankachan, S.V., Aluru, S.: The complexity of approximate pattern matching on de Bruijn graphs. In: Pe'er, I. (ed.) Research in Computational Molecular Biology - 26th Annual International Conference, RECOMB 2022, San Diego, CA, USA, 22–25 May 2022, Proceedings. Lecture Notes in Computer Science, vol. 13278, pp. 263–278. Springer (2022)
10. Navarro, G.: Improved approximate pattern matching on hypertext. Theor. Comput. Sci. **237**(1–2), 455–463 (2000)
11. Sirén, J.: Indexing variation graphs. In: Fekete, S.P., Ramachandran, V. (eds.) Proceedings of the Ninteenth Workshop on Algorithm Engineering and Experiments, ALENEX 2017, Barcelona, Spain, Hotel Porta Fira, 17–18 January 2017, pp. 13–27. SIAM (2017)
12. Williams, R.: A new algorithm for optimal 2-constraint satisfaction and its implications. Theor. Comput. Sci. **348**(2–3), 357–365 (2005)

Plastic: An Easy to Use and Modular Tool for Benchmarking Tumor Phylogeny Reconstruction Pipelines

Akshay Juyal[1], Zahra Tayebi[1], Alexander Zelikovsky[1], Mauricio Soto-Gomez[3], Simone Ciccolella[2], Gianluca Della Vedova[2], and Murray Patterson[1]()

[1] Georgia State University, Atlanta, GA, USA
mpatterson30@gsu.edu
[2] University of Milano-Bicocca, Milan, Italy
[3] University of Milano, Milan, Italy

Abstract. Several tools have been proposed to infer tumor phylogenies from single-cell sequencing data, but a systematic way of comparing them is still lacking. One of the reasons is that the decomposition of the complete procedure is not settled. We propose a description of the procedure in separate steps and we provide a tool that is able to combine those steps with the final goal of comparing the tools attacking each distinct step. Moreover, we have implemented an easy-to-use graphical application that can run the entire phylogeny inference pipeline, therefore facilitating the use of different tools.

Keywords: Single-cell sequencing · Tumor phylogeny reconstruction · Benchmarking

1 Introduction

Recent developments in sequencing technologies has fueled an increasing interest in algorithmic tools for the evolutionary history reconstruction problem, particularly in the case of tumors, known as cancer progressions in this context. Understanding this evolutionary history can help clinicians understand the cell heterogeneity in the tumors of various types of cancer and, more importantly, gives insights on how to devise therapeutic strategies [28,38].

Evolutionary research has traditionally relied on bulk sequencing methods for inferring tumor phylogenies. The widespread availability and affordability of next-generation sequencing (NGS) data used in this problem have led to a large number of tools [5,13,16,19,23,26,29,33,35,37,40,41]. Despite their clear advantages, the use of this kind of data conveys some fundamental drawbacks. The main one is that samples are sequenced in groups, which usually contain a mixture of both healthy and cancerous cells, which may belong to different clones; that is, groups of cells having a specific set of mutations. Therefore, in bulk sequencing data, only the relative proportion of cells with any given mutation is observable, obscuring the details of the individual clonal evolution.

The introduction of single-cell sequencing (SCS) technologies promises to greatly reduce such uncertainty, by enabling thorough exploration of genomic diversity at the granularity of individual cells. Unfortunately, this methodology is not free of problems. Firstly, SCS is still much more expensive than bulk sequencing, hence limiting its adoption in practice. Moreover, the technical difficulties associated with SCS make the quality of data obtained from this technique not on par with bulk sequencing [21]. For instance, some of the problems include: (1) *doublet cell captures*, that is, data originating from two cells instead of one; (2) *false negatives from allelic dropout*, that is, the presence of undetected mutations, and (3) *missing values* due to low coverage. Despite these obstacles, cost reduction and the development of novel methodologies [11] are continuously being developed to overcome these challenges, making SCS a promising alternative.

Motivated by this fact, numerous methods have been proposed to infer tumor phylogenies from SCS data [10,14,18,31,34,42,43] or by using a hybrid approach that combines both SCS and VAF (variant allele frequency) from bulk sequencing data [22,30,32].

Cancers are under strong evolutionary selective pressures, therefore going through a fast evolutionary process that results in mutations that are very rare to observe otherwise. For example, tumor cells can have large deletions [6]. The wide range of mutations that have to be considered greatly enlarges the search space, which explains why so many methods have been published (*e.g.*, TRaIT [30], SiFit [43], SASC [10], and SPhyR [14]), without a clear winner arising. SPhyR introduces the idea of pre-processing the input SCS matrix by performing a clustering method to reduce the problem dimensionality and ultimately its running time. This strategy was further developed in [9] by devising a clustering method tailored to SCS data—while SPhyR relied on the ubiquitous k-means [2,27] algorithm. Moreover, it can be expected that the reduction of the cost of SCS will result in a faster growth of the data availability respect to computational power [21]. Thus, pre-processing techniques allowing to reduce the dimension of the instances will become a crucial step in the near future.

Although tumor phylogeny inference is crystallizing into a sequence of well-established steps to obtain a complete tool that starts from an SCS dataset and ends with one or more potential phylogenies, the way in which these steps are combined remains largely ad-hoc, making the development of the complete pipeline a time-consuming process. For example, depending on the model and the parameters associated to a particular inference method, we can obtain different potential phylogenies from the same SCS dataset. This motivates the need for a method that allows to quantify the similarities of different phylogenies in the form of a validation tool. For instance, given a set of solutions for a specific SCS dataset, a large cluster with several highly similar phylogenies is likely to be more reliable since the underlying evolution is confirmed by multiple methods. In this direction, some methods to measure the distance between two tumor phylogenies have recently been proposed [3,4,7,12,15,17,20]. While these measures vary greatly in practical applicability, they all express the need for incorporating the evolutionary process into the definition of distance.

With the aim of making the analysis of cancer data more streamlined, we developed `plastic` (PipeLine Amalgamating Single-cell Tree Inference Components), an integrated and easy-to-use tool that provides a modular and systematic framework for integrating the different steps of the phylogenetic inference (including clustering, phylogeny inference, and comparison) [1]. The `plastic` tool is available under the MIT license, and available at https://github.com/plastic-phy/plastic.

Recently, some tools for simulating SCS datasets [25,36,39], have appeared. To remain current, we have hence added this functionality to the pipeline (see step a. of Fig. 1). This adds the ability to benchmark the above-mentioned tumor phylogeny reconstruction tools against ground-truth data, in addition to comparing their outputs. We have also enhanced the `plastic` tool with a graphical user interface (GUI) in the form of a Docker containerized web application that is easy to use and is publicly available.[1] Currently, our GUI tool incorporates the publicly available tools scDNA-sim, *celluloid*, and *SASC*, for the simulation, clustering, and inference steps, respectively.

2 Pipeline Anatomy

We have developed an easy-to-use graphical tool to run a complete pipeline to execute and evaluate cancer progression inference tools, called `plastic`, which integrates several steps, some of which are optional, depending on the desired outcome. The steps of such a pipeline supported by `plastic` is depicted in Fig. 1. A notable feature of `plastic` is that all steps share the same data structure, making easier to add more tools to a comparative analysis.

In particular, `plastic` provides an SCS matrix data structure that is enriched with some additional information that can be shared among the different steps, and a phylogeny tree structure that is used for communication between the phylogeny inference and the phylogeny comparison steps. Such structures are transparent to the user and contribute many additional functionalities without complications from the end-user perspective.

Furthermore, given the nature of `plastic`, we added some graphical capabilities, so that each step of the pipeline can be displayed within an interactive notebook or be exported to separate files. This has been added because exploring cancer simulation tools can be overwhelming, especially for those unfamiliar with computer administration and programming languages like Python or R. Many of these tools require the installation of various packages and libraries, often leading to compatibility issues across different operating systems. For instance, a tool developed on Linux or MacOS might encounter setup failures on Windows.

To address these challenges, we propose a user-friendly solution. Our tool simplifies the entire setup process with just one script which can be run on

---

[1] Available in the published version.

the system. It automates the installation of necessary packages and libraries, while also launching a web application accessible directly from the browser. This approach eliminates the hassle of manual installations and ensures seamless usage for researchers and practitioners alike.

We have developed a Docker containerized web application that seamlessly integrates the `plastic` pipeline with various simulators, with a current focus on the scDNA-sim [36]. Since we are distributing a Docker container, the only dependency is Docker[2] or podman[3], an open source drop-in replacement for Docker developed and maintained primarily by *Red Hat, Inc.*

Our application streamlines the user experience by eliminating the need for manual integration and installation of the pipeline components, which are managed by the application automatically. The application comprises both front- and back-end components, with the former built using `ReactJS` and `d3JS`, and the latter developed in `Python` with `Flask` with various internal helper scripts to communicate within the individual components of the pipeline, without the need for user interaction.

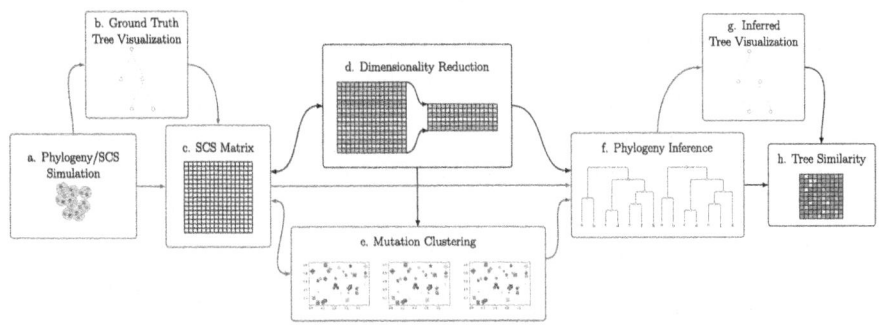

Fig. 1. The `plastic` framework. The steps highlighted in blue are currently supported by the GUI (in addition to the source available at https://github.com/plastic-phy). This figure is best viewed in color.

We now describe in detail each step of the pipeline supported by `plastic`, depicted in Fig. 1, with the corresponding screenshots of the GUI. The steps are

a. Simulation of a phylogeny and the associated input SCS matrix,
b. Ground truth phylogeny tree visualization,
c. Input SCS matrix,
d. Dimensionality reduction (cell clustering)
e. Mutation clustering,
f. Phylogeny inference,
g. Inferred tree visualization and interactive plotting, and

---

[2] https://www.docker.com.
[3] https://podman.io.

h. Tree similarity and comparison

Note that while steps d. Dimensionality reduction, and h. Tree similarity are available in `plastic` at https://github.com/plastic-phy, they are the subject of future work regarding implementation in the GUI. We describe all other steps with some screenshots of the execution of a minimal example which serves as the default example for users who use the tool from the Docker container.

a. Simulation of a phylogeny and the associated input SCS matrix

This step involves selecting an SCS tumor phylogeny simulator, and then calling it with its parameters. Figure 3 depicts the GUI screenshot of this step, more precisely when selecting and calling the scDNA-sim [36] simulator. This simulator models the accumulation of SNVs, and the duplication or deletion of chromosomal segments (copy-number aberrations, or CNAs), producing several files corresponding to the phylogenetic tree and mutational profiles generated. The phylogenetic tree will serve as the ground-truth tree for further benchmarking. The mutational profiles are amplified and then noise is added according to error rates (FP, FN and Drop rate) of SCS technologies to produce the input SCS matrix for downstream steps. While scDNA-sim [36] is fully integrated into `plastic`, CNAsim [39] is partially integrated, while the open-source nature of `plastic` allows for adaption of any future simulator.

Notice that this step is optional, but it must be used if we want to compare the results of the phylogeny inference step with a ground truth. If this step is not run, the pipeline computes cancer progressions starting from a given SCS matrix (step c. of Fig. 1, which would then be required as input).

b. Ground truth phylogeny tree visualization

This step is performed only if the optional simulation step is run, and it displays the (ground truth) phylogeny tree generated from the simulation. This step, depicted in Fig. 2, utilizes the files produced by the previous step to visualize the evolutionary tree using d3JS, providing a comprehensive graphical representation. The tool offers functionalities to adjust the orientation of the trees, zoom in and out, and collapse parent nodes for enhanced insights into evolutionary relationships. By incorporating evolutionary data, users can explore the evolutionary history of the entities represented in the tree, enriching the understanding of their genetic relationships and evolutionary patterns.

c. Input SCS matrix

This step of the pipeline is the starting point if the user provides an input SCS matrix, and does not use the simulator. Otherwise (in the case the user runs a simulation), this matrix is the result of the simulation step (step a. of Fig. 1).

d. Dimensionality reduction (cell clustering)

The cell clustering step, though optional, proves beneficial when handling large datasets, particularly in scenarios where generated data spans hundreds of cells. In fact, several tools silently fail when they are fed a dataset that is too large. SPhyR [14] is, to the best of our knowledge, the only tool that incorporates an explicit clustering step to overcome this problem.

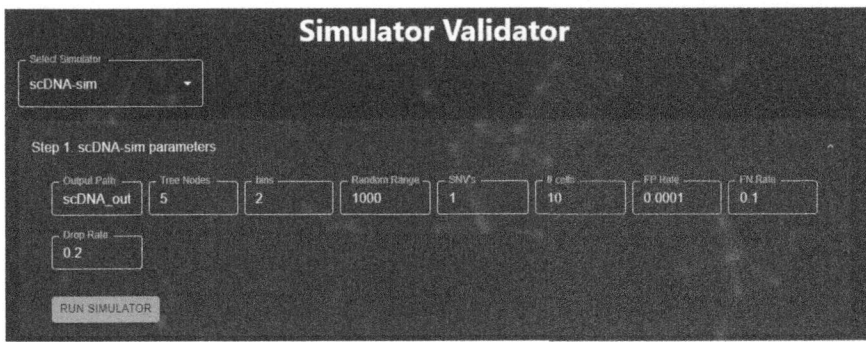

Fig. 2. Screenshot of an execution of the simulation step, in selecting (via the "Select Simulator" dropdown menu) and calling scDNA-sim [36] with the following (default) parameters (which can be changed) by clicking the "RUN SIMULATOR" button. A phylogenetic tree with 5 nodes ("Tree Nodes") and 10 cells ("# cells"), over 2 genomic "bins" is generated. The SNVs are indexed starting at 1000 ("Random Range"), with one SNV ("SNVs": 1) accumulating in each parent-node to child-node generation. To the SCS matrix resulting from this tree (bins, cells and SNVs), noise is added at "FP Rate": 0.0001, "FN Rate": 0.1 and "Drop Rate": 0.2. All files (the phylogeny and SCS matrix) are written to "Output Path": scDNA_out.

In plastic, we use Autoencoder (AE) and ridge regression (RR) for such cell clustering step—see details in [1]. This step is currently not implemented in the GUI, but it will be available in the full version of this paper.

e. Mutation clustering

The mutation clustering step, though optional, also proves beneficial when handling datasets of more than one thousand mutations, or simulated data on many genomic bins, which reflect the resolution of currently available datasets. Again, SPhyR [14] is, to the best of our knowledge, the only tool that incorporates an explicit clustering step.

A mutation clustering phase has two benefits:

(i) it shrinks the instance of the phylogeny inference step, and
(ii) it reduces the dimensionality of the matrix.

The first benefit reduces the running time of the tree inference step, which is the most computationally intensive part of the pipeline, while reducing the dimensionality improves the accuracy of the downstream steps.

The topmost pane of Fig. 4 depicts the (optional) Mutation Clustering step using *celluloid*. This clusters the mutations (columns) of the SCS matrix into (here, 5) super-mutations, and the downstream phylogeny inference is run on this clustered instance, where each supermutation (a column) represents a set of mutations from the original SCS matrix. While only *celluloid* is supported in the GUI, the open-source nature of this tool allows the addition of other mutation clustering methods.

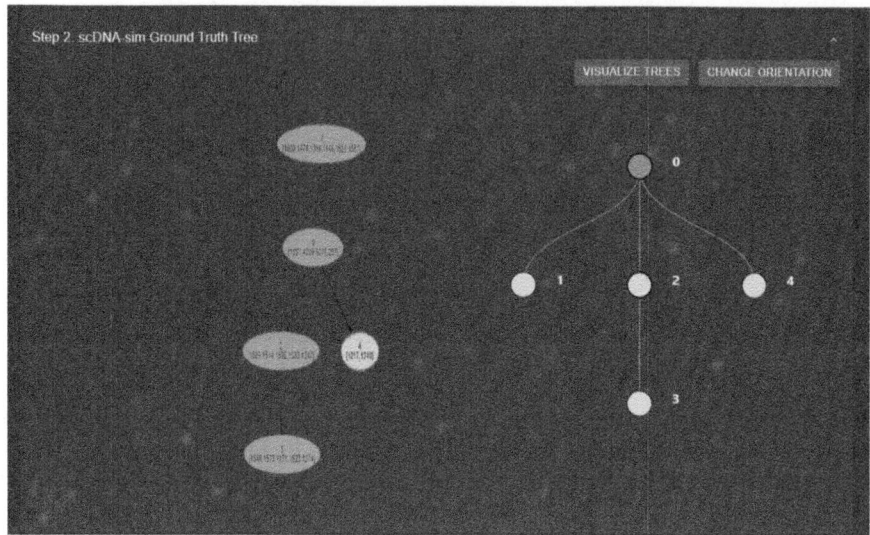

Fig. 3. Screenshot of an execution of the tree visualization step, in visualizing the ground truth tree produced by the simulator, by clicking the "VISUALIZE TREES" button. The tree on the right depicts the topology of this tree, rooted at node 0. The tree on the left provides this topology, along with more information in the form of the new mutations (SNVs) which appear at this node. Note that the (four) mutations at node 0 are germline mutations. The orientation of these trees can be changed by clicking the "CHANGE ORIENTATION" button, and the nodes of the tree on the left can be moved around with the mouse.

f. Phylogeny inference

This step computes a cancer progression from the SCS matrix computed in the previous steps, as depicted in the middle pane of Fig. 4. It is the most computationally expensive part of the pipeline. Currently, the phylogeny Inference step uses SASC [8] which is a robust framework based on Simulated Annealing for the inference of cancer progression from the single-cell sequencing data. The main objective of SASC is to overcome the limitations of the Infinite Sites Assumption by introducing a version of the k-Dollo parsimony model which allows the deletion of mutations from the evolutionary history of the tumor. We plan to add the possibility of using also SCiTE [18] and PhiSCS [24] in the future.

g. Inferred tree visualization and interactive plotting

This step allows to visualize the tree inferred by *SASC* (see the Phylogeny inference step), using its interactive SASCviz tool, as depicted in the bottom-most pane of Fig. 4.

h. Tree similarity and comparison

The tree comparison is an essential component of an evaluation among different cancer progression tools or to find a consensus among different inferred phylogenies. It can be used in two different ways, depending on the goal one

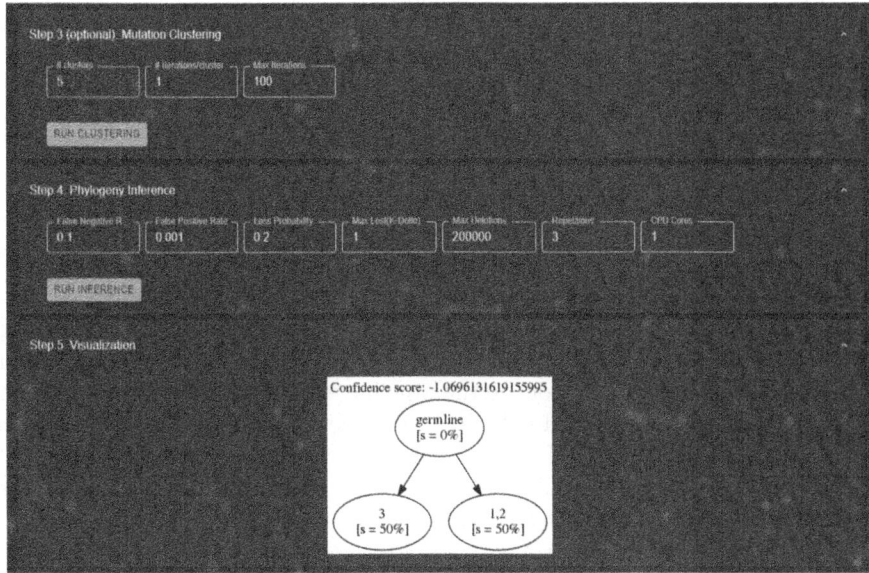

Fig. 4. Screenshot of an execution of the mutation clustering, phylogeny inference and tree visualization steps. In the (optional) Mutation Clustering step (topmost pane), the number of desired clusters ("# clusters": 5), *i.e.*, columns of the output SCS matrix is chosen, along with the number of iterations per cluster ("# iterations/cluster": 1) and the overall number of iterations ("Max Iterations": 100) of the clustering method, *celluloid*, are chosen, and the clustering is executed by clicking the "RUN CLUSTERING" button. In the Phylogeny Inference step (middle pane), the inference method *SASC* is parameterized with the expected FN, FP and Drop rates: "False Negative Rate": 0.1, "False Positive Rate": 0.0001, and "Loss Probability": 0.2, respectively. The additional parameters k (in the Dollo-k model, "Max Loss(K-Dollo)": 1), the maximum number of overall mutational losses ("Max Deletions": 200000), the maximum number of repetitions of *SASC* ("Repetitions": 3) and the number of "CPU Cores": 1 used are chosen, and then *SASC* is launched with "RUN INFERENCE". Finally the Visualization step (bottom-most pane), is run to visualize the tree inferred from the previous Phylogeny Inference step.

wants to achieve: (1) compare two predicted phylogenies, that is the results of two different executions of inference tools, or (2) compare a predicted phylogeny with a ground truth.

The first case is useful if a ground truth is not available, as it is common case in real-world scenarios, and we want to find likely outliers among the results obtained, by difference methods, or different results produced by the same method. The second case is much more relevant for benchmarking; in this case we want to measure how close the predicted solutions are to the ground truth.

To improve the quality of our evaluation, we have incorporated a recently proposed measure of tree similarity, MP3 [7]. While this is implemented in `plastic`, it will be added to the GUI in the full version of this paper. Moreover we plan to add some other measure of similarity, such as ancestor-descendant and different lineages, in the next releases of the tool.

3 Discussion

Given the large amount of SCS tumor data that are being published, there is a need for comparing and tuning the software tools that are used to compute the phylogenies representing cancer progressions, with an easy-to-use interface.

For this purpose we developed `plastic`, a command line tool to execute and compare pipelines to infer tumor phylogenies, and a companion GUI application to allow a user to run and customize the pipeline even without writing a Python program. Moreover, we have divided our pipeline into well separated steps, so that it is possible to benchmark each specific aspect of a more complete pipeline. The entire program has the only dependency of a Docker-compatible container management system making the setup more streamlined.

Future improvements for `plastic` would be to include more tools into the pipeline to extend the breadth of analyses available and to provide additional algorithmic alternatives for the same task. This would allow to use `plastic` as a complete benchmarking pipeline for comparing cancer progression inference tools. In fact, we are currently working on integrating SCITE [18] into `plastic`: this step requires the development of new bindings to transform the output produced by SCITE into the internal representation used by `plastic`.

Final improvements would be the integration of the more complex interaction of `plastic` in the GUI albeit requiring a bit more involvement of the final user, since they cannot be automatically streamlined as the rest.

Another possible direction, which is a bit outside of the current focus, is to allow different SCS matrices. This would allow to compare the effect of using different *upstream* tools for obtaining such matrices. In this case we do not have a ground truth, and we are not aware of simulators that are actually able to produce one in this context; for this reason we have not explicitly considered this extension in the design of `plastic`.

Acknowledgments. This project has received funding from the European Union's Horizon 2020 research and innovation programme under the Marie Skłodowska-Curie grant agreement No 872539 (Pangenome Graph Algorithms and Data Integration—PANGAIA). Mauricio Soto-Gomez gratefully acknowledges the financial support provided by the project "National Center for Gene Therapy and Drugs based on RNA Technology", PNRR-NextGenerationEU program [G43C22001320007].

References

1. Ali, S., Ciccolella, S., Lucarella, L., Vedova, G.D., Patterson, M.: Simpler and faster development of tumor phylogeny pipelines. J. Comput. Biol. **28**(11), 1142–1155 (2021)
2. Anderberg, M.: Cluster Analysis for Applications. Academic Press (1973)
3. Bernardini, G., Bonizzoni, P., Della Vedova, G., Patterson, M.: A rearrangement distance for fully-labelled trees. In: 30th Annual Symposium on Combinatorial Pattern Matching (CPM 2019) (2019)
4. Bernardini, G., Bonizzoni, P., Gawrychowski, P.: On two measures of distance between fully-labelled trees. In: 31st Annual Symposium on Combinatorial Pattern Matching (CPM 2020), vol. 161, pp. 6:1–6:16 (2020)
5. Bonizzoni, P., Ciccolella, S., Della Vedova, G., Soto Gomez, M.: Does relaxing the infinite sites assumption give better tumor phylogenies? An ILP-based comparative approach. In: IEEE/ACM TCBB, pp. 1–1 (2018)
6. Brown, D., Smeets, D., Székely, B., et al.: Phylogenetic analysis of metastatic progression in breast cancer using somatic mutations and copy number aberrations. Nat. Commun. **8**, 14944 (2017)
7. Ciccolella, S., Bernardini, G., Denti, L., Bonizzoni, P., Previtali, M., Vedova, G.D.: Triplet-based similarity score for fully multi-labeled trees with poly-occurring labels. Bioinformatics **37**, 178–187 (2020)
8. Ciccolella, S., Gomez, M.S., Patterson, M., Vedova, G.D., Hajirasouliha, I., Bonizzoni, P.: Inferring cancer progression from single cell sequencing while allowing loss of mutations. bioRxiv **268243** (2018). https://doi.org/10.1101/268243
9. Ciccolella, S., Patterson, M., Bonizzoni, P., Della Vedova, G.: Effective clustering for single cell sequencing cancer data. IEEE JBHI (2021)
10. Ciccolella, S., et al.: Inferring cancer progression from single-cell sequencing while allowing mutation losses. Bioinformatics **37**(3), 326–333 (2020)
11. DePasquale, E.A., Schnell, D.J., Camp, P.J.V., et al.: DoubletDecon: deconvoluting doublets from single-cell RNA-sequencing data. Cell Rep. **29**(6), 1718-1727.e8 (2019)
12. DiNardo, Z., Tomlinson, K., Ritz, A., Oesper, L.: Distance measures for tumor evolutionary trees. Bioinformatics **36**, 2090–2097(2019)
13. El-Kebir, M., Satas, G., Oesper, L., Raphael, B.: Inferring the mutational history of a tumor using multi-state perfect phylogeny mixtures. Cell Sys. **3**(1), 43–53 (2016)
14. El-Kebir, M.: SPhyR: tumor phylogeny estimation from single-cell sequencing data under loss and error. Bioinformatics **34**(17), i671–i679 (2018)
15. Govek, K., Sikes, C., Oesper, L.: A consensus approach to infer tumor evolutionary histories. In: Proceedings of the 2018 ACM International Conference on Bioinformatics, Computational Biology, and Health Informatics, pp. 63–72 (2018)
16. Hajirasouliha, I., Mahmoody, A., Raphael, B.J.: A combinatorial approach for analyzing intra-tumor heterogeneity from high-throughput sequencing data. Bioinformatics **30**(12), i78–i86 (2014)
17. Jahn, K., Beerenwinkel, N., Zhang, L.: The Bourque distances for mutation trees of cancers. Algs. for Mol. Biol. **16**(1), 9 (2021)
18. Jahn, K., Kuipers, J., Beerenwinkel, N.: Tree inference for single-cell data. Genome Biol. **17**(1), 86 (2016)
19. Jiao, W., Vembu, S., Deshwar, A.G., et al.: Inferring clonal evolution of tumors from single nucleotide somatic mutations. BMC Bioinformatics **15**(1), 35 (2014)

20. Karpov, N., Malikic, S., Rahman, M.K., Sahinalp, S.C.: A multi-labeled tree dissimilarity measure for comparing "clonal trees" of tumor progression. Algorithms Mol. Biol. **14**(1), 17 (2019)
21. Kharchenko, P.V.: The triumphs and limitations of computational methods for scRNA-seq. Nat. Methods **18**, 1–10 (2021)
22. Malikic, S., Jahn, K., Kuipers, J., Sahinalp, C., Beerenwinkel, N.: Integrative inference of subclonal tumour evolution from single-cell and bulk sequencing data. bioRxiv (2017)
23. Malikic, S., McPherson, A.W., Donmez, N., Sahinalp, C.S.: Clonality inference in multiple tumor samples using phylogeny. Bioinformatics **31**(9), 1349–1356 (2015)
24. Malikic, S., Mehrabadi, F.R., Ciccolella, S., et al.: PhISCS: a combinatorial approach for subperfect tumor phylogeny reconstruction via integrative use of single-cell and bulk sequencing data. Genome Res. **29**(11), 1860–1877 (2019)
25. Mallory, X.F., Nakhleh, L.: SimSCSnTree: a simulator of single-cell DNA sequencing data. Bioinformatics **38**(10), 2912–2914 (2022)
26. Marass, F., Mouliere, F., Yuan, K., Rosenfeld, N., Markowetz, F.: A phylogenetic latent feature model for clonal deconvolution. Ann. Appl. Stat. **10**(4), 2377–2404 (2016)
27. McQueen, J.: Some methods for classification and analysis of multivariate observations. In: the 5th Berkely Symposium on Mathematical Statistics and Probability, pp. 281–297 (1967)
28. Morissey, A., Garzia, L., Shih, D., et al.: Divergent clonal selection dominates medulloblastoma at recurrence. Nature **529**(7586), 351–357 (2015)
29. Popic, V., Salari, R., Hajirasouliha, I., et al.: Fast and scalable inference of multi-sample cancer lineages. Genome Biol. **16**(1), 91 (2015)
30. Ramazzotti, D., Graudenzi, A., De Sano, L., Antoniotti, M., Caravagna, G.: Learning mutational graphs of individual tumour evolution from single-cell and multi-region sequencing data. BMC Bioinformatics **20**(1), 1–13 (2019)
31. Ross, E.M., Markowetz, F.: Onconem: inferring tumor evolution from single-cell sequencing data. Genome Biol. **17**(1), 69 (2016)
32. Salehi, S., Steif, A., Roth, A., et al.: ddClone: joint statistical inference of clonal populations from single cell and bulk tumour sequencing data. Genome Biol. **18**(1), 44 (2017)
33. Satas, G., Raphael, B.J.: Tumor phylogeny inference using tree-constrained importance sampling. Bioinformatics 33(14), i152–i160 (2017)
34. Singer, J., Kuipers, J., Jahn, K., Beerenwinkel, N.: Single-cell mutation identification via phylogenetic inference. Nat. Commun. **9**(1), 5144 (2018)
35. Strino, F., Parisi, F., Micsinai, M., Kluger, Y.: TrAp: a tree approach for fingerprinting subclonal tumor composition. Nucleic Acids Res. **41**(17), e165–e165 (2013)
36. Tayebi, Z., Juyal, A., Zelikovsky, A., Patterson, M.: Simulating tumor evolution from SCDNA-SEQ as an accumulation of both SNVS and CNAS. In: ISBRA 2023, pp. 530–540 (10 2023)
37. Toosi, H., Moeini, A., Hajirasouliha, I.: BAMSE: Bayesian model selection for tumor phylogeny inference among multiple samples. BMC Bioinformatics **20**(11), 282 (2019)
38. Wang, J., Cazzato, E., Ladewig, E., et al.: Clonal evolution of glioblastoma under therapy. Nat. Genet. **48**(7), 768–776 (2016)
39. Weiner, S., Bansal, M.S.: CNAsim: improved simulation of single-cell copy number profiles and DNA-seq data from tumors. Bioinformatics **39**(7), btad434 (2023)

40. Wu, Y.: Accurate and efficient cell lineage tree inference from noisy single cell data: the maximum likelihood perfect phylogeny approach. Bioinformatics **36**(3), 742–750 (2019)
41. Yuan, K., Sakoparnig, T., Markowetz, F., Beerenwinkel, N.: Bitphylogeny: a probabilistic framework for reconstructing intra-tumor phylogenies. Genome Biol. **16**(1), 36 (2015)
42. Zafar, H., Navin, N., Chen, K., Nakhleh, L.: Siclonefit: Bayesian inference of population structure, genotype, and phylogeny of tumor clones from single-cell genome sequencing data. Genome Research (2019)
43. Zafar, H., Tzen, A., Navin, N., Chen, K., Nakhleh, L.: SiFit: inferring tumor trees from single-cell sequencing data under finite-sites models. Genome Biol. **18**(1), 178 (2017)

A 3D Deep Learning Architecture for Denoising Low-Dose CT Scans

Armen Kasparian[1(✉)], Guohua Cao[2], and Wu-chun Feng[1]

[1] Department of Computer Science, Virginia Tech, Blacksburg, USA
{armen,wfeng}@vt.edu
[2] School of Biomedical Engineering, ShanghaiTech University, Shanghai, China
caogh@shanghaitech.edu.cn

Abstract. Low-dose computed tomography (LDCT) scans reduce the radiation dose of computed tomography (CT) scans but come at the expense of image quality. Deep-learning (DL) image denoising techniques can enhance these LDCT images to match the quality of their regular-dose CT counterparts. To achieve better denoising performance than the current state of the art, we present a novel 3D DL architecture for LDCT image denoising called **3D-DDnet**. The architecture leverages the inter-slice correlation in volumetric CT scans to obtain better denoising performance and employs distributed data parallel (DDP) strategies along with transfer learning to achieve faster training. The DDP training strategy enables a scalable multi-GPU approach on Nvidia A100 GPUs, which allows the training of previously prohibitively large volumetric samples. Our results show that **3D-DDnet** achieves 10% better mean square error (MSE) on LDCT scans than its 2D predecessor (i.e., 2D-DDnet). In addition, the transfer learning in **3D-DDnet** leverages existing trained 2D models to "jump start" the weights and biases of our 3D DL model and reduces training time by 50% while maintaining accuracy.

Keywords: deep learning · distributed data parallel · transfer learning · computed tomography · image enhancement

1 Introduction

Radiologists rely on non-invasive CT scans to obtain images of internal organs for the diagnosis and treatment of patients [1]. Unfortunately, it exposes the patient to significant radiation. Fazel et al. found that 75.4% of patients' total effective radiation dose in the USA can be attributed to CT and nuclear medicine scans and used 50 millisieverts (mSv) as the "high" exposure range for radiation [2]. A standard chest CT scan exposes patients to 4.55 mSv of radiation. With the need to minimize radiation exposure, LDCT scans have emerged to reduce radiation exposure to a mere 0.5 mSv but at the expense image fidelity [3].

Both statistical methods and deep learning-based methods have been used to combat the noise and artifacts found in lower-quality LDCT scans. For example, statistical model-based iterative reconstruction (MBIR) enhances image quality and reduces noise in CT imaging Deep learning (DL) architectures like RED-CNN [4] and DDnet [5] learn features from standard-dose CT scans and use that information to denoise LDCT scans. These DL architectures focus on denoising these three-dimensional (3D) CT scans via two-dimensional (2D) slices as 3D denoising has been computationally impractical. Enabling the transition from today's 2D-slice denoising architectures to a 3D architecture requires training of a parameter space that is nearly 5× larger, i.e., 4,712,723 vs. 1,021,619.

Our 3D denoising architecture, 3D-DDnet, consists of two main components: (1) distributed data parallel training to improve performance and (2) data loader modifications to improve accuracy. The parallel training approach uses a volumetric representation of CT scans by selectively choosing the slices from the dataset. The number of slices selected for a volume can now be explored as a hyperparameter to fit into GPU VRAM. The increase in GPU VRAM size allows larger sample volumes to be offloaded from CPU to GPU for faster training.

2 Related Work

2D vs. 3D Deep-Learning Architectures. Existing 3D deep-learning (DL) architectures for CT imaging have explored segmentation, classification, and detection techniques versus their 2D counterparts. Avesta et al. compare 2D and 3D U-Net-based DL architectures for image segmentation in brain MRIs [6] and show that 3D models need 20× more memory than 2D models. The 3D models also converge to a more accurate state but at the expense of significantly more computational overhead due to the extra dimension. Our 3D-DDnet approach reduces this increased computational overhead of 3D approaches while improving the accuracy of state-of-the-art U-Net-based models.

Challenges with 3D Deep-Learning Networks. Singh et al. find that relative to classification, segmentation, and detection for 3D deep-learning networks, common challenges include increased difficulty in training and hyperparameter tuning and the inability to utilize smaller datasets [7]. Crespi et al. study the transition from 2D convolution neural networks (CNN) to 3D CNNs and find the larger dimension space problematic due to the lack of available datasets [8]. In contrast, our 3D-DDnet approach proposes a data-loading strategy that creates 3D volumes from existing 2D slice-based CT scan datasets. This strategy reduces the overhead of curating a custom dataset by introducing the ability to reuse previous datasets designed for 2D CT scan denoising.

Impact of Multislice Inputs on Accuracy. Multislice inputs for 3D CNN noise reduction have previously been explored on the accuracy front. Zhou et al. show that the improved accuracy of a 3D network comes at the expense of longer training time per epoch [9]. The 2D network takes 25 min to train per epoch while the 3D equivalent takes 270. In addition, due to hardware limitations, their data loader is limited to sampling 64×64 patches of the source 512×512 slices. An earlier implementation of 3D-DDnet [10] was also limited by hardware, resulting in the use of a sample patching and stitching technique.

3 Approach and Design of 3D-DDnet

Increasing the dimensions of an architecture from 2D to 3D dramatically increases the parameters of the network (in our case, from 1,021,619 to 4,712,723), allowing for more feature extraction but at the expense of more computation [11]. To alleviate these costs, we present an overhaul to the training strategy and enhancements to the data loader for our 3D-DDnet architecture.

3.1 Architecture

The 3D-DDnet architecture extends our previous "2D DenseNet and Deconvolutional neural network" (i.e., DDnet) [5], as shown in Fig. 1. This extended architecture generalizes more information from the source data by using the correlations found between slices and delivers better accuracy.

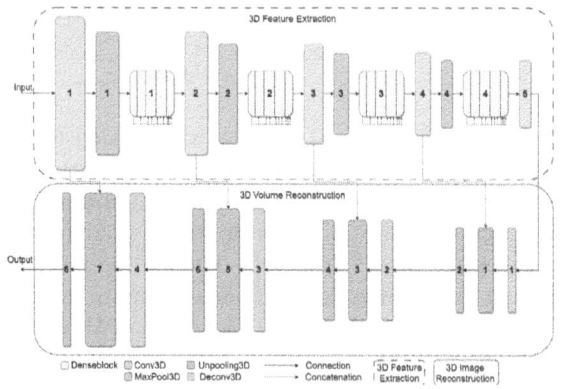

Fig. 1. Architecture of 3D-DDnet. Layer numbers correspond to Table 1.

The convolution layers in the upper half of 3D-DDnet now consist of 3D dense blocks and 3D convolutions. The 3D dense blocks contain internal 3D convolution layers that are followed by 3D max pooling layers, which work like their 2D counterparts but reduce the latent space by a factor of two in the x, y, and z dimensions of the sample volume.

Table 1 shows the input and output sizes of each network layer, where a volume consists of 32 slices. The architecture allows for the number of slices utilized in a sample to be a hyperparameter. This hyperparameter is then set at runtime, facilitating further research into the effects of different volume sizes.

3.2 Datasets

The data used to train and test 3D-DDnet came from three sources: (1) Mayo Clinic with 100 healthy CT scans at full and quarter dosage, (2) Lung Image Database Consortium (LIDC) with 722 CT scans, and (3) Medical Imaging Databank of the Valencia region (BIMCV) with 397 CT scans. If the dataset did *not* contain LDCT images, we ran the high-dose images through an algorithm to simulate LDCT scans. These numerically simulated low-dose images were then paired with the corresponding high-dose CT images to create a uniform dataset containing source and target images. All the CT images from the dataset are of size 512×512 and rotated to be the same orientation.

Table 1. Input/output sizes and filter sizes of the layers in 3D-DDnet

Layers	Output Size	Details
Conv3D 1	512 × 512 × 32 × 16	Filter size=7 × 7 × 7, Stride=1
MaxPool3D 1	256 × 256 × 16 × 16	Filter size=3 × 3 × 3, Stride=3
DenseBlock3D 1	256 × 256 × 16 × 80	Filter size=(5 × 5 × 5) × 4
Conv3D 2	256 × 256 × 16 × 16	Filter size=1 × 1 × 1, Stride=1
MaxPool3D 2	128 × 128 × 8 × 16	Filter size=3 × 3 × 3, Stride=3
DenseBlock3D 2	128 × 128 × 8 × 80	Filter size=(5 × 5 × 5) × 4
Conv3D 3	128 × 128 × 8 × 16	Filter size=1 × 1 × 1, Stride=1
MaxPool3D 3	64 × 64 × 4 × 16	Filter size=1 × 1 × 1, Stride=2
DenseBlock3D 3	64 × 64 × 4 × 80	Filter size=(5 × 5 × 5) × 4
Conv3D 4	64 × 64 × 4 × 16	Filter size=1 × 1 × 1, Stride=1
MaxPool3D 4	32 × 32 × 2 × 16	Filter size=3 × 3 × 3, Stride=2
DenseBlock3D 4	32 × 32 × 2 × 80	Filter size=(5 × 5 × 5) × 4
Conv3D 5	32 × 32 × 2 × 16	Filter size=1 × 1 × 1, Stride=1
Unpooling3D 1	64 × 64 × 4 × 16	Scale factor = 2
Deconv3D 1	64 × 64 × 4 × 32	Filter size=5 × 5 × 5, Stride=1
Deconv3D 2	64 × 64 × 4 × 16	Filter size=1 × 1 × 1, Stride=1
Unpooling3D 2	128 × 128 × 8 × 16	Scale factor = 2
Deconv3D 3	128 × 128 × 8 × 32	Filter size=5 × 5 × 5, Stride=1
Deconv3D 4	128 × 128 × 8 × 16	Filter size=1 × 1 × 1, Stride=1
Unpooling3D 3	256 × 256 × 16 × 16	Scale factor = 2
Deconv3D 5	256 × 256 × 16 × 32	Filter size=5 × 5 × 5, Stride=1
Deconv3D 6	256 × 256 × 16 × 16	Filter size=1 × 1 × 1, Stride=1
Unpooling3D 4	512 × 512 × 32 × 16	Scale factor = 2
Deconv3D 7	512 × 512 × 32 × 32	Filter size=5 × 5 × 5, Stride=1
Deconv3D 8	512 × 512 × 32 × 1	Filter size=1 × 1 × 1, Stride=1

3.3 Data Loaders

2D data loaders supply batches of image slices to models, with a batch size of one meaning one image per epoch for training. Larger batch sizes send more slices per epoch. 3D data loaders, on the other hand, send volumes-stacked sets of 2D slices-as single samples. Batches with 3D loaders contain multiple volumes processed separately, increasing computational demands. For instance, a 512×512 slice requires 1.5 GB of memory, while a 512×512×16 volume needs nearly 20 GB. This memory constraint makes volume slice selection

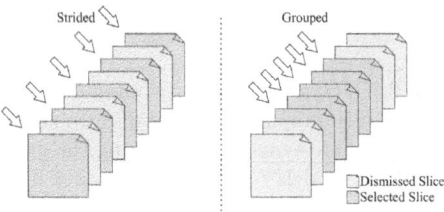

Fig. 2. Strided vs. grouped data loader. A selection of a five-slice volume from a seven-slice sample with a stride length of size two.

critical, determined by both desired accuracy and hardware capabilities. Our paper introduces and evaluates two data loader types (see Fig. 2): strided and grouped.

Strided Data Loader. The strided data loader selects slices at uniform intervals. The strided loader's advantage is its consistent slice selection across the entire CT scan, useful when crucial data is spread throughout. However, it uniformly values all slices, which might not be ideal. For example, the initial and final slices of LDCT may offer limited relevant data; yet, this loader picks slices uniformly, potentially including less valuable ones for training.

Center-Grouped Data Loader. The center-grouped data loader minimizes the distance between slices selected for the volume to help the model learn the correlation between slices in the third dimension. This data loader works by selecting a group of slices centered around the middle of the sample. The main benefit of this data loader is that the selected slices for the volume contain the highest correlation between them.

3.4 Training Strategy

Current datacenter GPUs offer up to 80GB of HMB2e VRAM, a significant upgrade over the previous 32GB of HBM2, thus supporting the loading of larger datasets onto GPUs and the shift from 2D slices to 3D volumetric data. For context, a single 512×512 slice occupies about 1.5GB, while 16-slice and 32-slice volumes require 20GB and 40GB, respectively, underscoring the demand for enhanced GPU capacity.

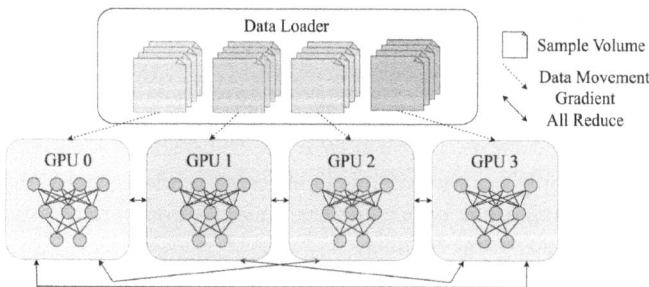

Fig. 3. Overview of the distributed data parallel architecture. The data loader individually sends unique batches of data to each GPU while each of the models loaded onto the GPUs are synchronized during the gradient, all-reduce step found in backpropagation. The models receive different data from the data loader but average the gradients per epoch, leaving the models synchronized.

Past generations of datacenter GPUs could not even load a single sample volume onto the GPU. With this hardware limitation now lifted, we need to be smarter with how we load the dataset onto the GPU. Since we are unable to load *multiple* sample volumes onto a single GPU, our approach to solve this problem uses a distributed data parallel (DDP) approach. Distributed data parallelism (DDP) enables data parallelism at the model level. It allows for a synchronized model on each device while different data is passed for training during each epoch. This approach, built into PyTorch, works by creating a replica of the model architecture on each GPU. For each epoch, each replica model running on its corresponding GPU receives a different batch of data to run through the forward pass. Gradients are then computed locally for each process. These local gradients are then synchronized during the backward pass with an all-reduce gradient synchronization that calculates the mean of all the gradients across all the processes. These average gradients are then distributed to the individual processes for the backward pass. After this backward pass completes, all the models across each of the individual processes are the same and prepared for the next epoch and batch of data. A visualization of how the data loader works in tandem with the synchronized models can be seen in Fig. 3.

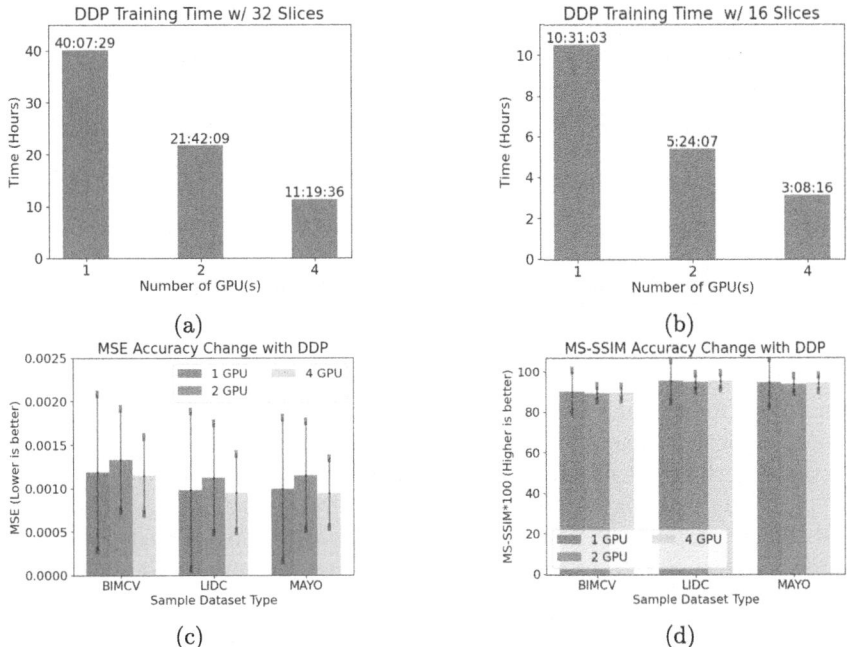

Fig. 4. Performance results with respect to accuracy and training time. (a/b) Show training time scaling with respect to the number of GPUs. (c/d) Show accuracy variance between datasets with respect to the number of GPUs.

Data Parallelism with Batches. When using the DDP training strategy, we need to understand how it affects the batch size and learning during training. With DDP, we modify the batch size to be an effective batch size that relates to the number of processes being run. This effective batch size is equivalent to the local batch size (the number of samples being processed per epoch per process) multiplied by the number of processes (GPUs) being utilized.

In our application with volumetric CT image enhancement, this now removes the VRAM limitation of the GPU and increases the previous maximum batch size from one to equal the number of GPUs available for training. It is important to note that this is only possible since we can load, at minimum, a single 3D volumetric sample onto the individual GPU. Then, by utilizing multiple GPUs, we can increase our effective batch size and scale with data parallelism.

The performance speedup shown in Fig. 4.(a/b) is due to the data parallelism leveraged by utilizing the DDP training algorithm. The use of DDP also facilitates reasonable, effective batch sizes that allow for faster training times while maintaining accuracy, as shown in the Fig. 4.(c/d).

When looking at Fig. 4.(c/d), we can see the variance bands continually decrease as more GPUs are utilized during training. Using more GPUs in distributed data parallel training reduces variance in model training by allowing for larger batch sizes and more consistent gradient estimates, leading to smoother, more stable updates and better generalization. This is achieved through parallel processing, averaging gradients across GPUs, and efficient utilization of computational resources.

3.5 Transfer Learning

Leveraging pre-existing 2D networks can offset the computational challenges of transitioning to 3D networks. Transfer learning applies the weights and biases of a pre-trained model to new settings, providing an advantage in training. Specifically for 3D networks, it enhances efficiency by strategically employing the 2D architecture's weights and biases. Numerous 2D models in the biomedical sector allow us to harness their computational investment to amplify the 3D counterparts through the reuse of older network structures.

Typically, in transfer learning, trained layers from a donor network are integrated directly into a recipient network, ensuring the pre-learned parameters initiate the recipient's training. However, this is feasible only when both architectures have matching dimensions. When dimensions differ, as in our study, an adapted transfer learning strategy is needed. Transferring from 2D to 3D involves embedding the weights of a 2D image kernel into a 3D volume kernel, as visualized in Fig. 5.

Unoccupied values in the 3D kernel are filled with zeros from the pretrained 2D kernel. We tested alternative fill methods, such as consistently using the 2D kernel for the entire 3D space or introducing random noise. All variations displayed comparable performance, with negligible differences in metrics like MSE and MS-SSIM.

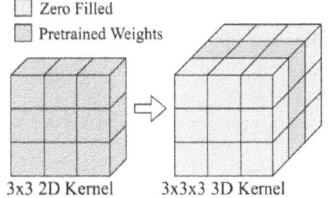

Fig. 5. Transfer learning diagram showing insert of 2D kernel into 3D kernel

4 Evaluation Metrics for the Loss Function

The loss function of the neural network dictates the metrics that backpropagation is trying to enhance. Mean squared error (MSE) is a direct pixel-to-pixel comparison of the differences between two images or volumes – see Eq. (1). The MSE equation iterates over N pixels, where X is the source image, and Y is the image after being denoised.

$$MSE(X,Y) = \frac{1}{N} \sum_{n=1}^{N} [X - Y]^2, \qquad (1)$$

The structural similarity (SSIM) index, shown in Eq. (2), measures the similarity between two images based on their structural information in the spatial domain [12]. The SSIM index considers luminance $l(X,Y)$, contrast $c(X,Y)$, and structure $s(X,Y)$ and returns a value between -1 and 1, where a value of 1 indicates perfect similarity. In Eq. 2, μ_x and μ_y are the mean intensities of windows, σ_x^2 and σ_y^2 represent the variances of windows, σ_{xy} is the covariance between windows which provides a measure of the strength and direction of the relationship between two sets of variables, $C1$ and $C2$ are constants added to stabilize for the dynamic range of the images where $C1 = (k_1 L)^2$ and $C2 = (k_2 L)^2$ where L is the dynamic range of the images and k_1 and k_2 are small constants (set to default values $k_1 = 0.01$ and $k_2 = 0.03$).

$$\text{SSIM}(X,Y) = \frac{(2\mu_x \mu_y + C1)}{(\mu_x^2 + \mu_y^2 + C1)} \times \frac{(2\sigma_{xy} + C2)}{(\sigma_x^2 + \sigma_y^2 + C2)} = l(X,Y) \times \text{cs}(X,Y) \qquad (2)$$

The multi-scale structural similarity index (MS-SSIM), defined in Eq. (3), is built upon the SSIM by images considering multiple scales (M and j) to better capture structural similarities across different levels of detail. MS-SSIM divides the image or volume into multiple smaller sub-regions and computes the SSIM value at different levels of image detail [13]. Like SSIM, MS-SSIM ranges from -1 to 1, with higher values indicating a higher similarity between the two samples.

α and β_j are weights for the luminance, contrast, and structure terms. These weights are set to one to equally weigh all the terms.

$$\text{MS-SSIM}(X,Y) = l_M^\alpha(X,Y) \cdot \prod_{j=1}^{M} cs_j^{\beta_j}(X,Y) \qquad (3)$$

Eq. (4) presents a composite loss function, $Loss(X,Y)$ that combines the higher-quality MS-SSIM with the MSE. This allows us to have a loss function where the MSE focuses on pixel-wise accuracy and the MS-SSIM emphasizes preserving structural and perceptual quality. Combining them offers a holistic approach, ensuring both pixel-level precision and high perceptual quality in the denoised images. In our loss function, a γ of 0.1 is selected to balance the magnitude of these two terms properly.

$$\text{Loss}(X,Y) = \text{MSE}(X,Y) + \gamma(1 - \text{MS-SSIM}(X,Y)), \qquad (4)$$

5 Results

Here we present both the quantitative results of 3D-DDnet with respect to accuracy and the qualitative results of 3D-DDnet with respect to the CT images and the difference maps between the original high-quality CT image and the denoised CT images produced by 2D-DDnet and 3D-DDnet. These results are run with the Adam optimizer and a learning rate of 0.001. An ablation study is done between the different optimizations which can be seen in the quantitative results below.

5.1 Quantitative Results

The baseline 2D-DDnet model sets the benchmark for our transition to 3D-DDnet. Training on a single GPU for about 8 h with a batch size of one, it achieves an average MSE of 0.003998 and an MS-SSIM of 0.788877. In contrast, our 3D-DDnet variants show improvements in both MSE and MS-SSIM, highlighting the benefits of the 3D approach.

Comparisons between using 32 and 16 slices per volume in 3D-DDnet (Fig. 6 and Table 2) reveal negligible differences (± 1) in performance, suggesting that using more than 16 slices may lead to unnecessary computational expense. Therefore, for optimal efficiency and accuracy, a 16-slice volume is recommended.

Speed enhancements are evident with the distributed data parallel strategy. Scaling from 1 to 2 GPUs nearly doubles the speed, while increasing from 2 to 4 GPUs offers a 1.75× speedup.

To verify that parallel architecture doesn't compromise the model's stability, we examined loss curves for single and multi-GPU setups (Fig. 7). The loss consistently decreases over 50 epochs, indicating stable training. Notably, transfer learning provides a substantial acceleration in learning, allowing the model to rapidly approach optimal loss values. In instances where surpassing the 2D model's performance is the goal, training can be shorter. At 15 epochs, the 3D model already outperforms its 2D counterpart. The loss function plateaus at 50 epochs (Fig. 7(c)), signifying that the network has reached its optimal point.

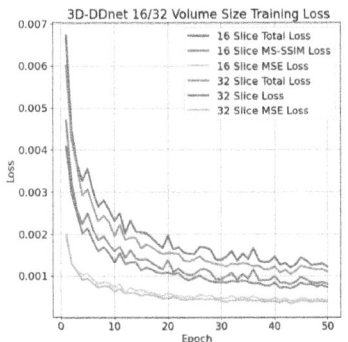

Fig. 6. Training loss similarities between the 16- and 32-slice volume data loader

Our results show that there is very little difference between the utilization of 32 slices per volume compared to the 16 slices per volume in both training (Fig. 6) and testing. From Table 2, it is apparent that while holding all hyperparameters the same but the number of slices per volume there is only a ±1% difference which is within the margin of error between the two models. This leads to the conclusion that the utilization of more slices in the volume can be considered wasted computation time as the training time is almost double. For maximum parallel efficiency while maintaining accuracy, selecting 16 slices is critical.

Table 2. Quantitative Results showing the training time, average MSE, and average MS-SSIM of testing. The architecture section contains GDL (Grouped Data Loader), SDL (Strided Data Loader), and TL (Transfer Learning) optimizations present. Batch Size = GPU Count

Architecture	Epochs	Slice Count	Batch Size	Training Time (H:M:s)	Average MSE	Average MS-SSIM
2D-DDnet	50	N/A	1	8:44:32	0.0039±0.0015	0.7888±0.0857
3D-DDnet-SDL	50	32	2	21:47:50	0.0022±0.0014	0.9165±0.0455
3D-DDnet-GDL	50	32	2	21:42:09	0.0012±0.0009	0.9233±0.0779
3D-DDnet-TL	15	32	2	1:56:13	0.0014±0.0008	0.9201±0.0596
3D-DDnet-TL-GDL	50	16	1	10:31:03	0.0010±0.0006	**0.9339±0.0572**
3D-DDnet-TL-GDL	50	16	2	5:24:07	0.0011±0.0006	0.9258 ±0.0603
3D-DDnet-TL-GDL	50	16	4	**3:08:17**	0.0010±0.0006	0.9303±0.0667
3D-DDnet-TL-GDL	50	32	2	21:43:28	0.0021±0.0080	0.9246±0.0634
3D-DDnet-TL-GDL	50	32	4	11:19:36	0.0011±0.0009	0.9257±0.0713
3D-DDnet-TL-GDL	25	32	2	10:52:32	0.0016±0.0017	0.9247±0.0866
3D-DDnet-TL-GDL	25	32	4	5:40:45	0.0015±0.0013	0.9076±0.0807

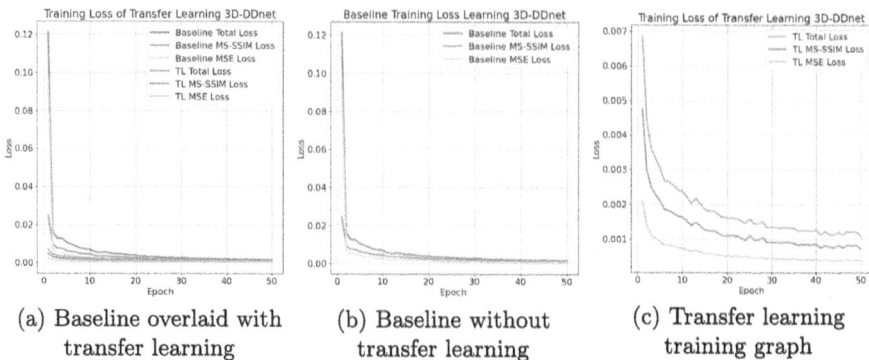

(a) Baseline overlaid with transfer learning

(b) Baseline without transfer learning

(c) Transfer learning training graph

Fig. 7. Graph (a) shows the order of magnitude difference in the loss function between utilizing transfer learning and not. Graph (b) shows the loss function without transfer learning, and graph (c) shows the loss function with transfer learning. Both graphs (b) and (c) show that transfer learning does not affect the ability of the network to optimize the loss function.

5.2 Qualitative Results

The difference maps in Fig. 8 also give an insight into what the network focuses on changing. At the same time, the 2D network is more focused on minor changes that pertain to the streaking. The 3D network focuses on the lung tissue, with the difference map showing the most changes in that area.

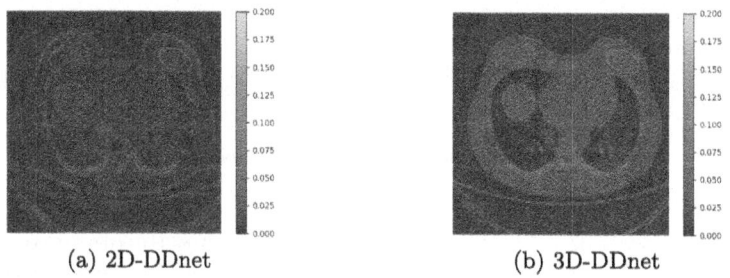

(a) 2D-DDnet

(b) 3D-DDnet

Fig. 8. Difference maps of the same randomly selected sample CT slice

The direct comparison between the 2D and 3D reconstructed images shows how the two architectures differ in what they aim to enhance and denoise. Looking at the 2D reconstructed image in Fig. 9(c), much of the streaking is removed from the source LDCT slice found in Fig. 9(a). While this may seem positive initially, many features found in the target CT slice shown in Fig. 9(b) are lost in this process. Looking at the 3D reconstructed image in Fig. 9(d), we can see that the streaking is overall reduced from the LDCT scan. With this reduction comes the retention of features, which can be seen on the top portion of the scan.

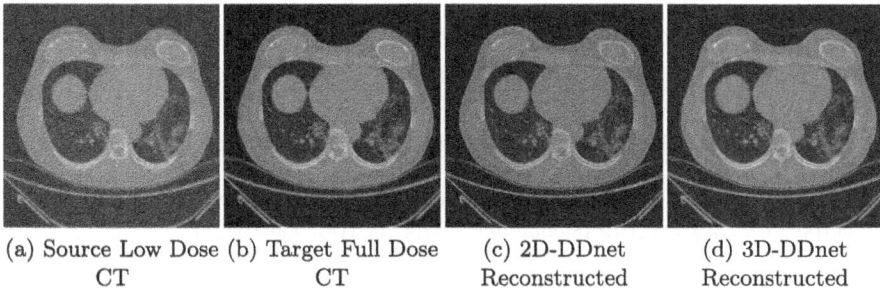

(a) Source Low Dose CT (b) Target Full Dose CT (c) 2D-DDnet Reconstructed (d) 3D-DDnet Reconstructed

Fig. 9. Comparison between the LDCT, target slice, 2D reconstructed slice, and 3D reconstructed slice. The sample CT slice was randomly selected from the dataset. 3D-DDnet can maintain features that are lost in the 2D reconstruction. Retaining these features comes at the cost of some streaking and artifacts still present in the 3D reconstruction.

6 Conclusion

This work presents a method for adopting 3D architectures in biomedical image denoising, demonstrating effective training of larger models on single-node multi-GPU systems using distributed data parallelism. We adapt existing 2D data loaders and datasets for 3D modeling and employ transfer learning for efficient reuse of datasets and pretrained models.

While focused on LDCT scan denoising, our **3D-DDnet** architecture is versatile, applicable to fields like image segmentation and classification, and shows potential in MRI imaging. This study indicates the need for scaling to multi-node setups for larger datasets due to the correlation between batch size and GPU utilization.

The shift to 3D introduces hyperparameter sensitivity, addressed through manual tuning in this study. Future research could explore automated optimization methods. Overall, **3D-DDnet** effectively outperforms its 2D predecessor, highlighting the advantages of 3D models in biomedical imaging.

Acknowledgments. This research was supported in part by NSF IIS-2027607 and NSF CCF-2031215 as well as Dr. Cynthia McCollough, the Mayo Clinic, and the American Association of Physicists in Medicine and grants EB017095 and EB017185 from the NIBIB, for providing the Mayo Clinic dataset. We acknowledge Garvit Goel for his initial work on the 2D model and its evolution to the 3D model as well as Advanced Research Computing at Virginia Tech for providing the computational resources and technical support that contributed to the results in this paper.

References

1. Hsieh, J.: Computed Tomography: Principles, Design, Artifacts, and Recent Advances, vol. 1 (2015)
2. Fazel, R., et al.: Exposure to low-dose ionizing radiation from medical imaging procedures. N. Engl. J. Med. **361**, 849–857 (2009)
3. Zhou, Y., et al.: Radiation dose levels in chest computed tomography scans of coronavirus disease 2019 pneumonia. Medicine **100**, e26692 (2021)
4. Chen, H., Zhang, Y., Kalra, M.K., Lin, F., Chen, Y., Liao, P., Zhou, J., Wang, G.: Low-dose CT with a residual encoder-decoder convolutional neural network. IEEE Trans. Med. Imaging **36**(12), 2524–2535 (2017)
5. Zhang, Z., Liang, X., Dong, X., Xie, Y., Cao, G.: A sparse-view CT reconstruction method based on combination of DenseNet and deconvolution. IEEE Trans. Med. Imaging **37**(6), 1407–1417 (2018)
6. Avesta, A., Hossain, S., Lin, M., Aboian, M., Krumholz, H.M., Aneja, S.: Comparing 3D, 2.5D, and 2D approaches to brain image segmentation (2022)
7. Singh, S.P., Wang, L., Gupta, S., Goli, H., Padmanabhan, P., Gulyás, B.: 3D deep learning on medical images: a review (2020)
8. Crespi, L., Loiacono, D., Sartori, P.: Are 3D better than 2D convolutional neural networks for medical imaging semantic segmentation? In: International Joint Conference on Neural Networks (IJCNN), vol. 2022, pp. 1–8 (2022)
9. Zhou, Z., Huber, N.R., Inoue, A., McCollough, C.H., Yu, L.: Multislice input for 2D and 3D residual convolutional neural network noise reduction in CT. J. Med. Imaging **10**(1), 014003 (2023)
10. Goel, G., Gondhalekar, A., Qi, J., Zhang, Z., Cao, G., Feng, W.: Computecovid19+: Accelerating Covid-19 diagnosis and monitoring via high-performance deep learning on CT images. In: Proceedings of the 50th International Conference on Parallel Processing. Association for Computing Machinery (2021)
11. Szegedy, C., et al.: Going deeper with convolutions. In: IEEE Conference on Computer Vision and Pattern Recognition (CVPR), vol. 2015, pp. 1–9 (2015)
12. Wang, Z., Bovik, A., Sheikh, H., Simoncelli, E.: Image quality assessment: from error visibility to structural similarity. IEEE Trans. Image Process. **13**(4), 600–612 (2004)
13. Wang, Z., Simoncelli, E.P., Bovik, A.C.: Multiscale structural similarity for image quality assessment. In: 37th Asilomar Conference on Signals, Systems, and Computers, 2003, vol. 2, 2003, pp. 1398–1402 (2003)

A Simple and Interpretable Deep Learning Model for Diagnosing Pneumonia from Chest X-Ray Images

Lucas Otavio Leme Silva, Karine Marques Hara,
Pedro Henrique Mendes de Paula, Alexandre Rossi Paschoal,
and Fabricio Martins Lopes(✉)

Universidade Tecnológica Federal do Paraná (UTFPR) Câmpus Cornélio Procópio,
Departamento Acadêmico de Computação (DACOM), Avenue Alberto Carazzai,
1640 - CEP 86300-000 Cornélio Procópio, PR, Brazil
lucasotavio@alunos.utfpr.edu.br, fabricio@utfpr.edu.br

Abstract. Pneumonia is an infectious disease that has afflicted humanity for centuries. Its origins can be diverse, such as bacterial, viral, fungal, or chemical agents. It is one disease that causes the most deaths among children and adults worldwide. There are many ways to treat pneumonia, however, it is a fact that the sooner it is detected, the greater the chances of successful treatment. Therefore, it is right to think that developing ways to make the diagnosis faster, and facilitating early treatment, is of general interest. Therefore, the present work presents a neural network for the analysis of chest radiographs for the diagnosis of pneumonia. The proposed method showed remarkable results when compared to similar methods in the literature. In addition, the proposed method presents a more transparent diagnosis through relevance aggregation, highlighting the regions of images that were recognized by the neural network to perform the diagnosis, also contributing to interpretable results.

Keywords: Pneumonia · pattern recognition · deep learning · relevance aggregation · explainable AI

1 Introduction

Pneumonia is an acute respiratory infection, caused by viruses, bacteria, fungi, or chemical reagents, it causes pleural effusion, a condition in which fluids fill the lung, causing difficulty breathing. The disease was responsible for the deaths of over 700,000 children in 2019 alone [1]. Delayed diagnosis of pneumonia causes the disease to worsen, which leads to increased mortality [2]. In severe cases of pneumonia, adequate medical infrastructure is necessary to treat the disease, requiring hospitalization and the use of mechanical ventilation. Pneumonia is one of the most deadly diseases caused in the lungs [20] and remains one of the leading causes of infant mortality worldwide, causing millions of deaths [3].

Computer vision and pattern recognition allow very important approaches to solve complex problems. Classical computer vision approaches consider feature extraction, classification algorithms and feature selection methods in order to generate and identify features that are relevant for class identification [4–10]. In fact, machine learning approaches provide important contributions in the medical research field, such as reducing the time and costs associated with discoveries and diagnoses [11].

On the other hand, recently the deep learning models, specifically convolutional neural networks (CNNs), are frequently used for the detection and classification of images and many other applications [12]. In particular, deep learning approaches have been successfully applied to the identification of pneumonia [13–15]. More specifically, the chest radiography is the most commonly used form of pneumonia diagnosis. However, the analysis of this exam has a subjective variability [16]. Thus, it is important to improve the machine learning methods for the diagnosis of pneumonia.

In this context, the CheXNet method was proposed by considering a 121-layer convolutional neural network, adopting the "Chest X-ray14" dataset [18]. The adopted dataset has 112,120 x-ray images of the front view of the chest, and with labels of 14 diseases. The CheXNet algorithm achieved an AUC of 76.8%. Compared to other current works, such as [23], which also uses a convolutional neural network, and adopted a database of 3,883 images, presented an AUC of 96.8%. Another important contribution was recently proposed by considering the same database ("Chest X-ray14") and based on a set of deep learning models to perform the prediction, namely: GoogLeNet, ResNet-18, and DenseNet-121 [19]. Recently, four CNN's models: simple CNN, VGG16, VGG19 and InceptionV3, were proposed to detect pneumonia from the chest x-ray images [20]. However, considering a set of models also increases its complexity and computational cost.

Although deep learning approaches are widely used, the lack of understanding of the results is a general limitation in health applications and particularly in diagnosis [21]. Besides, the development of more sophisticated models to assist in the radiograph,s diagnosis can lead to a gain in the correct cases. However, may cause a decrease in the interpretability of its results. To overcome this limitation, it is proposed relevance aggregation, which aims to provide the interpretation of the results [17,22].

In this way, this study presents a simplified deep learning model, maintaining high accuracy and sensitivity in the automatic detection of childhood pneumonia, also contributing to the health area by presenting interpretable results.

2 Materials and Methods

2.1 Materials

The adopted dataset [23] in this study is publicly accessible on Kaggle: https://www.kaggle.com/datasets/paultimothymooney/chest-xray-pneumonia. The dataset contains 5,856 x-ray images, with different gray-scale dimensions, in which it is divided between people with pneumonia and healthy people. Among healthy people, the dataset has 1,583 images, while for people with pneumonia, the dataset contains 4,273. The chest X-ray images (anterior-posterior) were selected from pediatric patients aged one to five years from the women's and children's medical center in Guangzhou, China, and the chest X-ray examinations were interpreted to confirm the diagnosis by two medical specialists. The details of the adopted image dataset are available in Table 1.

The adopted image dataset [23] presents three groups of image samples: training, testing, and validation. Thus, this organization was kept to allow a straightforward comparison of the results with other approaches that adopt the same dataset.

Table 1. Chest x-ray images adopted in this study.

Dataset	# Healthy	# Pneumonia	# Total
training samples	1,341	3,875	5,216
test samples	234	390	624
validation samples	8	8	16
Total	1,583	4,273	5,856

2.2 Methods

An overview of the method proposed in this study is available in Fig. 1.

Data Pre-processing. Once the dataset is defined, data preprocessing is proposed in order to perform the re-scaling of all the x-ray images using the MinMax rescale. The rescaling in this study is inspired by the Logistic Regression [25] and is important because it reduces the effects of illumination differences between images, and since the image is on a gray scale, the pixels range from intensity 0 (black) to 255 (white), so with the rescaling of the data the neural network converges faster from 0 to 1 than from 0 to 255.

Another common technique adopted in the pre-processing of the images was the histogram equalization [26]. This technique comprises adjusting the distribution of intensity values of the histogram to stress details not previously visible, increasing the contrast of the image. The calculation to perform histogram equalization can be seen in the Eq. 1. The results of applying this technique can be seen in Fig. 2.

$$g = (L-1) * \sum_{j=0}^{L-1} \frac{n_j}{n} \quad (1)$$

where: n_j is the number of pixels with intensity j, L is the number of possible intensity values, n is the total number of pixels.

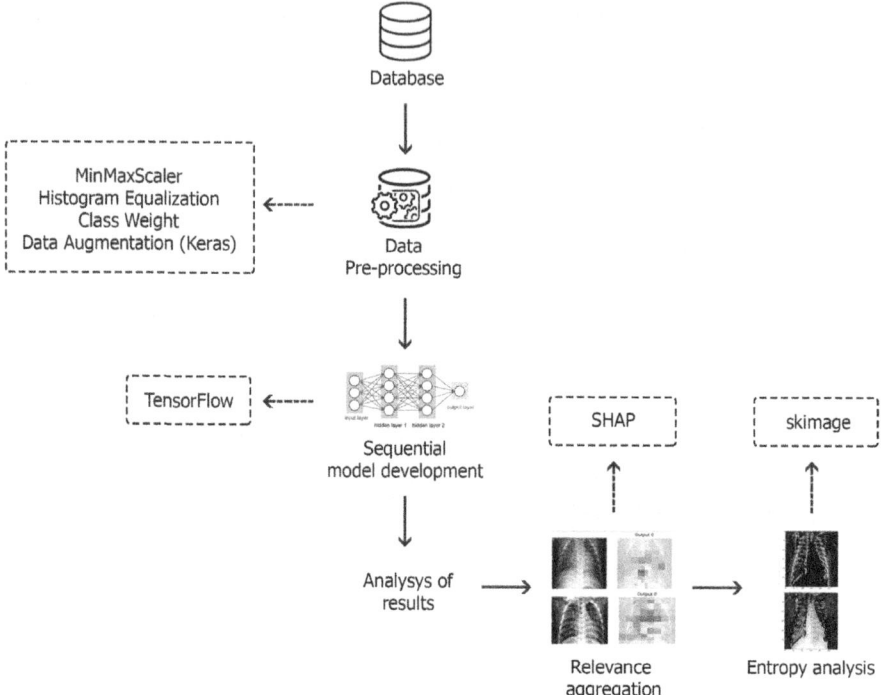

Fig. 1. Overview of the steps of the proposed method.

Still considering the pre-processing step, it is possible to notice in Table 1 that there is an unbalance in the number of samples when comparing the classes with pneumonia and healthy, in which there is a much larger number of samples with pneumonia than healthy. To balance the classes, a weight is adopted for each class, so the method considers the weight ($class_{weight}$) of the error of each sample ($n_{samples}$) in relation to its class ($n_{occurrences}$), as defined in Eq. 2.

$$class_{weight} = \frac{n_{samples}}{n_{classes} - n_{occurrences}} \qquad (2)$$

In order to test the robustness of the proposed approach, Keras Image Data Generator was adopted [27]. This technique consists of applying different transformations to original images that result in several transformed copies of the same image. Each copy, however, is different from the other in certain aspects, depending on the augmentation techniques applied, such as displacement, rotation, inversion, among others. Applying these small variations to the original image does not change the target class, but only provides a new perspective on capturing the object in real life. The parameters used in Keras can be seen in the Table 2.

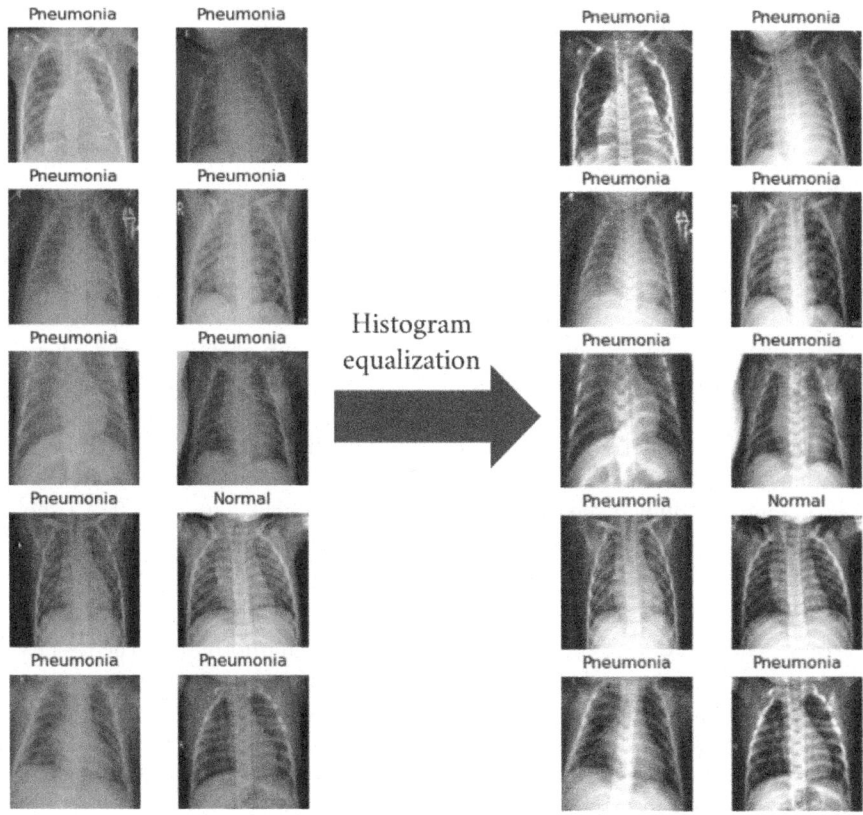

Fig. 2. Examples of images before (left) and after (right) the application of histogram equalization.

Table 2. Parameters used in Keras image data generator

Parameters	Values
rotation range	30
zoom range	0.2
width shift range	0.1
height shift range	0.1
horizontal flip	True

Sequential Model Development. Regarding the development of the neural network, a sequential model was implemented, which allows inserting layers of a neural network in series, in which the output of the first layer serves as input to the second, and so on. This model used only 5 convolution layers, Batch Normalization and Max Pooling, 2 dropout layers, 1 flatten layer and 2 dense layers that had the sigmoid and RELU functions as activation. In the end, the model has 499,777 parameters, of which 499,297 are trainable and 480 are untrainable.

The parameters used for the Adam optimizer, for the ReduceLROnPlateau callback and for compiling and training the model can be seen in the Table 3. The ReduceLROnPlateau callback aims to reduce the learning rate, helping to reduce model overfitting.

Relevance Aggregation. The relevance aggregation is an important step in this study, which gives the interpretability on the results. The main idea is to identify which are the features that the neural network considers important for the classification, and the analysis of these features can show the most relevant areas for the detection of pneumonia in the images [22]. In this study, to carry out the relevance aggregation, the SHAP [28] was adopted. It is based on the value of Shapley, a method to calculate the contribution of each player to the result of a game. In generalizing this idea, the pixels of the image were considered as players and the image classification as the result of the game. Section 3 shows the images resulting from this study with masks in the regions identified as important in the classification of pneumonia.

Entropy Analysis. The entropy is commonly used in information theory based on the Shannon entropy, which applies to quantify the amount of information in a source [29]. One way to apply entropy is using [30] as an objective evaluation method for image segmentation. Entropy can also be used for the analysis of x-ray images [31]. Therefore, in this study, entropy in x-ray images with pneumonia was adopted in order to evaluate the regions of interest that bring more information about the diagnosis of pneumonia. The library skimage [32] was adopted. In addition, the image binarization technique was adopted to visualize the areas of the x-ray that contained the highest levels of entropy.

3 Results and Discussion

In order to evaluate the proposed method and compare it with similar methods in the literature, the dataset presented in Table 1 was considered. 5,216 chest X-rays images were adopted for the training step and the tests were performed used the test dataset, which had 624 images.

Analysis of the results was divided into 2 steps. The first was to evaluate the training results by adopting the precision, recall, accuracy and loss. Figure 3 present the quantitative results of the adopted metrics. It is possible to notice that proposed method achieved an accuracy of 0.9450, loss of 0.1471, precision of 0.8513 and a recall of 0.9523. Besides, the proposed method achieved high assertiveness and rapid convergence after only a few training epochs.

The second step was to analyze the results on the test dataset. Thus, the accuracy, precision and recall were performed. The results on the test data pointed out an accuracy of 0.9071, loss of 0.2365, precision of 0.8826 and a recall of 0.8675.

A Simple and Interpretable Deep Learning Model for Diagnosing Pneumonia

Table 3. Parameters used for Adam, ReduceLROnPlateau and Training.

	Parameter	Value
Adam optimizer	Learning Rate	0.00017
	Beta 1	0.9
	Beta 2	0.999
	Epsilon	1e-07
	amsgrad	False
ReduceLROnPlateau	monitor	accuracy
	patience	2
	verbose	1
	factor	0.3
	min-lr	0.000001
Training	Loss Function	Binary Crossentropy
	Optimizer	Adam
	Assertiveness Metrics	Accuracy/Precision/Recall
	Weights for classes	0.673..(Pneumonia)/1.944...(Healthy)
	Callbacks	ReduceLROnPlateau

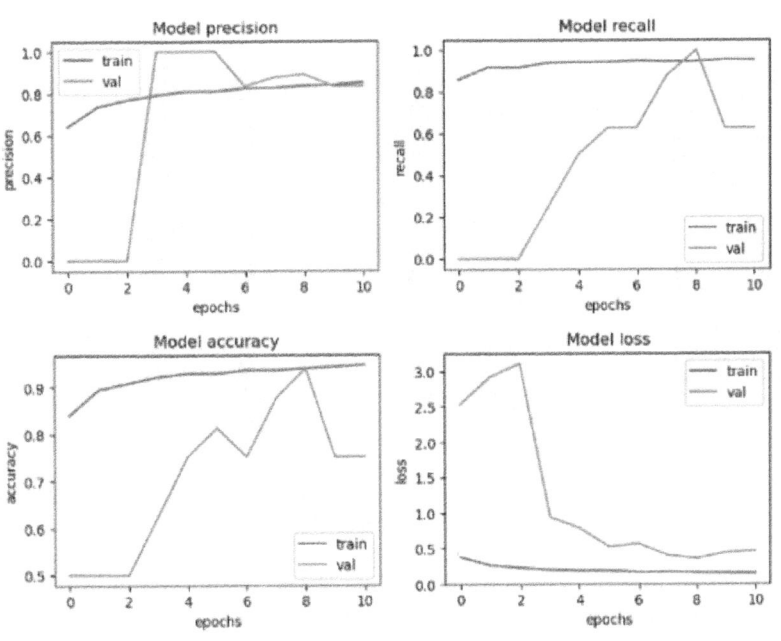

Fig. 3. Training results of the proposed method considering the number of epochs.

The comparative results to similar methods available in literature for the diagnosis of pneumonia using x-ray images can be seen in Table 4, which contains the values of the metrics got during the experiments and the number of x-ray images. Besides assertiveness results, an important contribution of this study is its interpretability by adopting the relevance aggregation. Thus, the contribution of the pixels of the image and the outcome of the pneumonia. Figure 4 presents 9 x-ray images as examples, in which the relevance aggregation was performed. The red pixels in the image represent the areas which increase the probability of the pneumonia class being predicted, while the blue pixels decrease the probability of the class being predicted.

Table 4. Comparison of results with similar methods in the literature.

Method	Number of X-ray images	Accuracy	AUC
Kermany et al. [23]	5.232	92.8%	93.2%
Rajpurkar et al. [18]	112.120	—	76.8%
xRayAID [24]	26.684	87,9%	86,9%
Stephen [13]	5,856	93%	—
Proposed method	5,856	90.71%	89.91%

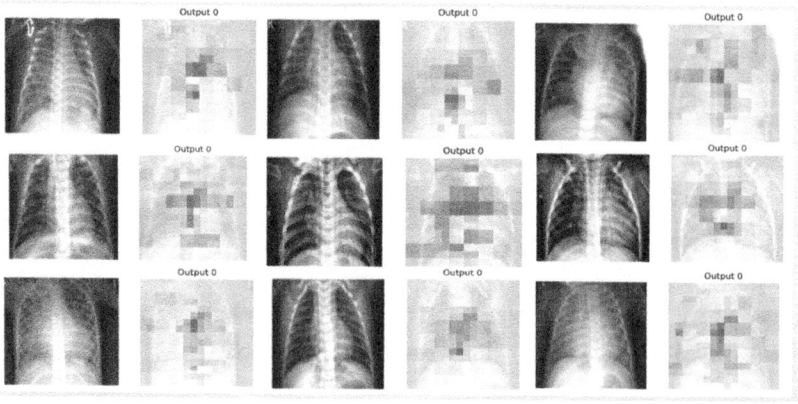

Fig. 4. Examples of image from the proposed method with relevance aggregation.

In order to improve the interpretability of the results, an entropy analysis was performed on x-ray images with pneumonia, and as a result, bring more information about the diagnosis of pneumonia. Experiments were performed in order to analyze the pixel neighborhood radius for the entropy analysis. Figure 5 shows examples of results changing the considered radius, from the value 1 to 9.

It can be seen that as we increase the radius of the disk, the image becomes more blurred. Therefore a disk radius value has been selected that is a good balance between sharpening and blurring. The experiments carried out showed that the radius value 4 produced the suitable results.

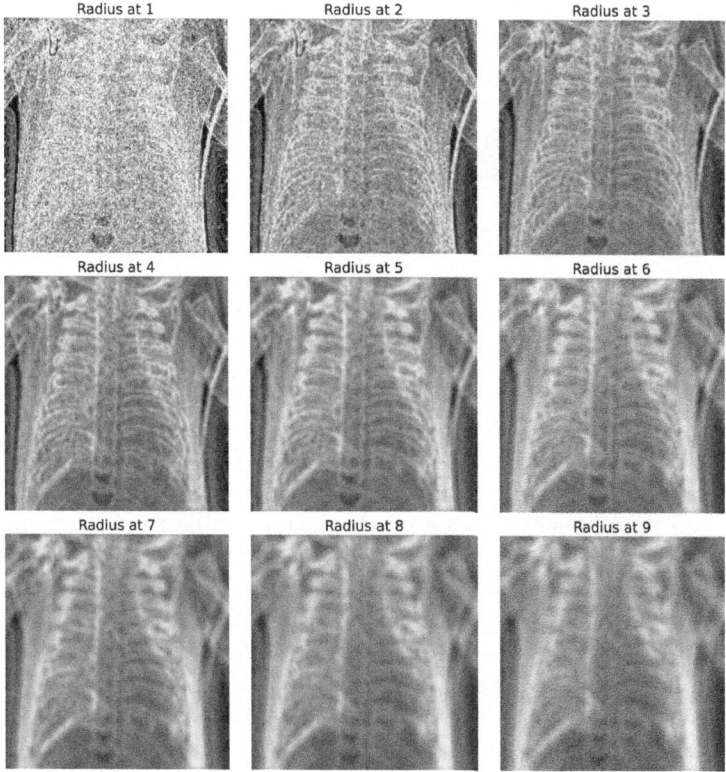

Fig. 5. Entropy analysis from X-ray images with different radius values.

The last step is the image binarization technique, in order to visualize the areas of the x-ray that contained the highest entropy levels. In this step, the images with different threshold levels are produced in order to find regions with high entropy values. Thus, it was identified that the best results were produced with the threshold value 0.8. As a result, by considering the binary images, it was possible to identify the relevant areas for the automatic diagnosis of pneumonia. Figure 6 shows an example output, the image on the left shows in the x-ray the regions that have entropy values greater than 0.8 considering a radius of 4, which are represented by the areas in white, using the same approach, the image on the right shows the areas of entropy with greater value than 0.8 in black color. In this way, the proposed method allows to identify the regions of interest of the images were important for the diagnosis of pneumonia.

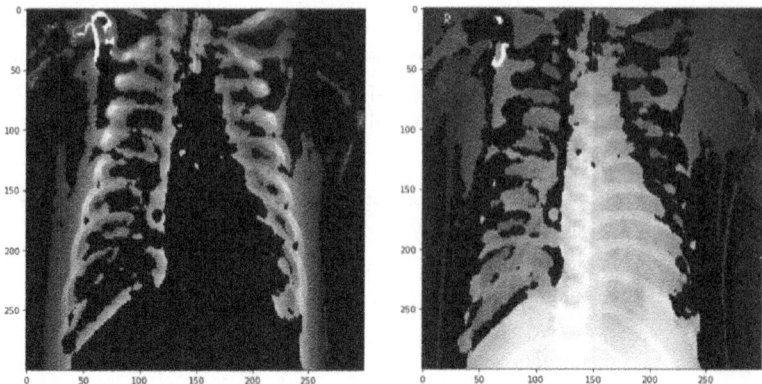

Fig. 6. Entropy analysis to identify the region of interest in the diagnosis of pneumonia.

4 Conclusion

The greatest infectious cause of the death of children in the world is pneumonia. Early detection of the disease is fundamental in treating the patient correctly and avoid severe cases of the disease, which can lead to death. There are several ways to identify if a person has the disease. The most used is the chest X-ray. However, the assertiveness depends on the doctor's subjective variability.

Given this problem, this study presents an artificial intelligence, based on deep learning, that classifies chest X-ray images into two classes, with pneumonia or healthy. Using a simpler model, the proposed approach got 90.71% accuracy, 88.26% precision, and 86.75% recall. Other studies in the same area have got similar performances, however these models are much more complex and computationally demanding. Moreover, the proposed approach presents as a contribution to interpreting the results, through relevance aggregation. As a result, images are produced, showing the regions of interest that are important for the diagnosis of pneumonia automatically.

This study used only radiographic images of children aged 1 to 5 years from a hospital in China. Because different x-rays can produce diverse results, in future work, it is projected to expand and use different image datasets that consider different age groups and ethnicities, in order to produce a more general learning model.

Acknowledgments. This study was financed by the Fundação Araucária (Grant number 035/2019, 138/2021 and NAPI - Bioinformática) and CNPq (Grant number 440412/2022-6 and 408312/2023-8).

References

1. Pneumonia in Children. https://www.who.int/news-room/fact-sheets/detail/pneumonia. Accessed 29 May 2023
2. Luna, C.M.: Appropriateness and delay to initiate therapy in ventilator-associated pneumonia. Eur. Respir. J. **27**, 158–64 (2006)
3. Mohanan, M., Vera-Hernández, M., Das, V., et al.: The know-do gap in quality of health care for childhood diarrhea and pneumonia in rural India. JAMA Pediatr. **169**(4), 349–357 (2015)
4. Pinto, S.C.D., Mena-Chalco, J.P., Lopes, F.M., Velho, L., Cesar, R.M.: 3D facial expression analysis by using 2D and 3D wavelet transforms. In: 2011 18th IEEE International Conference on Image Processing. IEEE (2011)
5. Brilhador, A., Colonhezi, T.P., Bugatti, P.H., Lopes, F.M.: Combining texture and shape descriptors for bioimages classification: a case of study in ImageCLEF dataset. In: Ruiz-Shulcloper, J., Sanniti di Baja, G. (eds.) CIARP 2013. LNCS, vol. 8258, pp. 431–438. Springer, Heidelberg (2013). https://doi.org/10.1007/978-3-642-41822-8_54
6. Brilhador, A., Serrarens, D.A., Lopes, F.M.: A computer vision approach for automatic measurement of the inter-plant spacing. In: CIARP 2013. LNCS, vol. 8258, pp. 219–227. Springer, Cham (2015). https://doi.org/10.1007/978-3-319-25751-8_27
7. de Lima, G.V.L., Castilho, T.R., Bugatti, P.H., Saito, P.T.M., Lopes, F.M.: A complex network-based approach to the analysis and classification of images. In: CIARP 2013. LNCS, vol. 8258, pp. 322–330. Springer, Cham (2015). https://doi.org/10.1007/978-3-319-25751-8_39
8. de Lima, G.V., et al.: Classification of texture based on bag-of-visual-words through complex networks. Expert Syst. Appl. **133**, 215–224 (2019)
9. de Souza Piotto, J.G., Lopes, F.M.: A feature extraction approach based on LBP operator and complex networks for face recognition. In: Tavares, J.M.R.S., Papa, J.P., González Hidalgo, M. (eds.) CIARP 2021. LNCS, vol. 12702, pp. 440–450. Springer, Cham (2021). https://doi.org/10.1007/978-3-030-93420-0_41
10. de Souza Piotto, J.G., Lopes, F.M.: Combining SURF descriptor and complex networks for face recognitio. In: 2016 9th International Congress on Image and Signal Processing, BioMedical Engineering and Informatics (CISP-BMEI). Datong, China, vol. 2016, pp. 275–279 (2016)
11. Garg, A., Mago, V.: Role of machine learning in medical research: a survey. Comput. Sci. Rev. **40**, 100370 (2021)
12. Shinde, Pramila P., Shah, S.: A review of machine learning and deep learning applications. In: 2018 Fourth International Conference on Computing Communication Control and Automation (ICCUBEA). IEEE (2018)
13. Stephen, O., Sain, M., Maduh, U.J., Jeong, D.U.: An efficient deep learning approach to pneumonia classification in healthcare. J. Healthc. Eng. **2019**, 4180949 (2019)
14. Gabruseva, T., Poplavskiy, D., Kalinin, A.: Deep learning for automatic pneumonia detection. Proceedings of the IEEE/CVF Conference on Computer Vision and Pattern Recognition Workshops (2020)
15. Li, Y., et al.: Accuracy of deep learning for automated detection of pneumonia using chest X-Ray images: a systematic review and meta-analysis. Comput. Biol. Med. **123**, 103898 (2020)

16. Williams, G.J., et al.: Variability and accuracy in interpretation of consolidation on chest radiography for diagnosing pneumonia in children under 5 years of age. Pediatr. Pulmonol. **48**(12), 1195–1200 (2013)
17. Gomes, J.M., Lopes, F.M.: Interpretability with relevance aggregation in neural networks for absenteeism prediction'. In: 2022 IEEE-EMBS International Conference on Biomedical and Health Informatics, Greece, 2022, pp. 01-04 (2022)
18. Rajpurkar, P., et al.: ChexNet: radiologist-level pneumonia detection on chest x-rays with deep learning. arXiv preprint arXiv:1711.05225 (2017)
19. Kundu, R., Das, R., Geem, Z.W., Han, G.T., Sarkar, R.: Pneumonia detection in chest X-ray images using an ensemble of deep learning models. PLoS ONE **16**(9), e0256630 (2021)
20. Labhane, G., et al.: Detection of pediatric pneumonia from chest X-Ray images using CNN and transfer learning. In: 3rd Internationa Conference on Emerging Technology in Computer Engineering: Machine Learning and Internet of Things, India, 2020, pp. 85–92 (2020)
21. Stiglic, G., et al.: Interpretability of machine learning-based prediction models in healthcare. Wiley Interdiscip. Rev. Data Min. Knowl. Discov. **10**(5), e1379 (2020)
22. Grisci, B.I., et al.: Relevance aggregation for neural networks interpretability and knowledge discovery on tabular data. Inf. Sci. **559**, 111–129 (2021)
23. Kermany, D.S., et al.: Identifying medical diagnoses and treatable diseases by image-based deep learning. Cell **172**(5), 1122–1131.e9 (2018)
24. Trevisan, V., Rodrigues, D., Rezende, E.: xRayAID detecting pneumonia using artificial intelligence. In: Anais do XXI Simpósio Brasileiro de Computação Aplicada à Saúde, pp. 1–12. Porto Alegre (2021)
25. Tomz, M., King, G., Zeng, L.: ReLogit: rare events logistic regression. J. Stat. Softw. **8**(2), 1–27 (2003)
26. Manpreet Kaur, Jasdeep Kaur and Jappreet Kaur, "Survey of Contrast Enhancement Techniques based on Histogram Equalization" International Journal of Advanced Computer Science and Applications(IJACSA), 2(7), 2011
27. Chollet, F. & others, 2015. Keras. https://github.com/fchollet/keras
28. Lundberg, Scott M., and Su-In Lee. "A Unified Approach to Interpreting Model Predictions." Advances in Neural Information Processing Systems, edited by I. Guyon et al., vol. 30, Curran Associates, Inc., 2017
29. Shannon, C.E.: A mathematical theory of communication. The Bell System Technical Journal **27**, 379–423 (1948)
30. Hui Zhang, Jason E. Fritts, and Sally A. Goldman "An entropy-based objective evaluation method for image segmentation", Proc. SPIE 5307, Storage and Retrieval Methods and Applications for Multimedia 2004, (18 December 2003)
31. D. Abin, SD Thepade, H. Mankar, S. Raut e A. Yadav, "Blending of Contrast Enhancement Techniques for Chest X-Ray Pneumonia Images", 2022 International Conference on Electronics and Renewable Systems, Tuticorin, 981-985
32. Stéfan van der Walt et. al. scikit-image: Image processing in Python. PeerJ 2:e453 (2014) https://doi.org/10.7717/peerj.453

FedDP: Secure Federated Learning with Differential Privacy for Disease Prediction

Bin Li, Hongchang Gao, and Xinghua Shi[✉]

Department of Computer and Information Sciences, College of Science and Technology, Temple University, Philadelphia, PA 19122, USA
mindyshi@temple.edu

Abstract. Integrative analysis of distributed biomedical data is essential for maximizing knowledge discovery, accelerating medical breakthroughs, and improving patient care through collaborative research and practices. However, it is challenging to share and aggregate biomedical data distributed among multiple institutions or computing resources due to various concerns including data privacy, security, and confidentiality. The federated Learning (FL) framework can effectively enable multiple institutions to jointly perform machine learning by training a robust model without sharing local data to satisfy the requirement of user privacy protection as well as data security. However, conventional FL methods are exposed to the risk of gradient leakage and cannot be directly applied to genomic data since they cannot address the unique challenges of data imbalance typically seen in biomedicine. To provide secure and efficient disease prediction based on biomedical data distributed across multiple parties, we propose an FL framework enhanced with differential privacy (FedDP) on trained model parameters. The key idea of FedDP is to deploy differential privacy on intermediate gradients that are computed and transmitted by optimizers from local parties. In addition, the unique weighted min-max loss in FedDP is deployed to address the challenge of fair prediction on highly imbalanced datasets. Our experiments on label-imbalanced datasets for cancer prediction demonstrate that FedDP provides a powerful tool to implement and evaluate various strategies in support of privacy preservation and model performance guarantee to overcome data imbalance.

Keywords: Federated Learning · Differential Privacy · Data Imbalance

1 Introduction

Disease prediction and personalized medicine have become increasingly important in the field of healthcare. Cancer patients, in particular, can benefit from the analysis of their genomic and molecular data to determine tumor subtypes, pathological stages, and personalized treatment options. Artificial intelligence (AI) algorithms can be employed to calculate disease risk scores, classify tumors, develop treatment strategies, and predict clinical outcomes based on genetic and molecular data [35]. However, the successful implementation of these AI models often relies on access to large-scale datasets that are distributed across multiple institutions, healthcare systems, and geographic regions [25]. Data sharing and aggregation are challenging due to privacy concerns and technical constraints, hindering the development of robust and secure AI systems for biomedical research and clinical practices [14,28,29].

The field of AI and machine learning (ML) has made significant advancements including in the medical field regarding disease prediction and diagnosis [9]. However, the effectiveness of AI/ML algorithms often depends on the amount of data that is used to train these algorithms. A critical issue at the intersection of ML and medicine is to collect and analyze different types of data at a large scale to enable precision medicine with consideration of the security and privacy of participating individuals. [10] Therefore, new computing frameworks need to be explored to provide secure and privacy-preserving analysis of medical data where data and computing can be distributed across multiple institutions and diverse infrastructures. Federated Learning (FL) is such a computing framework that can effectively help multiple institutions to perform data analysis and ML modeling under the requirements of user privacy protection, data security, and government regulations [12,31]. Within the framework of FL, generic approaches [28] have been proposed for privacy-preserving training of ML models in an N-party setting that employs multiparty lattice-based cryptography.

To enhance the FL performance on privacy preservation, some works focused on secured aggregation algorithms for federated processing among multiple parties based on homomorphic encryption (HE) [11,22,33]. Although the FL with HE improves the security against potential attacks, it is not appropriate for complicated models due to its limitation in the costs of encryption and decryption [20,33]. Thus, some studies have employed differential privacy (DP) to improve the privacy preservation of FL [16,23]. The model based on both FL and DP has significant privacy protection capabilities in the direction of medical diagnosis. The main idea is to share model parameters with DP [36], not original data, aggregating them after local training steps, eventually leading to a shared global model with secure training on data distributed at multiple parties [26].

In addition, biomedical data are highly imbalanced and typically distributed in different institutions, limiting the possibility of training well-performing FL models to benefit healthcare decision-making [13,15]. Specifically, those FL methods typically optimize the accuracy-induced objective function, e.g., the cross-entropy loss function, whose incapability of handling imbalanced data has been confirmed by numerous studies [8,18]. Hence, it is imperative to develop new FL approaches to address the issue of imbalanced data distribution while preserving data privacy as demonstrated in security analysis [6].

In this study, we propose a federated learning framework incorporated with differential privacy (FedDP) to provide differentially private federated learning for disease and pathological stages prediction, with optimization strategies deployed to handle imbalanced data. In FedDP, differential privacy was deployed on model parameters during the transmission instead of local datasets. Meantime, we used a weighted min-max optimizer to assist in enhancing the capacity of secure federated learning on imbalanced data which is commonly found in biomedical datasets. Our empirical results on various datasets have shown that FedDP is efficient in support of privacy-preserving cancer prediction, even when data are imbalanced among different parties. The contribution of this study is summarized below.

First, we investigate the shortcomings of current federated models and develop a strategy to strike a trade-off between the security and the performance of federated models for disease prediction. We implemented such a strategy in a newly developed FedDP framework to support secure and privacy-preserving disease prediction. FedDP is a privacy-preserving FL model in that model gradients are shared with (ϵ, δ)-DP (instead of direct sharing of the original data), aggregated after local training steps, and finally lead to a securely-shared optimized global model.

Second, The proposed FedDp employs the Area under the ROC Curve (AUC)-induced min-max loss function rather than optimizing the accuracy-induced loss (e.g., cross-entropy), which can consider both precision and recall metrics to learn a comprehensive model across imbalanced data.

Third, we demonstrate the proposed FedDP by applying it to two cancer datasets from The Cancer Genome Atlas (TCGA) project for cancer prediction. We evaluate the performance of the FedDP framework in two classification scenarios for breast cancer and kidney cancer. Experimental results show that FedDP performs well under the strong privacy budget in both two cases as well as dealing with the label-imbalance problem. Hence, these results demonstrate the generality and effectiveness of the FedDP framework for disease prediction on label-imbalanced biomedical datasets.

The remainder of this paper is organized as follows. We first describe the details of FedDP in Sect. 2. Then we present the empirical evaluation results in Sect. 3. Finally, we conclude the paper in Sect. 4 with future directions.

2 Framework of FedDP

In this section, we present our privacy-preserving federated learning framework that combines FL with DP (FedDP) for disease prediction.

2.1 Federated Learning Scheme

In this study, we assumed that the data source and type are transparent to each party, meaning local data held by each party are gene expression sequences of the same disease. We built a horizontal FL framework to satisfy the assumption. As simplified in Fig. 1 (a), we illustrate a simplistic setting where this exemplar FL system is composed of one centralized server and three parties deployed at different devices. Each federated round is conducted following four steps:

- Initialize the models and parameters for each party and the server.
- Each party trains the model separately in certain epochs (local hyperparameter) based on the current model parameters, then extracts new local model parameters.
- Add DP noises to local parameters then upload to the central server.
- The global optimizer of the server conducts an aggregation method to compute the global parameters. Aggregated parameters are broadcast from the server to the parties at the end of each federated round. Afterward, each party updates its local models with the received parameters.

We employed the (ϵ, δ)-DP mechanism for securing the intermediate parameters in the transmission process. Weights from each local deep learning network will not be transmitted in a federated process, instead, gradients computed by local optimizers are selected as intermediate parameters with DP noises. Specifically, for arbitrary party C_i holding local data x_i, Gaussian noises $\mathcal{N}(0, \frac{\delta_P^2}{\epsilon^2})$ were added to its model parameters $\nabla W(x_i)$ to satisfy the DP condition:

$$Pr[\nabla W'(x_i)] \leq e^\epsilon Pr[\nabla W'(x_i')] + \delta \qquad (1)$$

where $\nabla W'(x_i)$ is secured parameters calculated by

$$\nabla W'(x_i) = \nabla W(x_i) + \mathcal{N}(0, \frac{\delta^2}{\epsilon^2}) \qquad (2)$$

We call this framework FedDP since it is a secured FL framework that combines gradients-transmitted FL and DP.

It should be noted that there are two kinds of optimizers in FedDP: local optimizers and one server optimizer. Local optimizers are only used to compute local model updates at each party. DP is added to the gradients to be transmitted from local optimizers to ensure the ability of privacy-preserving. The server optimizer is responsible for computing global gradients by aggregating local gradients from parties. From this perspective, this distributed architecture ensures that each party does not need to share their data with others.

Training of FedDP. In FedDP, global gradients were aggregated by averaging all gradients learned from local parties. Multiple parties could set different aggregating weights regarding their relative data sizes. The proposed procedure protects the per-sample privacy of each participating party dataset. Differential privacy is added by parties individually where the dataset sizes can be very different. Depending on the setting of the sampling strategy, the sampling ratio q can be different from party to party. Thus the privacy loss is variant by the local dataset. Each party should keep a record of the spent privacy along each update and stop its update to the central server once the predefined privacy threshold is reached.

2.2 Model Architecture

It is admitted that the efficiency of centralized federated learning is significantly limited by the bandwidth of the server since the whole upload process from parties to the server is synchronous. Therefore, deploying complex deep learning models to this one-to-many structure would increase the number of parameters, resulting in increased communication overhead of the whole network. Most FL-based studies [5,19,34] have used simple models for training, including linear models, random forests, and support vector machines (SVMs). Hence, we prefer to deploy simple models for parties to ensure efficiency and communication overhead. Specifically, we built a two-layered CNN model for local training. Since the input data are a two-dimensional matrix, we constructed two Cov1D layers in which the size is $32 \times 2000 \times 16$. We observed that the classification task is prone to overfitting because of the large number of input features. Thus, we reduced the number of effective weights by deploying L1 regularization at the end of each convolution layer. In addition, we constructed the logistic regression model with L1 regularization (lasso) as a comparison.

2.3 Optimization

The conventional accuracy-optimized cross-entropy loss function is indeed applicable to binary and multi-class classification problems. However, it is challenging to learn a well-performing classifier with imbalanced biomedical data. The model is prone to run into a dead-end in which all test samples are predicted as the major category, in this way, the loss function no longer drops because the accuracy is very high (very low precision or recall) at a certain point. Therefore, recent studies were focused on optimizing the AUC which is a reasonable metric for imbalanced-based-data classification tasks. The main feature of AUC is that it aggregates across different threshold values for binary prediction, separating the issues of threshold setting from predictive power. Meanwhile, AUC considers both precision and recall metrics to learn a comprehensive model across imbalanced data. Thus, it is rational to directly optimize the AUC score, rather than the cross-entropy loss function. Specifically, Ying et al. [32] developed the following minimax loss function for the AUC maximization problem:

$$\min\max \mathfrak{L}(w, \hat{w}_1, \hat{w}_2, \theta; a, b) = (1-p)(f(w;a) - \hat{w}_1)^2 \mathbb{K}_{[b=1]} \\ + p(f(w;a) - w_2)^2 \mathbb{K}_{[b=0]} - p(1-p)\theta^2 \\ + 2(1+\theta)(pf(w;a))\mathbb{K}_{b=0} - (1-p)f(w;a)\mathbb{K}_{[b=1]}) \quad (3)$$

where $\hat{w} \in \mathbb{R}^d$ denotes the weights of model f, $\hat{w}_1 \in \mathbb{R}$, $\hat{w}_2 \in \mathbb{R}$, $\theta \in \mathbb{R}$ are additional parameters to compute AUC score, a is the feature of samples and b represents the label. p is the rate of positive samples to all samples and $\mathbb{K}$ is an indicator function. This loss was optimized by ADAM SGD methods [17].

3 Data and Experiments

In this section, we present our experimental details and result analysis. We illustrate the details of the imbalanced biomedical datasets we used in this research. Then, we show the fine-tuned hyperparameter settings and analyze the performance of FedDP for cancer prediction tasks. We evaluate the capacity to deal with the imbalanced problem of datasets based on FedDP.

3.1 Datasets

To investigate the contribution of privacy preservation for a federated learning framework, we evaluated FedDP with two different models on imbalanced datasets. The Cancer Genome Atlas (TCGA) [2], which is a project overseen by the National Cancer Institute and the National Human Genome Research Institute to apply high-throughput genome analysis techniques to help people have a better understanding of cancer as well as improve the ability to prevent, diagnose and treat cancer [30].

To evaluate FedDP for the cancer prediction tasks, we utilized one breast cancer dataset [1], BC-TCGA that labels the sample as two categories, and one kidney cancer dataset, TCGA-Kidney. **BC-TCGA**: is a binary labeled dataset containing 61 normal samples and 529 breast-tumor samples. Each sample has gene expression quantification in 17,814 genes. **TCGA-Kidney**: a dataset of all 1,024 kidney-tumor samples with different pathological stages, integrated as 513 early stages (stage I), 126 mid stages (stage II), and 350 advanced stages (stage III and IV) based on categorical standards from National Cancer Institute [3]. Each sample has expression values of 60,487 genes. All of the expression quantification in two datasets comes from RNA Sequencing data, generated by gene level and exon level quantification in Fragments Per Kilobase of transcript per Million mapped reads (FPKM). Then normalized into Upper Quartile normalized FPKM (FPKM-UQ) formation.

Additionally, two datasets enable us to test FedDP to explore the potential performance of FedDP for binary prediction tasks while multi-class prediction tasks could be tested on the kidney dataset. In addition, since these two datasets have different label imbalance rates that approximately 9:1 (tumor to normal) for the BC-TCGA dataset and 4:1:3 (early to mid to advanced) for the TCGA-Kidney dataset, we can further evaluate the capacity of FedDP for addressing the imbalance problem with different imbalance ratios.

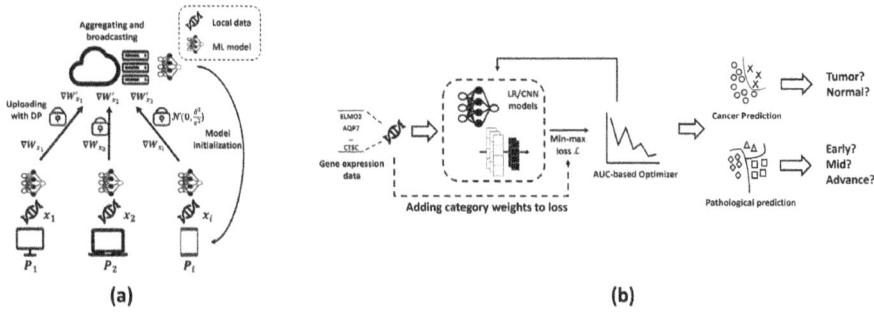

Fig. 1. (a). An illustration of a secured federated framework with DP. Models and data are kept at local parties without sharing. Local parameters secured by DP are transmitted to the central server, and the server updates global parameters and coordinates communications with local parties. (b). An illustration of prediction tasks for imbalanced biomedical datasets. Two classification tasks were performed by FedDP. Binary-labeled breast cancer prediction using BC-TCGA gene expression data. Multi-labeled kidney cancer prediction using TCGA-Kidney gene expression data. The Min-max loss was employed to address the label-imbalance problem and improve accuracy.

3.2 Evaluation Metrics

Our solutions are based on the conventional (ϵ,δ)-DP model implemented in Tensorflow [4]. The term ϵ, called privacy budget, is the metric to determine if a system has a strict privacy-preserving ability under DP. Theoretically, the smaller epsilon a system has, the more stringent privacy protection there will be. ϵ is estimated by the Dual and Central Limit Theorem (CLT) [7,21] and is influenced by the factors below:

- Sampling rate. The sampling rate is represented as the ratio of batch size to sample size. Theoretically, a lower sampling rate generates lower ϵ if other factors are constant, which means better privacy-preserving capabilities.
- Noise multiplier: Noise multiplier is the ratio of the standard deviation of the Gaussian noise to the l2-sensitivity of the function to which it is added. This governs the amount of noise added during training. Generally, more noise results in better privacy and lower utility. The density of the Gaussian noises added in terms of μ is directly influenced by the noise multiplier via

$$\mu = \frac{\sqrt{2mq}}{q} \cdot \sqrt{e^{\frac{1}{\lambda^2}} \mathcal{F}(\frac{3}{2\lambda}) + 3\mathcal{F}(-\frac{1}{2\lambda}) - 2} \qquad (4)$$

where m is the number of epochs and q is the sampling rate, $\mathcal{F}$ is the cumulative distribution function of Gaussian distribution.
- Delta(δ): Delta bounds the probability of an arbitrary change in model behavior. We can usually set this to a very small number (1e-7 or so) without compromising utility. A rule of thumb is to set it to be less than the inverse of the training data size. (We set delta to 1e-3).

Table 1. Performance of FedDP on BC-TCGA (D_1) and TCGA-Kidney(D_2). Each row represents the accuracy under certain hyperparameter settings. There are two kinds of models deployed in FedDP: logistic regression and two-layered CNN. Min-max loss function has been deployed for both cases. The trade-off taken between accuracy and privacy budget epsilon should be taken to obtain the best performance.

Noise Multiplier	Delta	Batch Size	Privacy Budget ϵ_{d1}	Privacy Budget ϵ_{d2}	Acc			
					D_1,LR	D_2,LR	D_1,CNN	D_2,CNN
1.5	1e-3	32	0.84	1.91	0.95	0.34	0.98	0.41
1.3	**1e-3**	**32**	**1.11**	**2.35**	**0.96**	**0.52**	**0.98**	**0.65**
1.1	1e-3	64	2.39	4.15	0.98	0.59	0.99	0.71
1.1	1e-3	32	1.6	3.11	0.98	0.56	0.99	0.67
1.0	1e-3	32	1.94	3.69	0.99	0.56	0.99	0.68
1.0	1e-3	32	3.11	4.89	0.99	0.62	0.99	0.72
0	0	32	None	None	0.99	0.63	0.99	0.72

3.3 Experimental Results

The whole process of FL is simulated on one high-performance clustering server that has three NVIDIA Tesla V100 GPUs. We evaluate the performance of the proposed algorithm under a varying (ϵ,δ)-DP privacy budget. The grid-search method is performed to fine-tune the parameters in experiments. The result under each set of parameters was averaged 5 times.

Fine-Tuned Results. Several optimal parameter settings for the two models on two different datasets are shown in Table 1. Our experimental results show that the smaller the batch size is and the greater noise multiplier is, the smaller the privacy budget is, which means the higher degree of privacy-preserving capability. Moreover, results demonstrate that increasing batch sizes will lead to higher accuracy but low privacy security. Thus, the accuracy of disease prediction in FedDP is positively related to the noise multiplier and δ. Consequently, reducing batch sizes means that more time is consumed during training. Thus, it is vital to make a trade-off between privacy budget and accuracy which are chosen as evaluation metrics throughout the whole parameter tuning process.

Prediction Vs. Privacy. To explore how DP impacts the accuracy of the model, we tuned the hyperparameters to find the best setting for ϵ by conducting a grid search. The results indicate the model performance deteriorates with the decrease of the privacy budget ϵ. Specifically, for the classification of the BC-TCGA dataset based on the logistic regression model (LR), to maintain higher accuracy rather than privacy, the learning rate of DP-SGD on the server is appropriate at 0.003, and the number of epochs per round on each party is 15, and the noise multiplier is set to 1.0. Based on this setting, we have achieved a 0.99 accuracy under the DP criteria that ϵ is 3.11 and δ is 1e-3. Moreover, if we increased the batch size for the training dataset. There would be a higher privacy budget after calculation, which means all procedures will be secured under strict privacy criteria. Certainly, the performance reached max without any DP added to the transmission. However, this will be exposed to the risk of privacy leakage. On the contrary, the best privacy budget ϵ we obtained is 0.84, which means the whole training process is under very strict privacy preservation. This also sacrifices the accuracy in test processing, which has 0.98 accuracy. In fact, this result is relatively promising due to the complexity of the BC-TCGA dataset.

It is straightforward to find the binary classification boundary for the BC-TCGA dataset. However, for multi-label classification on the TCGA-Kidney dataset, it is more difficult to find a classification boundary. For example, the LR model was not able to reach an accuracy greater than 0.65. The CNN model could reach an accuracy of 0.72 without any privacy preservation. We have thus investigated possible reasons for the model's incapability to achieve a high accuracy rate. From the perspective of the dataset, TCGA-Kidney is "fatter" than BC-TCGA, that is, compared with the BC-TCGA dataset (17,814 features and 590 samples), the number of features in TCGA-Kidney has significantly increased, reaching 60,843 gene expression values per sample while the sample size has not increased a lot (1,024 samples). This leads us to use stricter regularization and improve the randomness of the model to prevent overfitting. With consideration of security, we did not perform effective feature selection to reduce redundant features because this will inform the server about the data details of each client and violate local data privacy. Hence, our results are based on raw feature data without prepossessing.

We evaluated the privacy-preserving ability of FedDP under different privacy budgets with the best hyperparameter setting (shown in **Table** 1). Figure 2 shows the curves of accuracy versus privacy for two classification tasks using the logistic regression model and the two-layered CNN model with the min-max loss function. For the breast cancer prediction task, we finally could achieve a 1.0 accuracy under the DP criteria that ϵ is 5.38 and δ is 1e-3. Moreover, we can gain an even stricter privacy budget of 0.98 with 0.95 accuracy based on the CNN model. For the prediction of kidney cancer pathological stages, our method achieved 0.65 accuracy with privacy budget $\epsilon = 2.35$ in case of balancing the security preservation and model utility.

Addressing Label Imbalance. The imbalanced dataset (i.e., $\frac{pos}{neg} > 9$) is randomly divided into three subsets, each allocated to one of three parties. It is crucial as it introduces variability and imbalance in the data available to each party. We could further explore if all parties collaboratively optimize the loss function to learn the best model parameter against imbalance. We have tested and plotted the performance of two prediction tasks using the same CNN model with two different loss functions (min-max loss and standard cross entropy loss function as a comparison) under the two certain privacy budgets, as shown in Fig. 3. Results of the BC-TCGA dataset and TCGA-Kidney dataset are shown as two panels, Fig. 3 (a) and Fig. 3(b). Each panel contains three columns.

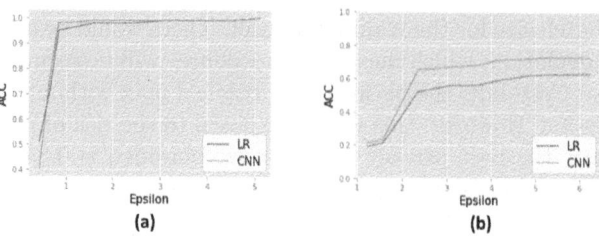

Fig. 2. Testing accuracy of logistic regression (blue) and CNN (orange) models with different privacy budget epsilon values under the data-center setting with three parties: (a) Results on BC-TCGA dataset. (b) Results on TCGA-Kidney datasets. Min-Max loss is used for both models. (Color figure online)

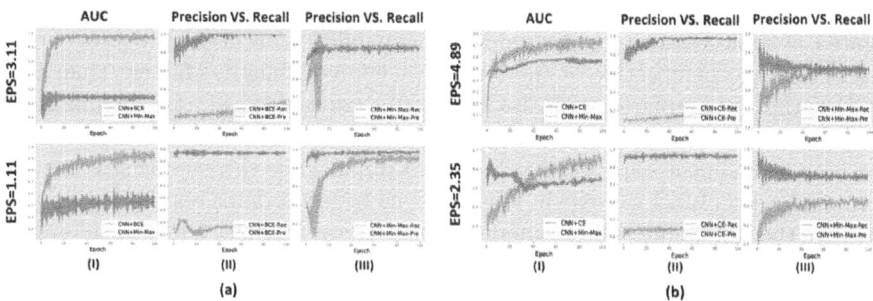

Fig. 3. Cancer prediction results of the min-max optimization for CNN models under two different privacy budgets in case of the data-center setting with three parties: (a) results of breast cancer prediction using BC-TCGA datasets. (b). results of pathological stages prediction for kidney cancer using TCGA-Kidney dataset. (I) AUC scores of the min-max optimization (Orange) and a regular binary cross-entropy (BCE) optimization (Blue) on the same CNN model over 100 epochs. Light-shaded areas reflect the dispersion of AUC scores for the three parties. (II) The precision (orange) and recall (blue) score of the CNN model with BCE loss over 100 epochs. (III) Precision (orange) and recall (blue) score with min-max loss function over 100 epochs. (Color figure online)

Column (I) shows the classification performance (AUC) of min-max loss compared with conventional binary cross-entropy (BCE) loss. AUC results of two datasets show that min-max loss could improve the classification performance while using cross-entropy loss could only achieve around 0.5 AUC score, tending to be a random-guess model. Column (II) shows that the precision is low and cannot gain much learning during the training process using the cross-entropy loss, meaning this loss function cannot address the imbalance problem because precision represents how accurate the prediction in the positive sample is. Low precision indicates there is a large amount of false positive predictions, the model is prone to predict all samples as positive. This explains why the AUC score of the model with cross-entropy loss is around 0.5 since it didn't learn any relation between features and labels due to the high-imbalanced data. Column (III) shows weighted min-max could overcome the data-imbalanced problem for two imbalanced datasets, the precision score is improved obviously during the training process. The shaded area reflects the fluctuation of learning among all parties since the data-imbalanced ratio of local parties is different.

In conclusion, by exploring the trade-off between accuracy and privacy budgets, we first illustrated that the CNN model has a better capability to protect privacy compared with the linear logistic regression model. Then, we demonstrated the min-max loss function is more appropriate than the standard cross-entropy loss function for two classification tasks from the perspective of solving the label unbalancing problem that happens in most biomedical datasets.

4 Conclusion

In this paper, we propose a privacy-preserving FL framework, FedDP, that integrates DP strategies into an FL schema, coupled with new strategies for efficient optimization and communications on modeling imbalanced data. We have evaluated FedDP for two classification prediction tasks: binary-labeled prediction of breast cancer and multiple-labeled prediction of pathological stages in kidney cancer. Experimental results demonstrate that the CNN with min-max loss achieved a good trade-off between accuracy and privacy budget as well as the consideration of time consumption. The min-max loss has significantly improved the low accuracy problem caused by the label imbalance from gene expression datasets. We test our FedDP with an imbalanced data load for each party. The model works well for a stringent privacy budget and the accuracy finally converges. We believe frameworks like FedDP will facilitate the adoption of new analytic and predictive modeling for disease prediction and management, dealing with complicated scenarios where data might be imbalanced and distributed across diverse infrastructures and multiple institutions. Although FedDP is a robust and flexible framework for secure disease prediction, there are several directions we'd like to explore in future studies. For example, the current implementations of FedDP require relatively high bandwidth for communications between the server and each party. As in any FL framework, the throughput capacity of the central server is the bottleneck and thus this bottleneck needs to be mitigated to support large-scale versions of the FedDP framework. In the meanwhile, as more complicated deep models are included in FedDP, more hyperparameters including weights as well as gradients, need to be transferred into the FedDP framework and thus strategies will be explored to reduce communication costs in FedDP. Compression methods could also be applied during communications to reduce the size of gradients so that FedDP can be applied in low-bandwidth and large-scale networks. We will explore several types of compression methods that can be applied in our FedDP including top-k gradient sparsification [27] and gradient quantization [24].

References

1. Bc-TCGA dataset. http://dx.doi.org/10.17632/v3cc2p38hb.1#file-c6b0b7d3-a63d-4ec7-87d5-ac2b81f38e00
2. The cancer genome atlas (TCGA) program. https://www.genome.gov/Funded-Programs-Projects/Cancer-Genome-Atlas
3. National cancer institute. https://www.cancer.gov/about-cancer/diagnosis-staging/staging
4. Tensorflow privacy. https://github.com/tensorflow/privacy
5. Ahmed, H., Hamad, S., Shedeed, H.A., Hussein, A.S.: Enhanced deep learning model for personalized cancer treatment. IEEE Access **10**, 106050–106058 (2022)
6. Al Aziz, M.M., Anjum, M.M., Mohammed, N., Jiang, X.: Generalized genomic data sharing for differentially private federated learning. J. Biomed. Inf. **132**, 104113 (2022)
7. Balle, B., Barthe, G., Gaboardi, M.: Privacy amplification by subsampling: Tight analyses via couplings and divergences. In: Advances in Neural Information Processing Systems, vol. 31 (2018)
8. Beguier, C., Terrail, J.O.d., Meah, I., Andreux, M., Tramel, E.W.: Differentially private federated learning for cancer prediction. arXiv preprint arXiv:2101.02997 (2021)
9. Bentley, A.R., Callier, S., Rotimi, C.N.: Diversity and inclusion in genomic research: why the uneven progress? J. Community Genet. **8**(4), 255–266 (2017). https://doi.org/10.1007/s12687-017-0316-6
10. Chowdhury, A., Kassem, H., Padoy, N., Umeton, R., Karargyris, A.: A review of medical federated learning: applications in oncology and cancer research. In: Crimi, A., Bakas, S. (eds.) International MICCAI Brainlesion Workshop, pp. 3–24. Springer, Cham (2022). https://doi.org/10.1007/978-3-031-08999-2_1
11. Fang, H., Qian, Q.: Privacy preserving machine learning with homomorphic encryption and federated learning. Future Internet **13**(4), 94 (2021)
12. Geyer, R., Klein, T., Nabi, M.: Differentially private federated learning: a client level perspective (2017)
13. Huai, M., Wang, D., Miao, C., Xu, J., Zhang, A.: Pairwise learning with differential privacy guarantees. In: Proceedings of the AAAI Conference on Artificial Intelligence, vol. 34, pp. 694–701 (2020)
14. Islam, M., Haque, M., Iqbal, H., Hasan, M., Hasan, M., Kabir, M.N., et al.: Breast cancer prediction: a comparative study using machine learning techniques. SN Comput. Sci. **1**(5), 1–14 (2020)
15. Jain, D., Singh, V.: Feature selection and classification systems for chronic disease prediction: a review. Egyptian Inf. J. **19**(3), 179–189 (2018)
16. Khanna, A., Schaffer, V., Gürsoy, G., Gerstein, M.: Privacy-preserving model training for disease prediction using federated learning with differential privacy. In: 2022 44th Annual International Conference of the IEEE Engineering in Medicine & Biology Society (EMBC), pp. 1358–1361. IEEE (2022)
17. Kingma, D.P., Ba, J.: Adam: a method for stochastic optimization. arXiv preprint arXiv:1412.6980 (2014)
18. Liu, M., Yuan, Z., Ying, Y., Yang, T.: Stochastic AUC maximization with deep neural networks. arXiv preprint arXiv:1908.10831 (2019)
19. Liu, Y., et al.: Federated forest. IEEE Trans. Big Data **8**, 843–854 (2020)
20. Ma, J., Naas, S.A., Sigg, S., Lyu, X.: Privacy-preserving federated learning based on multi-key homomorphic encryption. Int. J. Intell. Syst. **37**(9), 5880–5901 (2022)

21. Mironov, I., Talwar, K., Zhang, L.: R\'enyi differential privacy of the sampled gaussian mechanism. arXiv preprint arXiv:1908.10530 (2019)
22. Park, J., Lim, H.: Privacy-preserving federated learning using homomorphic encryption. Appl. Sci. **12**(2), 734 (2022)
23. Ramalingam, V., Dandapath, A., Raja, M.K.: Heart disease prediction using machine learning techniques: a survey. Int. J. Eng. Technol. **7**(2.8), 684–687 (2018)
24. Reisizadeh, A., Mokhtari, A., Hassani, H., Jadbabaie, A., Pedarsani, R.: FedPAQ: a communication-efficient federated learning method with periodic averaging and quantization. In: International Conference on Artificial Intelligence and Statistics, pp. 2021–2031. PMLR (2020)
25. Rong, G., Mendez, A., Assi, E.B., Zhao, B., Sawan, M.: Artificial intelligence in healthcare: review and prediction case studies. Engineering **6**(3), 291–301 (2020)
26. Sarkar, E., Chielle, E., Gursoy, G., Chen, L., Gerstein, M., Maniatakos, M.: Scalable privacy-preserving cancer type prediction with homomorphic encryption. arXiv preprint arXiv:2204.05496 (2022)
27. Sattler, F., Wiedemann, S., Müller, K.R., Samek, W.: Robust and communication-efficient federated learning from Non-IID data. IEEE Trans. Neural Netw. Learn. Syst. **31**(9), 3400–3413 (2019)
28. Sav, S., Pyrgelis, A., Troncoso-Pastoriza, J.R., Froelicher, D., Bossuat, J.P., Sousa, J.S., Hubaux, J.P.: POSEIDON: privacy-preserving federated neural network learning. arXiv preprint arXiv:2009.00349 (2020)
29. Uddin, S., Khan, A., Hossain, M.E., Moni, M.A.: Comparing different supervised machine learning algorithms for disease prediction. BMC Med. Inform. Decis. Mak. **19**(1), 1–16 (2019)
30. Wang, Z., Jensen, M.A., Zenklusen, J.C.: A practical guide to the cancer genome atlas (TCGA). In: Mathé, E., Davis, S. (eds.) Statistical Genomics. MMB, vol. 1418, pp. 111–141. Springer, New York (2016). https://doi.org/10.1007/978-1-4939-3578-9_6
31. Yang, Q., Liu, Y., Chen, T., Tong, Y.: Federated machine learning: concept and applications. ACM Trans. Intell. Syst. Technol. **10**(2), 1–19 (2019)
32. Ying, Y., Wen, L., Lyu, S.: Stochastic online AUC maximization. In: Advances in Neural Information Processing Systems, vol. 29 (2016)
33. Zhang, L., Xu, J., Vijayakumar, P., Sharma, P.K., Ghosh, U.: Homomorphic encryption-based privacy-preserving federated learning in IoT-enabled healthcare system. IEEE Trans. Netw. Sci. Eng. **10**, 2864–2880 (2022)
34. Zhang, W., et al.: Blockchain-based federated learning for device failure detection in industrial IoT. IEEE Internet Things J. **8**(7), 5926–5937 (2020)
35. Zheng, N., et al.: Predicting Covid-19 in China using hybrid AI model. IEEE Tans. Cybern. **50**(7), 2891–2904 (2020)
36. Zhu, X., Wang, J., Hong, Z., Xiao, J.: Empirical studies of institutional federated learning for natural language processing. In: Findings of the Association for Computational Linguistics: EMNLP 2020 (2020)

Computational Tumor Progression Analysis via Seriation Based Trajectory Inference

Marmar R. Moussa, Charles H. Street, and Sriram Boddeda

School of Computer Science, University of Oklahoma, Norman, OK, USA
marmar.moussa@ou.edu

Abstract. Precise lineage or evolution path determination play a crucial role in discerning the dynamic developmental or temporal progression patterns observed in single cell RNA-Seq data. In this work, we present a novel computational approach for progression pattern inference of normal or tumor cell populations that are actively progressing along a dynamic pathway in single cell resolution. This is achieved via ordering the cellular transcriptional profiles identifying the progression of cell populations along differentiation, signaling, or tumor evolution paths. Here, we developed a seriation-based progression pattern inference method using optimally reordered hierarchies and provide advanced principal-curves-based visualization of the inferred paths in three dimensional latent space representation of scRNA-Seq data. Additionally, we present novel metrics for evaluating the reconstructed order and identified pathways and evaluate our approach using real single cell transcriptomics datasets.

1 Introduction and Background

Reconstructing developmental or temporal dynamics from single cell omics data can be approached with various strategies, each with its own strengths and weaknesses. For instance, RNA 'velocity' methods have been used for modeling dynamics of cell states from splicing information, however assumptions like splicing rate and modeling of up and down regulation via highly parameterized differential equation models hinder the interpretability and prevent meaningful results and often requires expensive and signal-altering manipulations and pre-processing of the data, preventing this strategy to be widely used. For single cell RNA-Seq analyses, it has become a staple step in the recommended scRNA-Seq analysis workflows to perform a trajectory or pseudo-time analysis - interestingly whether or not RNA dynamics are forecasted in the data. The vast majority of trajectory inference methods that emerged in this space rely on Minimum Spanning Tree (MST) algorithms for creating a pseudo-timeline for cells' differentiation or development. While conveniently implemented in popular scRNA-Seq analysis packages, trajectory analysis based on MST face various issues and limitations associated with their practical application.

These methods are often sensitive to noise and outliers and struggle with sparse sampling, particularly in early development stages or rare cell types; they have difficulty handling branching trajectories; they often produce trajectories lacking biological interpretability, necessitating additional domain knowledge and validation; they involve parameter sensitivity, needing careful tuning; and can be computationally intensive for large datasets. Most importantly, MST-based methods assume linear trajectories, which do not capture the predominantly nonlinear dynamics of biological processes and pathways. Examples include cell differentiation or development, tumor microenvironment dynamics, signalling pathways, or progression from a pre-cancerous state, etc., which do not necessarily exhibit linear progression patterns. Finally, existing methods (see [12] for detailed bench-marking of trajectory inference methods) assume all cells within a dataset fit into one single progression pattern or trajectory (branched or not) that all genomic expressions of all cells are contributing to. This assumption does not allow cells to be found active in more than one signalling pathway or process (e.g. immune cells can be active in both signalling and maturation pathways, tumor cells can be activated along a proliferation path, but simultaneously active in a cancer-related signalling pathway and so forth).

To address these limitations, especially the last two assumptions of linearity and single pathway activity, we present a novel computational approach that uses seriation-based optimal ordering for the reconstruction of cell order to identify relationships of cell populations, such as precursor-successor or activated-inactivated, within and across various samples (with underlying longitudinal aspects, e.g. various tumor stages). Our method aims at reconstructing the progression mechanisms and dynamics unique to a specific biological pathway, but allows for multiple pathways activation patterns within cell populations. We also present an interactive implementation and novel advanced visualizations for the analysis steps defined in our approach including 3-D paths using Principal Curves. Additionally, we present novel metrics (Progression Pattern Index and Label Change Index) for analysing and evaluating the reconstructed order or identified pathways. We evaluate our scalable approach using real single cell transcriptomics datasets from various platforms. We apply our method towards inferring progression dynamics in scRNA-Seq datasets of cancer, COVID19 and other samples, showcasing the insights provided by this analysis in capturing and identifying pathway progression patterns and the related active gene programs.

2 Method Details

2.1 Order Reconstruction

For cell order reconstruction, our algorithm is based on identifying the optimal order for the leafs of a hierarchy (or dendrogram) representing the data, where leafs are the individual cells in scRNA-Seq data, an approach we previously applied successfully for cell cycle order inference in [10]. We first obtain the cells' representation in latent space using principal components followed by projection in 3D-t-SNE space. This ensures that the variability in the data is

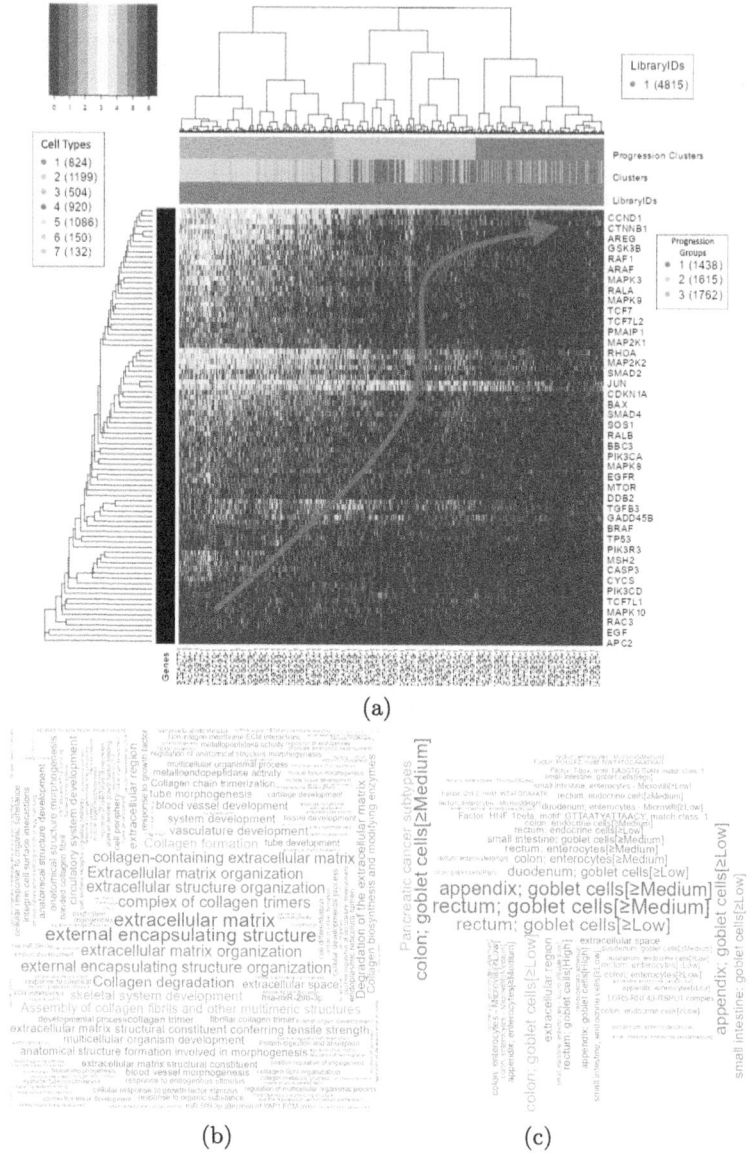

Fig. 1. Heatmap of ordered transcripomics data form the scRNA-Seq data from Colon Cancer sample of FFPE human specimen. The heatmap's dendrogram (a) reflects the order reconstruction by our algorithm using the gene programs from the colon cancer progression pathway as defined by KEGG pathways database. Two top color bars show the Progression Groups or CoPs in the first/top color bar - 3 CoPs (orange, light blue and red) - followed by cell type clusters (Cluster 1 to 7) in the second color bar of the heatmap. Figure panels (b) and (c) show the GO-term enrichment of Cell Types 5 (yellow, ECM) and 2 (light blue, Goblet Cells) respectively, which are mostly found in the most activated CoP (top bar orange group 3)

captured and that the locality of the original high-dimensionality (gene) space is well preserved [10][9]. We then apply the seriation/order reconstruction step; typically, to infer an optimal leaf ordering for n cells, the dendrogram produced by any rooted binary tree producing algorithm (e.g. agglomorative hierarchical clustering or graph-based clustering, etc.) has $n-1$ internal nodes and 2^{n-1} possible leaf orderings to be considered. Performing additional leaf-node reordering is equivalent to minimizing the length of a Hamiltonian path [2]. However, for our method, we use the Optimal Leaf Ordering (OLO) algorithm based on [2] that produces a leaf cell ordering which minimizes sum of distances between adjacent leaves in $O(n^3)$. The practical implementation in our method achieves further improved complexity since the cells' pairwise distances (we use cosine in our results) are pre-computed for clustering steps.

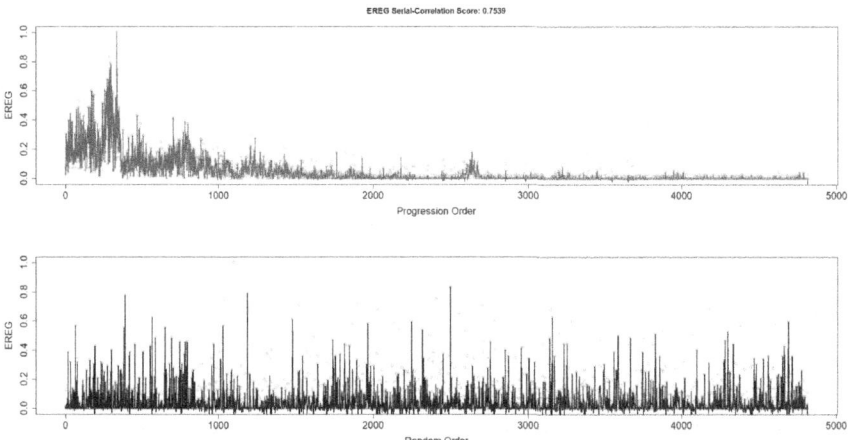

Fig. 2. Order reconstruction vs. random gene expression pattern. This Figure shows the effect of applying the re-constructed order on one gene, e.g. EREG gene expression values ordered (c, upper panel in red) show a clear pattern of multimodality recapturing the expression levels along the progression path, while the random order (c, lower panel in black) displays no apparent pattern

We base the cells' distances in this strategy on gene modules implicated in the differentiation or signaling process to be examined only (i.e. not all genes are used but only those from the pathway annotation with detectable expression). Gene groups participating in pathways' cascades are defined by knowledge-bases and can be obtained as annotated gene lists (e.g. pathway maps from KEGG Pathways Database [6] which we use in our results). Consequently, the order reconstructed is a representation of the dynamic activation or trajectory of the progression mechanism for each pathway under consideration and this strategy can accommodate multiple pathways analysis. The reconstructed hierarchy also provides clustering or grouping of reordered cells into progression groups or Clusters of Progression (CoP) that typically represent the progression stages.

Figure 1 illustrates in detail the results of our ordering method as applied to the Human Colon Cancer dataset[4] from 10x Genomics [1] against the 'Colorectal Cancer Pathway' form [6]. In this dataset, the inferred order starts by mostly cells from cell type cluster 5 (yellow, second color bar), enriched in extracellular matrix gene modules which indeed plays a significant role in the progression of colon cancer pathway situating this cell type/population within the most active CoP of the identified path (corresponding to CoP 3 (orange in top color bar)), followed by mostly goblet cells population coded by light blue (Cell type Cluster 2) cells in the second color bar. This indeed matches known activation patterns of colon cancer progression where strong evidence suggests the involvement of ECM remodeling and Goblet Cells enrichment in pre-cancerous states progression [3].

Furthermore, the effect of the inferred order can be strongly observed on individual genes' pattern of expression levels; in Fig. 2, for selected gene 'EREG', a clear pattern arises from the order reconstructed (red panel) as opposed to the random order panel (black). This non-parametric data-driven approach allows for patterns to be freely learned from the dataset as opposed to assuming a certain pre-defined model of expression that needs to fit all datasets regardless of the underlying dynamics.

2.2 Progression Pattern Index

To assess the obtained progression patterns (cell order + gene contribution to given order) we propose novel metrics based on the serial correlation that we define as Pathway Progression Index (PPI) metric as follows in Eq. 1:

$$Med\left(|s_{ord}(g_i) - \frac{1}{R}\sum_{j=1}^{R} s_{rand_j}(g_i)| \ : \ i = 1,\ldots,N\right) \qquad (1)$$

where N is number of annotated genes relevant to the progression pathway, $s_{ord}(g_i)$ denotes the first-order serial correlation score of gene i with respect to the given cell order, and $s_{rand_j}(g_i)$, $j = 1,\ldots,R$, denotes the score of gene i with respect to bootstrapping R times randomized cell orders ($R > 50$ in our results). This metric can be calculated per ordered CoP or progression group.

If the majority of genes are not active (i.e. the majority of genes have no significance/less than 0.05 PPI score) in the currently considered pathway, the cell population is deemed inactive for this pathway and not considered any further.

In Fig. 3 we show the boxplots aggregating the individual gene scores and illustrating the PPI score per group as the median line for each of the boxplots of the three progression groups/CoPs of the Colon Cancer dataset. Group 1 (red) has the lowest PPI (0.0415) indicating the majority of genes have a random pattern and therefore is classified not-active, while Group 3 clearly scores highest, with PPI score of 0.3238 and significance (t-test p-value) of 1.045e-09 when compared to the rest.

When there is considerable high activity for the majority of the genes of the examined pathway in a cell population, we can further tune the PPI into a pathway activity score to be calculated as in Eq. 2,

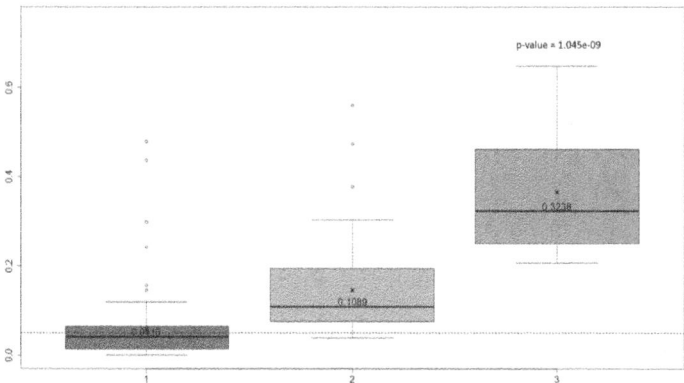

Fig. 3. Boxplots representing the PPIs of the Human Colon Cancer dataset with the three identified CoPs, 1 non-active (red), 2 moderate (blue) and 3 (orange) highest (Color figure online)

$$Med\ (GM_{High}(|s_{ord}(g_i) - \frac{1}{R}\sum_{j=1}^{R} s_{rand_j}(g_i)|\ :\ i = 1,\ldots,N)) \qquad (2)$$

where GM_{High} denotes the inclusion of only the genes classified as the generalized gaussian mixture model(GMM) with high first-order serial correlation. In other words, when the majority of the genes are significantly active in the considered pathway, the pathway activity can be confidently quantified based on the group of activated genes of that pathway. In detail, the gene scores contributing to the PPI index (i.e. before applying the median from Eq. 1) are analyzed using GMM (see Fig. 4 for visualization of this analysis). The model best fitting with the maximum Baysian Information Criterion(BIC) is selected to predict the genes' classes into at least one non-active gene class (lowest mean class) and one or more active gene classes (moderate to high activity classes). The pathway activity is then calculated using all but the non-active genes class making the PPI more precise per pathway and per active cell population. Additionally, and since the gene lists obtained from public databases are prior knowledge not dataset-specific, this data-driven strategy ensures tailoring gene lists to the data and allows for flexibility in pathway gene detection within each individual dataset, i.e. some genes from the pre-defined pathway are allowed to not participate in the identified order/activity for the specific dataset tested.

From an individual gene perspective, this score denotes whether the gene displays a pattern or not (without assuming a specific distribution model or pattern) given the ordered cell population. Each gene is free to play different roles given different pathways and this approach indeed enables this flexibility as shown for instance for the Human Colon Dataset in Figure (5.

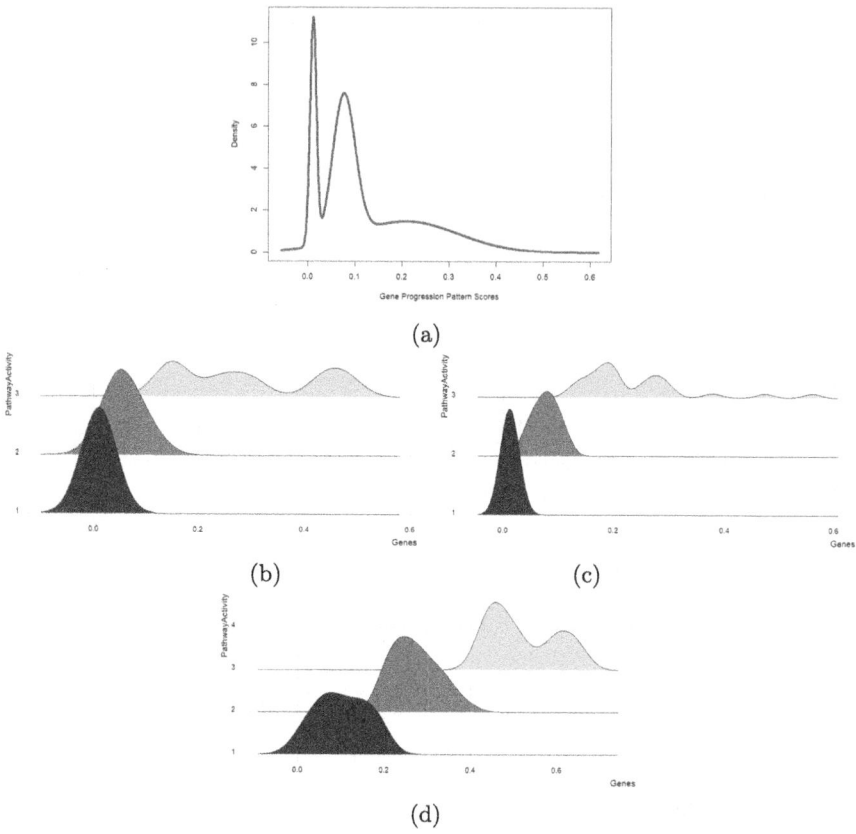

Fig. 4. PPI Score GMM-based multi-model analysis. In (a) we show a general example of the density distribution of the PPI score for all genes of a pathway, one can recognize multiple 'peaks' corresponding to multiple 'models' within the scores distribution. In (b,c, and d) we show the GMM analysis of the three CoPs (1,2 and 3 respectively) from the Colon Cancer dataset, each contains at least three gene models representing low mean PPI score (dark blue), medium activity genes (green) and highly active/high PPI scores gene group (yellow). In (d) the most active cell group, we can see that even the lowest class (dark blue) has a mean > 0

2.3 Label Change Index

For further assessing the inferred order, if ground truth is known for a dataset in the form of cell labels that correspond to certain phases or processes within a pathway, for instance cell cycle phases within cell cycle pathway or different known stages in carcinogenics, the evaluation for the labeled datasets can benefit from the 'Label Change Index' or LCI. We propose this metric adapted from [8] which measures how frequent an experimentally determined single cell labels changes along the identified progression order. Ideally, a perfect ordered set would change labels only between the phases or processes, using phases of

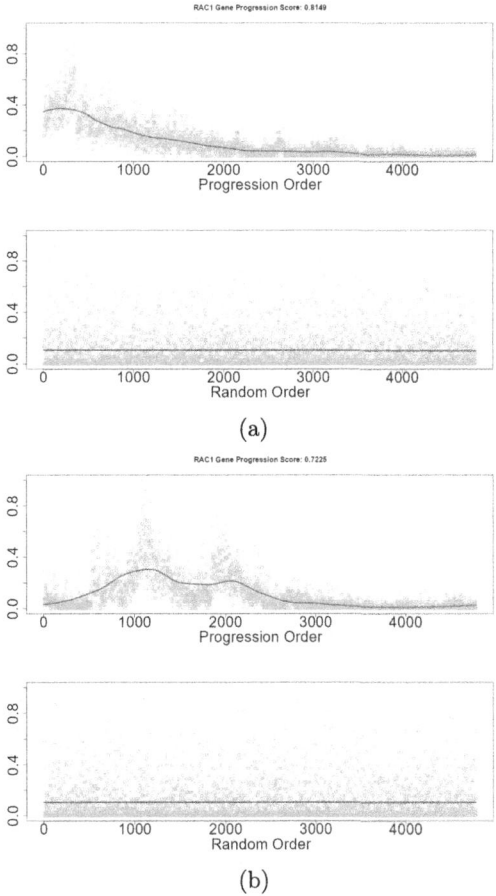

Fig. 5. Order reconstruction featureing Gene RAC1 in (a) Colon Cancer Pathway, and (b) Irritable Bowl Syndrome(IBS) pathway

proliferation as example, an ordered set of G1, S and G2M cells would include only two label changes along the pathway progression. Thus, we define the Label Change Index per CoP as:

$$LCI = 1 - (\sum LC - (n-1))/(N - n) \qquad (3)$$

where LC means the sum of the label changes between two adjacent cells in the given order, n is the number of identified groups and N is the total number of ordered cells. A perfect series or order per CoP would have change index value of 1, while the worst where $\sum LC = N - 1$ (i.e. all labels were changed) would have a value of 0.

To apply this metric, we use two labeled datasets, first, hESC from [11], a small dataset of 247 cells after QC, which is labeled by cell cycle phases, G1, G2 and S phases. Second, we combined three different single cell datasets from [1] constructed of Breast Cancer cells, Skin Cancer cells and healthy kidney cells. The inferred order should fairly group each of the cell populations together and in progressing ordered groups without label changing or interspersed patterns.

Table 1. LCI scored per each of the progression ordered groups in each dataset

	Group 1	Group 2	Group 3
Breast Cancer Data	0.9736	0.8024	0.9898
hESC Data	0.9730	0.9551	0.65

Results of utilizing the Label Change Index form analysis of the combined cancer dataset and hESC are shown in Table 1 and can be visually appreciated in Fig. 6.

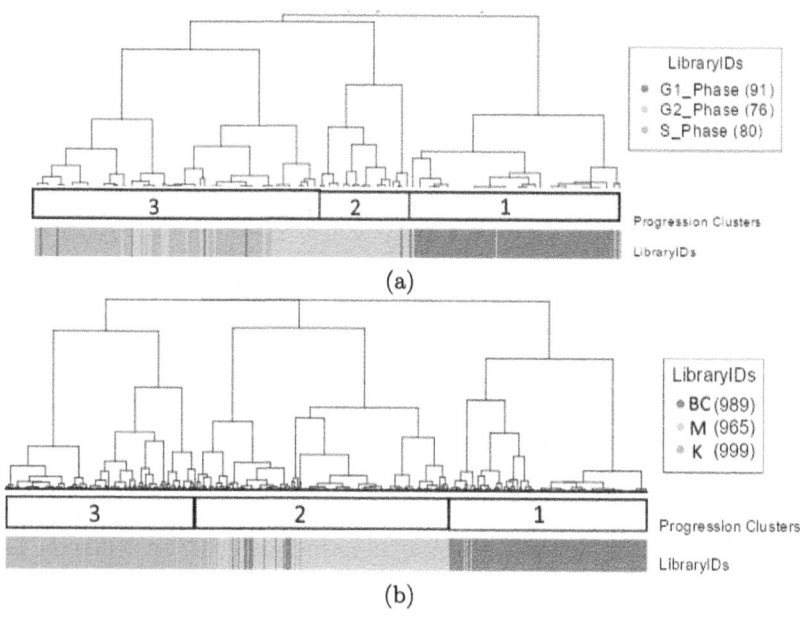

Fig. 6. Dendrogram of the reconstructed order for the hESC (a) and the combined Cancer (b) datasets

2.4 Principal Curves 3-D Order Visualization in Latent Space

A principal curve is a smooth curve passing through the middle of a multidimensional dataset, providing a (visual) non-linear summary of the data. The algorithm for constructing principal curves starts with a prior (ordered) summary of the data [5]. We utilize this algorithm to fitting a principal curve to our data in multiple dimensions given the learned order from our pathway progression method inference as the prior summary. Figure 7 shows the learned orders of the hESC in (a, 45 degree rotation) and (b, 120 degree rotation) and the Human Colon Cancer dataset in (c and d) fitted into a principal curve visualization in three dimensional t-SNE space providing an elegant visualization of the inferred order showing the succession of the progression groups or CoPs for each of the tested datasets.

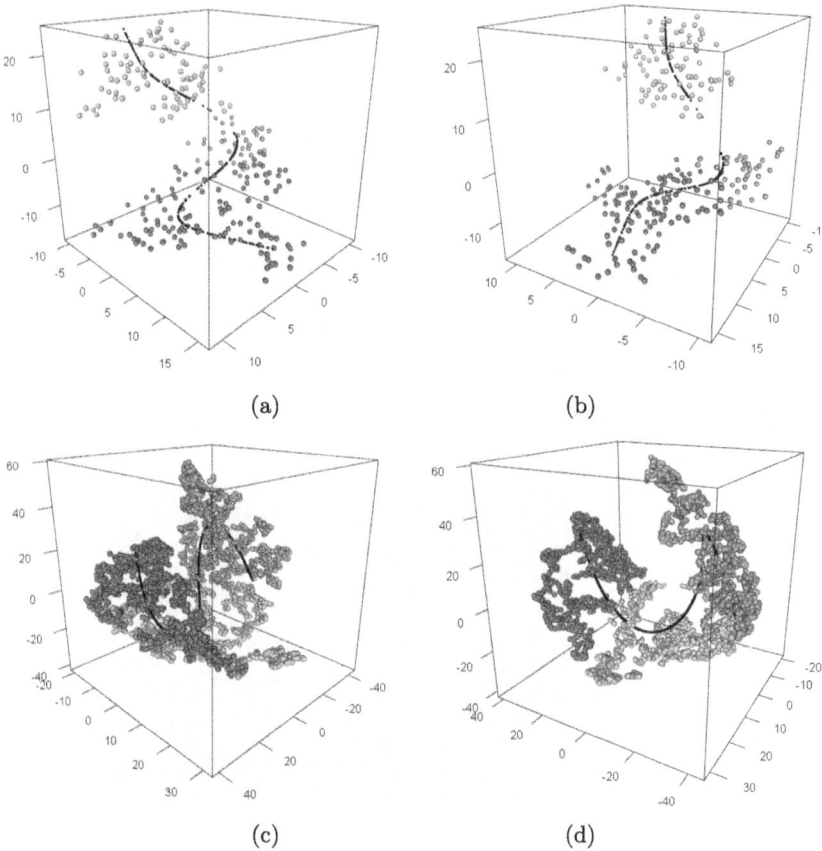

Fig. 7. Order 3D reconstruction showing principal curves based paths in black for hESC (a,b) and Human Colon Cancer (c,d)

3 Discussion

Our proposed approach presents a comprehensive analysis method for progression order reconstruction and pathway activity quantification that utilizes optimal leaf ordering algorithm and identifies clusters of progression learned from single cell RNA-Seq data. Indeed, the proposed method is able to first, infer a cell order along a progression pathway defined by gene modules from prior knowledge bases.

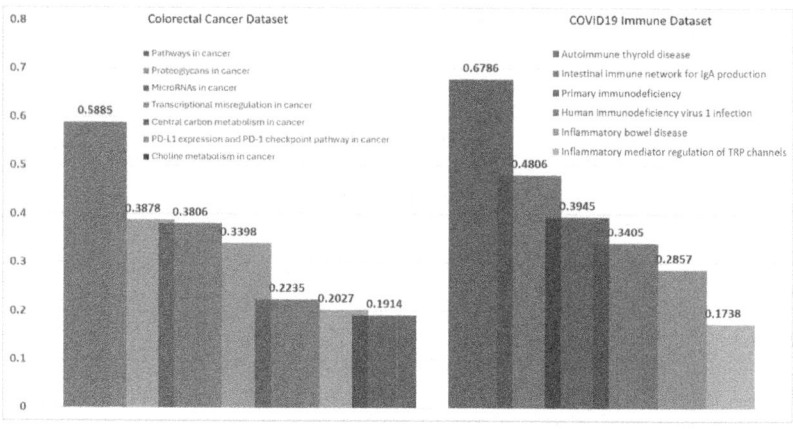

Fig. 8. Pathway Progression Index of select pathways

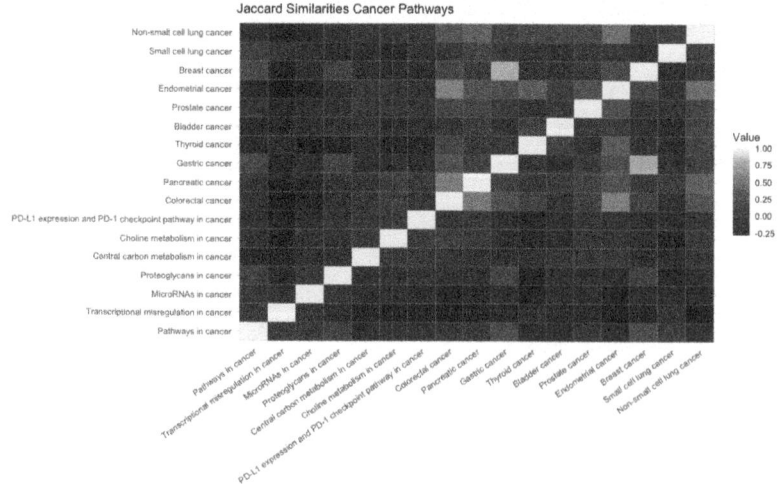

Fig. 9. Jaccard pairwise similarity scores of selected cancer-related pathways

We identify clusters of progressions, or progression ordered groups of cells where the detected gene signals and the PPI scores indicate how much activity each cell group or cluster of Progression is contributing. Not only can our method identify cell populations that are actively progressing along a given pathway, but we can use the PPI score to examine and quantify the activity for each potential pathway in a dataset. We show this use case in Fig. 8. Two datasets are tested, first, the Human Colon Cancer dataset in the left panel, and second a Covid19 immune cells dataset from [7]. The bar plots indicate the most active pathways in each dataset sorted by PPI scores (maximum per pathway in all groups). The relevance of the identified pathways using our method is clear from the pathways description active in each dataset. Typically, for basic science application, only relevant pathways need to be tested in each dataset; it is expected that pathways with mostly overlapping gene lists would produce similar PPI scores. In Fig. 9 we show the Jaccard similarity coefficient -a statistic used for gauging the similarity/diversity of sample sets- of a selected number of cancer related pathways. For the cancer types pathways, we observe a higher similarity than in cancer processes pathways. When testing pathways with lower similarities, the PPI score is particularly useful for picking out the most active pathway of the pathway set being tested.

Acknowledgements. This work was supported by the following awards: NSF 2341725; NIH NCI K25CA270079; and OU-BIC2.0. **Code and Data Availability** Code and processed data are available from the authors upon reasonable requests and will be available on github repository at https://github.com/moussa-sc-lab.

References

1. Cell RangerTM R Kit Tutorial: secondary analysis on 10x Genomics Single Cell 3' RNA-Seq PBMC Data. http://s3-us-west-2.amazonaws.com/10x.files/code/cellrangerrkit-PBMC-vignette-knitr-1.1.0.pdf
2. Bar-Joseph, Z., Gifford, D.K., Jaakkola, T.S.: Fast optimal leaf ordering for hierarchical clustering. Bioinformatics **17**(suppl_1), S22–S29 (2001)
3. Chen, B., et al et al.: Differential pre-malignant programs and microenvironment chart distinct paths to malignancy in human colorectal polyps. Cell **184**(26), 6262–6280 (2021)
4. 10x Genomics: Visium cytassist gene expression libraries of post-xenium human colon cancer FFPE using the human whole transcriptome probe set 2 replicate 1 (2024). https://www.10xgenomics.com/datasets/
5. Hastie, T., Stuetzle, W.: Principal curves. J. Am. Stat. Assoc. **84**(406), 502–516 (1989)
6. Kanehisa, M., Sato, Y., Kawashima, M., Furumichi, M., Tanabe, M.: KEGG: Kyoto encyclopedia of genes and genomes. https://www.kegg.jp/ (2024). Accessed 28 July 2024
7. Liao, M., et al.: Single-cell landscape of bronchoalveolar immune cells in patients with Covid-19. Nat. Med. **26**(6), 842–844 (2020)
8. Liu, Z., et al.: Reconstructing cell cycle pseudo time-series via single-cell transcriptome data. Nat. Commun. **8**(1), 22 (2017)
9. van der Maaten, L., Hinton, G.: Visualizing data using t-SNE. J. Mach. Learn. Res. **9**, 2579–2605 (2008)

10. Moussa, M., Măndoiu, I.I.: Computational cell cycle analysis of single cell RNA-Seq data. In: Jha, S.K., Măndoiu, I., Rajasekaran, S., Skums, P., Zelikovsky, A. (eds.) ICCABS 2020. LNCS, vol. 12686, pp. 71–87. Springer, Cham (2021). https://doi.org/10.1007/978-3-030-79290-9_7
11. Moussa, M., Măndoiu, I.I.: SC1: a tool for interactive web-based single-cell RNA-Seq data analysis. J. Comput. Biol. **28**, 820–841 (2021)
12. Saelens, W., Cannoodt, R., Todorov, H., Saeys, Y.: A comparison of single-cell trajectory inference methods. Nat. Biotechnol. **37**(5), 547–554 (2019)

Multilayer Network Analysis of Brain Signals for Detecting Alzheimer's Disease

Sean M. Nguyen[1], Mohammad Amin Basiri[2], and Sina Khanmohammadi[1,2]([✉])

[1] School of Computer Science, University of Oklahoma, Norman, OK 73019, USA
[2] Data Science and Analytics Institute, University of Oklahoma, Norman, OK 73019, USA
sinakhan@ou.edu

Abstract. Human neuroimaging datasets provide rich multi-scale spatiotemporal information about the state of the brain. Most current methods, such as spectral analysis, focus on a single facet of these datasets and do not take full advantage of the inherent spatiotemporal information. Here, we consider a multilayer cross-frequency functional connectivity analysis to capture the complex spatiotemporal features of neural datasets at multiple scales and show that such features could potentially provide a better description of the neural activity. We demonstrate the effectiveness of this approach by applying the proposed method to capture disruptions of cross-frequency brain connections in Alzheimer's patients. More specifically, we compared the multi-scale features extracted from electroencephalogram (EEG) data with traditional features in a machine learning framework to distinguish Alzheimer's patients from control subjects. Our results show that such multi-scale features improve the prediction accuracy when compared to traditional feature extraction methods in EEG analysis.

Keywords: EEG · Spatiotemporal Patterns · Multilayer Networks · Multi-Scale Features

1 Introduction

Alzheimer's disease (AD) stands as the most prevalent cause of dementia in the elderly population, which impacts about forty percent of individuals over the age of eighty [1]. In the early stages of AD, there is an observable difficulty in learning and memory, especially with declarative memory [2]. As the Alzheimer's disease progresses, individuals may experience worsening language difficulties [2] along with cognitive deficits extending into areas like judgment, logical reasoning, and organization [3]. On average, the life expectancy for patients with AD is about five to eight years following clinical diagnosis [4]. Early detection of AD is crucial to implement treatment strategies that mitigate the symptoms. However, a definitive diagnosis of Alzheimer's disease (AD) requires a post-mortem analysis of brain tissue [5].

Among various modalities used for early detection of Alzheimer's disease, electroencephalography (EEG) has garnered attention due to its wide availability, relative affordability, and non-invasiveness. Numerous studies have explored EEG irregularities associated with AD, employing both linear and non-linear analysis techniques [6–8]. The most prominent feature observed is a decrease in coherence between different brain regions [9]. However, most of these methods, such as spectral analysis, are focused on a particular property (i.e., frequency) and do not take full advantage of the rich spatiotemporal information in the data [10].

Here, we propose a multilayer cross-frequency network model that could adequately capture the spatiotemporal features of the neuroimaging dataset maintaining simplicity and computational efficiency. We demonstrate the effectiveness of the proposed approach by extracting spatiotemporal features of resting state EEG data from Alzheimer's patients and show that such features could provide complementary information to the conventional features used in neuroimaging studies.

2 Materials and Methods

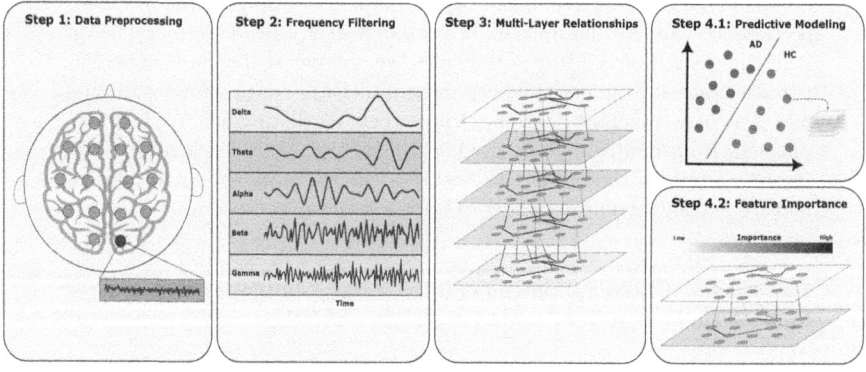

Fig. 1. Multilayer cross-frequency data analysis framework. (**1**) Resting state electroencephalogram (EEG) dataset from Alzheimer's patients is obtained [11] and preprocessed by removing the first and last 10 percent of the data points. (**2**) This dataset is filtered into five typical EEG frequency bands, each representing different aspects of brain function. Together, they form the foundation of the multilayer cross-frequency model. (**3**) All signals are aggregated from the different channels and frequencies, and then the Pearson correlation between all signal pairs is calculated. This will result in a super adjacency matrix with inter and intra-frequency relationships. (**4.1**) We then use Principal Component Analysis (PCA) to reduce the dimensionality of the data and use these features in the 3-Nearest Neighbors algorithm for classification and prediction. (**4.2**) We also use a logistic regression model to determine the relative importance of each feature group.

2.1 Data Preprocessing

We are utilizing the EEG dataset from [11], which includes data from 11 healthy controls and 12 patients with Alzheimer's disease with an average age of 73. The resting state data from each participant were recorded for an average of five minutes using 16 channels. The data consists of 663 trials with an average of 30 epochs per patient. For each epoch, we dropped ten percent of the samples from the beginning and end of the data to account for the EEG montage noise.

2.2 Frequency Filtering

In the second step of our framework (Fig. 1.2), we apply a finite impulse response (FIR) filter to extract the five typical EEG frequency bands of Delta (0.01–4 Hz), Theta (4–8 Hz), Alpha (8–12 Hz), Beta (12–30 Hz), and Gamma (30–49 Hz). Each frequency band constitutes different aspects of brain function [12], and together form the basis of our multilayer cross-frequency functional connectivity analysis.

2.3 Multilayer Relationships

After filtering the signals into different frequency ranges, we constructed the multilayer cross-frequency functional networks. This is done by aggregating all the signals from different channels and frequencies, followed by calculating the Pearson correlation between all signal pairs to create one super adjacency matrix that covers inter-frequency and intra-frequency relationships (Fig. 1.3).

2.4 Model Validation

We utilize two different evaluation methods to validate our proposed model. First, we considered a predictive machine learning framework to understand the discriminative power of the proposed multilayer network features as a whole, followed by a feature importance analysis to examine the relevant importance of each feature.

2.4.1 Predictive Modeling

To further account for the noise in the data, we used the Principal Component Analysis (PCA) to reduce the number of features to the minimum number that explains 95 percent of the variance in the data. Once the dimensionality of the data was reduced, we applied the k-nearest neighbors algorithm with three neighbors ($k = 3$) to classify the extracted features into Alzheimer's vs. healthy control groups. The model's performance was evaluated using a repeated 5-fold cross-validation with ten repetitions.

2.4.2 Feature Importance

To understand the relative importance of features, we have also applied a logistic regression model where the model's coefficients represent each feature's relative importance. These analyses were necessary to understand whether the cross-frequency features in our multilayer network framework have any added value.

3 Results

3.1 Multilayer Cross-Frequency Adjacency Matrix

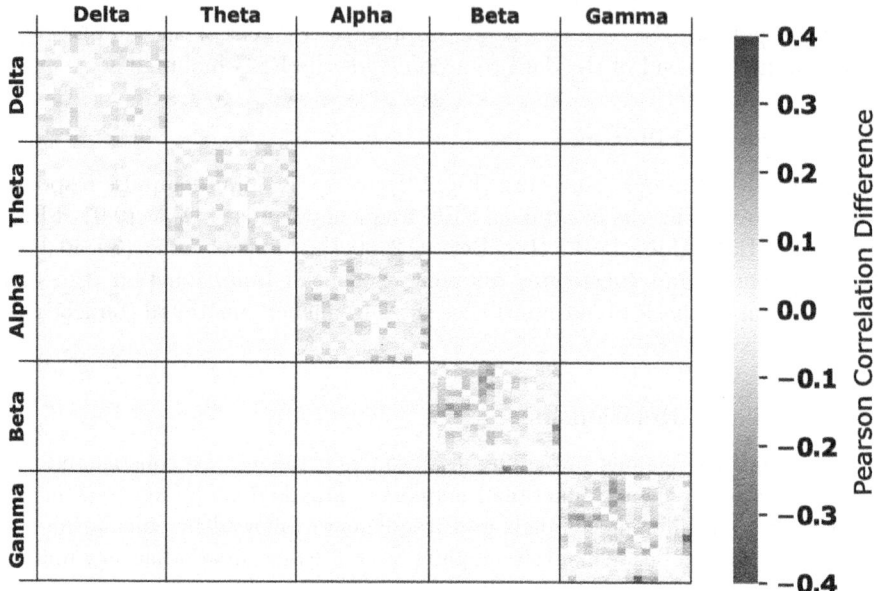

Fig. 2. Super-adjacency heatmap of mean differences between healthy and Alzheimer's patients. Each frequency region contains 16 channels, and interceptions between regions represent inter-frequency relationships. Every single square or intersection between a pair of channels represents the mean difference in their Pearson correlation coefficients between AD and control patients. Areas of darker color represent a stronger difference of correlation within those particular features.

Figure 2 represents our super-adjacency matrix of mean differences between healthy and Alzheimer's patients, visualized as a heatmap. The diagonal sections represent the intra-frequency relationships, and the areas outside the diagonal represent the inter-frequency relationships. As it can be seen in Fig. 2, the inter-frequency correlations are much smaller than the intra-frequency correlations, but nevertheless, we will see in the following sections that they provide valuable information in terms of classifying signals into Alzheimer's patients vs. healthy controls.

3.2 Model Accuracy Comparison

Figure 3 represents the results of our predictive modeling discussed in Sect. 3.1. The Figure includes results from applying the k-nearest neighbors algorithm to raw data, single-layer network features, and the proposed multilayer network features. The multilayer cross-frequency features provide significantly better results compared to the Delta, Theta, and Alpha bands, but they were not significantly higher than the Beta and Gamma bands. This was expected, as we discuss in Sect. 4.

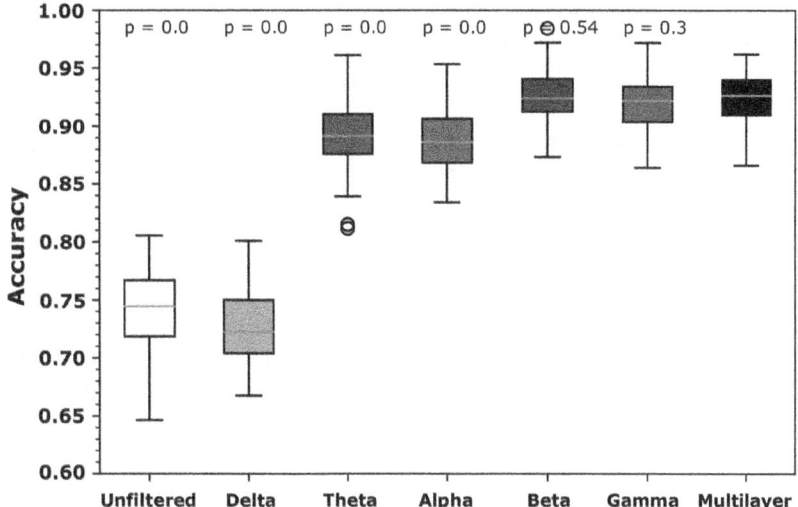

Fig. 3. Predictive model accuracy comparison. The "Unfiltered" model represents our data before any frequency filtering, whereas the "Delta", "Theta", "Alpha", "Beta", and "Gamma" represent single-layer features. The "Multilayer" model represents our proposed multilayer cross-frequency model.

3.3 Feature Importance

As discussed previously, we also compared the relative importance of each group of inter-frequency and intra-frequency features in terms of their contribution (coefficient value) in a logistic regression model. Figure 4 compares the distribution of relative importance for each group of features, where the green triangle represents the mean. As seen from the figure, all the cross-frequency features had nonzero feature importance, providing valuable information regarding the classification task.

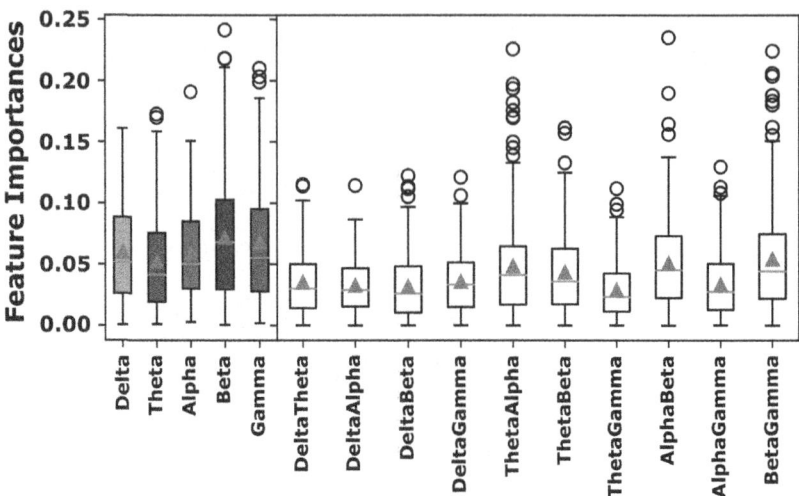

Fig. 4. Feature importance distribution comparison. Each boxplot represents the distribution of feature importance in each frequency region. The first five boxplots represent single-layer modeling, whereas the rest are features within various inter-frequency relationships.

3.4 Comparison Against Other Methods

We have also compared our model to four different EEG processing methods used for AD detection. Table 1 represents the comparison between our proposed method and the other methods using the dataset utilized in our study. The results in Table 1 are based on the predictive modeling approach outlined in Sect. 2.4.1. Each method was evaluated using five different performance metrics including Accuracy, F1-Score, sensitivity, precision, and specificity. Accuracy is the ratio between the correctly classified samples and the total number of samples in the evaluation dataset [13]. Sensitivity denotes the rate of correctly classified positive samples [13], and precision measures the accuracy of the positive predictions [13]. Specificity denotes the rate of correctly classified negative samples [13], and the F1 score is the harmonic mean of precision and sensitivity, meaning that it penalizes extreme values of either measure [13].

The last four rows in the table represent our multilayer method, evaluated using different dependency measures. The "Pearson Correlation" takes the Pearson correlation coefficients of the entire EEG signal, which was also utilized in Sect. 3.2. The "Signal Envelope" takes the Pearson correlation coefficients of the signal envelope of the signal, whereas the "Phase Synchrony Index (PSI)" calculates the phase-locking value between all pairs of channels. "Coherence", on the other hand, computes the frequency domain equivalent of the cross-correlation. Our results suggest that the "Pearson Correlation" approach provided the best results in terms of Accuracy, F1-Score, and Sensitivity, whereas the PSI method provided the best results for precision and specificity.

Table 1. Comparison of the proposed multilayer method performance versus others

Method	Accuracy	F1-Score	Sensitivity	Precision	Specificity
Fourier [14]	0.789 ± 0.033	0.828 ± 0.031	0.844 ± 0.043	0.814 ± 0.043	0.706 ± 0.062
Wavelet [14]	0.778 ± 0.029	0.821 ± 0.026	0.850 ± 0.038	0.797 ± 0.042	0.671 ± 0.059
Cross Frequency Coupling [15]	0.73 ± 0.039	0.774 ± 0.037	0.772 ± 0.054	0.78 ± 0.052	0.667 ± 0.079
Spectral Analysis [16]	0.778 ± 0.036	0.818 ± 0.035	0.836 ± 0.045	0.803 ± 0.043	0.69 ± 0.053
Pearson Correlation	**0.927 ± 0.025**	**0.917 ± 0.023**	**0.924 ± 0.033**	0.911 ± 0.037	0.863 ± 0.055
Signal Envelope	0.901 ± 0.029	0.888 ± 0.026	0.902 ± 0.038	0.876 ± 0.036	0.805 ± 0.057
Phase Synchrony Index (PSI)	0.898 ± 0.02	0.915 ± 0.018	0.92 ± 0.026	**0.912 ± 0.028**	**0.865 ± 0.039**
Coherence	0.878 ± 0.025	0.9 ± 0.021	0.913 ± 0.03	0.888 ± 0.035	0.825 ± 0.052

4 Discussion and Conclusions

The predictive model results in Fig. 3 represent a significant difference between the final multilayer model and the others, except for the beta band and gamma band single-layer models. This is aligned with previous studies that suggest enhanced high-frequency brain activity in patients with Alzheimer's disease [17]. Nonetheless, we don't usually know the dominant frequency a priori. Hence, considering all the inter-frequency and intra-frequency relationships in multilayer network models could be beneficial in practical applications. The importance of such cross-frequency information is further validated in results shown in Fig. 4, where all the inter-frequency features have nonzero feature importance.

According to our comparison results in Table 1, our proposed multilayer method performed better than the four other EEG feature extraction methods. Specifically, the Pearson correlation and phase synchrony values under our multilayer method provided significantly better results than conventional EEG analysis methods. It is worth noting that there have been a few recent studies that considered the multilayer functional connectivity networks in Alzheimer patients [18,19]. However, the multilayer networks in these papers have been purely defined in terms of coherence/cross-spectrum, which as shown in Table 1 provide inferior results compared to Pearson correlation based multilayer model. A possible explanation of this is that the coherence is more susceptible to noise and spurious phase relationships introduced by common referencing in EEG data [20].

We also would like to take a moment and discuss some of the limitations and potential future directions of this study. First, as shown in Fig. 3, the performance of delta features is relatively lower than the others. We believe such performance could be attributed to the difficulties of recording and analyzing low-frequency EEG signals as outlined in [21]. Hence, in our future work, we will consider methods to mitigate low-frequency noise defects. Second, we have not considered the interpretability of the obtained features. However, the applicability of the proposed method in real clinical settings will depend on the interpretability of the features. In our future work, we will consider modifications of the proposed method to generate more interpretable features that could provide information about different cognitive functions.

In conclusion, we explored the possibility of using multilayer cross-frequency network features to capture spatiotemporal information from EEG signals. We showed the effectiveness of such framework by using the extracted features in a predictive model to distinguish Alzheimer's vs. healthy control subjects. We hope these results provide some insights and enable further investigations of underlying spatiotemporal disruptions of brain dynamics in Alzheimer's patients.

References

1. Price, D.L.: Aging of the brain and dementia of the Alzheimer type. Principles Neural Sci. **17**, e12802 (2001)
2. Förstl, H., Kurz, A.: Clinical features of Alzheimer's disease. Eur. Arch. Psychiatry Clin. Neurosci. **249**, 288–290 (1999)
3. Bianchetti, A., Trabucchi, M.: Clinical aspects of Alzheimer's disease. Aging Clin. Exp. Res. **13**, 221–230 (2001)
4. Bracco, L., et al.: Factors affecting course and survival in Alzheimer's disease: a 9-year longitudinal study. Arch. Neurol. **51**(12), 1213–1219 (1994)
5. Weller, J., Budson, A.: Current understanding of Alzheimer's disease diagnosis and treatment. F1000Research **7**, F1000 (2018)
6. Jeong, J.: EEG dynamics in patients with Alzheimer's disease. Clin. Neurophysiol. **115**(7), 1490–1505 (2004)
7. Jin, S.-H., Jeong, J., Jeong, D.-G., Kim, D.-J., Kim, S.Y.: Nonlinear dynamics of the EEG separated by independent component analysis after sound and light stimulation. Biol. Cybern. **86**, 395–401 (2002)
8. Jonkman, E.: The role of the electroencephalogram in the diagnosis of dementia of the Alzheimer type: an attempt at technology assessment. Clin. Neurophysiol. **27**(3), 211–219 (1997)
9. Dunkin, J.J., Leuchter, A.F., Newton, T.F., Cook, I.A.: Reduced EEG coherence in dementia: state or trait marker? Biol. Psychiat. **35**(11), 870–879 (1994)
10. Bogéa Ribeiro, L., da Silva Filho, M.: Systematic review on EEG analysis to diagnose and treat autism by evaluating functional connectivity and spectral power. Neuropsychiatric Dis. Treatment **19**, pp. 415–424 (2023)
11. Escudero, J., Abásolo, D., Hornero, R., Espino, P., López, M.: Analysis of electroencephalograms in Alzheimer's disease patients with multiscale entropy. Physiol. Meas. **27**(11), 1091 (2006)
12. Abhang, P.A., Gawali, B.W., Mehrotra, S.C.: Introduction to EEG-and Speech-based Emotion Recognition. Academic Press (2016)
13. Hicks, S.A., et al.: On evaluation metrics for medical applications of artificial intelligence. Sci. Rep. **12**(1), 5979 (2022)
14. Fiscon, G., Weitschek, E., De Cola, M.C., Felici, G., Bertolazzi, P.: An integrated approach based on EEG signals processing combined with supervised methods to classify Alzheimer's disease patients. In: 2018 IEEE International Conference on Bioinformatics and Biomedicine (BIBM), pp. 2750–2752. IEEE (2018)
15. Yu, H., et al.: Variation of functional brain connectivity in epileptic seizures: an EEG analysis with cross-frequency phase synchronization. Cogn. Neurodyn. **14**, 35–49 (2020)
16. Lehmann, C., et al.: Application and comparison of classification algorithms for recognition of Alzheimer's disease in electrical brain activity (EEG). J. Neurosci. Methods **161**(2), 342–350 (2007)

17. Wang, J., Fang, Y., Wang, X., Yang, H., Yu, X., Wang, H.: Enhanced gamma activity and cross-frequency interaction of resting-state electroencephalographic oscillations in patients with Alzheimer's disease. Front. Aging Neurosci. **9**, 243 (2017)
18. Klepl, D., He, F., Wu, M., Blackburn, D.J., Sarrigiannis, P.G.: Cross-frequency multilayer network analysis with bispectrum-based functional connectivity: a study of alzheimer's disease. Neuroscience **521**, 77–88 (2023)
19. Guillon, J., et al.: Loss of brain inter-frequency hubs in alzheimer's disease. Sci. Rep. 7(1), 10879 (2017)
20. Bastos, A.M., Schoffelen, J.-M.: A tutorial review of functional connectivity analysis methods and their interpretational pitfalls. Front. Syst. Neurosci. **9**, 175 (2016)
21. Demanuele, C., James, C.J., Sonuga-Barke, E.J.: Distinguishing low frequency oscillations within the 1/f spectral behaviour of electromagnetic brain signals. Behav. Brain Funct. 3(1), 1–14 (2007)

DNA Methylation Based Subtype Classification of Breast Cancer

Sri Lakshmi Bhavani Pagolu, S. Suba, and Nita Parekh(✉)

Center for Computational Natural Sciences and Bioinformatics, International Institute of Information Technology, Hyderabad, Telangana 500032, India
{sri.lakshmi,suba.s}@research.iiit.ac.in, nita@iiit.ac.in

Abstract. Aberrant genome-wide DNA methylation patterns is common in cancers. Understanding how these affect the transcriptome can provide insights into subtype specific development and progression of tumorigenesis. In this study we carried out genome-wide analysis of DNA methylation and gene expression profiles in TCGA-BRCA breast cancer samples to propose a novel set of 35 methylation-based prognostic markers that may provide insights to molecular subtype specific disease stratification. Gene-set enrichment and pathway analysis of the predicted markers using MSigDB and DAVID revealed their role in mammary gland development pathway, various signaling pathways (ERBB2, NOTCH, etc.), and other cancer pathways, and show clear association with genes affected by hormone receptor status. We further show the discriminative power of the proposed DNA methylation signature in classifying breast cancer samples into three molecular subtypes, *viz.*, Luminal, HER2-enriched and Triple Negative. An accuracy of 94.12% and MCC of 0.87 is obtained in stratified 5-fold cross-validation for the three-class classification using SVM-RBF.

Keywords: DNA Methylation · Breast Cancer · Machine Learning · Subtype Classification

1 Introduction

Breast cancer is a heterogeneous disease and several subtypes have been identified with varied outcomes. Though some of the conventional measurements such as tumor grade and size, involvement of lymph node, and age of patient have been considered as important prognostic factors, the role of multi-omics data in breast cancer subtyping and therapeutic intervention is now undeniable. Based on the estrogen receptor (ER), progesterone receptor (PR), and human epidermal growth factor receptor 2 (HER2), three molecular subtypes of breast cancer have been identified: Luminal (ER+, PR±, HER2±), HER2-enriched (ER−, PR−, HER2+), and triple negative (ER−PR−, HER2−). Molecular subtyping is crucial for prognosis and treatment plan. For example, Luminal (LUM) tumors respond well to hormone therapy (tamoxifen or aromatase inhibitor) along with chemotherapy in some cases (Luminal B subtype), while HER2-enriched (HER2) and triple negative (TNBC) tumors have the worst outcomes for which hormone therapy has no benefit [1]. Various factors affect the transcriptome of tumor progression, of which epigenetic alterations, mainly DNA methylation changes are estimated to be observed in ~30% of breast cancers. Hypomethylation of oncogenes and metastatic genes and hypermethylation of tumor suppressors are shown to play a key role in cancer initiation, progression and metastasis. Because of the inherent reversibility of epigenetic states, there has been considerable interest in DNA methylation analysis with a focus on prevention and treatment [2].

Numerous earlier studies have explored the relationship between methylation patterns and molecular subtypes of breast cancer. However, identifying a small set of clinically relevant methylation markers has been a challenge because of large regions with aberrant methylation in cancer and encompassing numerous genes. Various studies have been conducted using supervised, unsupervised and deep learning models for subtype specific disease stratification. For feature selection, biological frameworks (using functional annotations, survival information, etc.), or statistical-based frameworks (t-test, chi-square test, etc.), or identified automatically by machine learning methods used [3–7]. For example, in DeepCC (2019), a supervised classification framework is proposed that uses a functional spectra representing gene enrichment scores for a subset of most variable genes (≥ 1000) [8]. In Zhang et al. (2018), Cox proportional risk regression models are used for identifying DNA methylation based prognostic markers and novel molecular subtypes of breast cancer using survival information. The discriminative ability of the proposed markers is confirmed using a Bayesian network classifier [9]. Two common strategies for integrating multi-omics data are used: directly concatenate features from different omics data for input to the classifier (integration in the input phase) and learning from individual classifiers from each data type sent to a final classifier for prediction (integration in the learning phase). For example, in DeepMO (2020), 5000 features are selected using chi-square test for each data type (mRNA, DNA methylation, Copy Number Variations) and the learned features from three encoding subnetworks are concatenated and sent to another classifier for prediction [10]. In Hierarchical Integration Deep Flexible Neural (HI-DFN) Forest model (2019), stacked autoencoder is used for learning from each multi-omics data (mRNA, miRNA, and DNA methylation data) and a hierarchical neural forest framework used for integrating all the learned representations into another autoencoder [11]. No biological knowledge is used in feature selection nor correlations between different data types considered during integration in these two studies. Also, the number of features is very large in many earlier studies, making them unsuitable as biomarkers. Multi-Omics Graph cOnvolutional NETworks (MOGONET, 2021) addresses this issue by first using Graph Convolutional Networks (GCN) for learning omics-specific features from each data type (mRNA, miRNA, DNA methylation) and then using View Correlation Discovery Network (VCDN) to identify cross-omics correlations for integration [12].

To address some of these issues, in this work we propose a knowledge-based framework for feature identification to obtain a small set of methylation markers. To assess the discriminative power of the proposed methylation markers, we considered six different ML models, *viz.*, SVM (with linear and Radial Basis Function (RBF) kernels), Logistic Regression (LR), Random Forest (RF), Linear Discriminant Analysis (LDA) classifier,

156   S. L. B. Pagolu et al.

and an Artificial neural network (ANN) to distinguish between LUM, HER2 and TNBC subtypes of breast cancer.

2 Materials and Methods

2.1 Data Acquisition

For this study, DNA methylation data from Illumina Infinium Human Methylation 450K array platform and gene expression data as raw sequence read counts from RNA sequencing (RNASeq) were obtained from The Cancer Genome Atlas (TCGA-BRCA dataset) [13], consisting of methylation levels of 485577 CpG sites and raw sequence read counts of 19962 protein coding genes. TCGA-BRCA dataset consists of 862 breast cancer patients (744 tumor, 84 normal), of which 538, 534 and 535 samples have the ER, PR and HER2 status labels, respectively. The three groups containing the respective receptor status labels were used for identifying DNA methylation-based features. Four datasets with PAM50 labels, GSE141338 ($n = 33$), GSE72245 ($n = 118$), GSE72251 ($n = 117$) and GSE84207 ($n = 252$) were downloaded from NCBI-GEO [14]. For identifying features for classification, the TCGA-BRCA was used, while for the classification task, a total number of 1260 BC samples, obtained by integrating TCGA-BRCA dataset with four NCBI GEO datasets, were used for training and validation (given in Supplementary Table 1).

2.2 Data Preprocessing

Preprocessing of methylation data was carried out using RnBeads to remove probes containing overlapping SNPs, context non-specific probes, cross-reactive probes, probes on sex chromosomes, and probes of low significance ($p \geq 0.05$) [15]. Probes containing more than 30% missing values were removed and for the remaining probes, imputation of missing values was performed by 'knn' method using *impute* package. Beta mixture quantile normalization (BMIQ) was used for type-bias correction and *methylumi.noob* for background intensity correction. Preprocessing of RNA-Seq raw read counts of 19,962 protein-coding genes was carried out using DESeq2 [16]. The preprocessing steps include removal of genes with less than 10 reads in total and median-of-ratios based normalization to adjust the counts within and across the samples.

2.3 Workflow for Identifying DNA Methylation-Based Subtype Specific Markers

The workflow to identify methylation-based markers for molecular subtyping of breast cancer is given in Fig. 1. We exploit here the role of three hormone receptors, namely, estrogen receptor (ER), progesterone receptor (PR), and human epidermal growth factor receptor 2 (HER2), in breast cancer subtyping. Three groups of TCGA-BRCA samples were constructed based on the receptor status of tumor samples and their adjacent normal: samples with ER (538), PR (534), and HER2 (535) receptor status labels, respectively. First, to filter CpG sites that may be relevant in tumorigenesis, sites that are differentially methylated between tumor and normal samples ($\Delta\beta$) are identified using the criteria

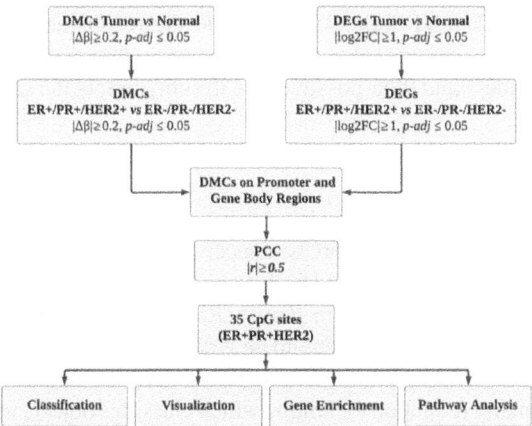

Fig. 1. Workflow for identifying subtype specific markers using TCGA-BRCA dataset. DMCs – Differentially Methylated CpG sites, DEGs – Differentially Expressed Genes, PCC - Pearson's correlation coefficient (r).

$|\Delta\beta| \geq 0.2$ and $p\text{-}adj \leq 0.05$ for each receptor group, determined using empirical Bayes static in '*limma*' package [18]. This resulted in 33730, 33902 and 33730 CpG sites for the three hormone receptor groups ER, PR and HER2, respectively. Next, considering the receptor status, CpG sites that are differentially methylated between samples with positive and negative receptor status within each group are identified using the same criteria. This resulted in 5266 (ER+ *vs* ER−), 1658 (PR+ *vs* PR−) and 794 (HER2+ *vs* HER2−) CpG sites for the three hormone status groups. Parallelly, a similar exercise was carried out with gene expression data to identify differentially expressed genes (DEGs). First, genes that are differentially expressed between tumor and normal samples are identified using the criteria, $|\log_2 \text{FoldChange}| \geq 1$, $p\text{-}adj \leq 0.05$, for each hormone group, determined using '*Wald*' test, to filter set of genes that may be of relevance in tumorigenesis. This resulted in 4737, 4738 and 4827 genes, for the three hormone receptor groups ER, PR and HER2, respectively. Next, using the same criteria, DEGs between positive and negative receptor status samples are identified in the three groups: 1648 (ER+ *vs* ER−), 1230 (PR+ *vs* PR−) and 754 (HER2+ *vs* HER2−). To identify methylation-based markers, we considered only those gene expression changes that may be associated with the regional DNA methylation status. For this, differentially methylated CpG sites (DMCs) on the promoter and gene body regions of the DEGs are filtered for the three groups: 575 (ER), 118 (PR) and 70 (HER2) CpG sites on 279 (ER), 83 (PR) and 43 (HER2) genes, respectively. Only those CpG sites whose methylation level (given by β) exhibit a correlation with the expression of the downstream gene are considered using the criteria $|r| \geq 0.5$, p-value ≤ 0.05 (using t-test), where r is Pearson's correlation coefficient. This resulted in 32 (ER), 12 (PR) and 3 (HER2) differentially methylated genic regions (DMRs). Considering only the most highly correlated CpG site for each DMR gene, a 35 methylation-based markers signature is proposed. To assess the discriminative power of these markers, the β value of these 35 CpG sites are used for classifying breast cancer samples into three subtypes, *viz.*, triple negative (TNBC),

HER2-enriched (HER2), and luminal (LUM) and the results are summarized below. Functional and pathway enrichment analysis of the genes associated with these markers was carried out to understand their possible role in subtype specific cancer initiation and progression.

2.4 Classification

For the multi-class classification of breast cancer samples (LUM, HER2 and TNBC), six machine learning models are considered: Support Vector Machine (SVM) ('Linear' kernel), SVM ('RBF' kernel), Random Forest (RF), Logistic Regression (LR), Linear Discriminant Analysis (LDA) classifier and an Artificial Neural Network (ANN). SVM and RF are supervised learning models, while LR is based on a linear regression model. LDA is a dimensionality reduction technique which classifies objects in a new embedded space by increasing variation between classes and decreasing within classes. ANN is a deep learning model that consists of many hidden layers between an input and output layer and back propagation through these layers adjusts the weights to correctly classify an object into the correct class.

2.5 Performance Evaluation and Visualization

Performance of the models was evaluated using stratified 5-fold cross-validation to ensure that the percentage of samples in different classes is preserved in every fold created for training and testing. For evaluating the performance of the ML models, the following metrics were computed: accuracy, precision, recall, F1-score, Mathew's correlation coefficient (MCC) and receiver operating characteristic - area under curve (ROC-AUC). For visualizing the clustering of samples, LDA plot was generated using the 35 CpG sites. Functional analysis of the differentially methylated genes corresponding to 35 DMCs was carried out using DAVID [17] and MSigDB [18].

3 Results

3.1 Evaluation of DNA Methylation-Based Markers for BC Subtype Classification

To assess their discriminative ability, the methylation levels (β value) of the 35 CpG sites is used as features in six widely used machine learning methods, SVM (with 'Linear' and 'RBF' kernels), RF, LR, LDA and ANN. Average values of the performance metrics from the stratified 5-fold cross-validation are summarized in Table 1. It may be noted that the performance of all the five models is comparable with the accuracy ranging from 92.62% to 94.12% and AUC values ≥ 0.94 for all the three classes. This clearly indicates the predictive power of proposed features in correctly classifying the samples. SVM-RBF model outperformed with best F1-scores (for all the three classes), accuracy and MCC values. Both Triple negative and HER2 classes had the best recall (0.96, 0.73) and F1-score (0.94, 0.77) with SVM-RBF, respectively. For the LUM class, the performance metrics are high across all models, while these are observed to be relatively low for the HER2 class. It is observed that the misclassified HER2 samples are labelled as LUM by all the models (confusion matrix in Fig. 2) The data imbalance (HER2 samples ~11%, while LUM ~70%), noisy labels, and that some Luminal B samples are HER2+ could be the probable reasons for this observation. Lower MCC values (0.84–0.87) are also indicative of the impact of data imbalance on the model's performance. To visualize the clustering of samples into three groups by the 35 CpG sites, LDA plot was generated as shown in Fig. 3. Three distinct clusters obtained indicate the reliability of the proposed methylation signature. A subset of 21 common features were identified from the 5-fold SVM-RBF models based on the feature importance scores. Functional analysis of these sites and corresponding genes (discussed below) indicates their potential role in the classification task. Ablation studies were performed using 30739 CpG sites (differentially methylated between Normal *vs* Tumor) and 5654 CpG sites (differentially methylated between ER+/PR+/HER2+ *vs* ER−/PR−/HER2−). SVM-RBF resulted in an accuracy of 93.81% and 94.60%, respectively for the two experiments, comparable to the accuracy obtained using 35 CpG sites (94.12%) (Supplementary file Sect. 2). Since machine learning models can identify best features for a classification task, this provides support to their relevance in distinguishing breast cancer subtypes.

Table 1. Performance of the ML models using stratified 5-fold cross validation on the integrated TCGA and GEO datasets.

Model	Class	Accuracy	Precision	Recall	F1-score	AUC	MCC
SVM-Linear	TNBC	93.73	0.91	0.94	0.93	**0.99**	0.86
	HER2		0.84	0.70	0.76	0.94	
	LUM		**0.96**	0.97	0.96	**0.98**	
SVM-RBF	TNBC	**94.12**	0.92	**0.96**	**0.94**	**0.99**	**0.87**
	HER2		0.83	**0.73**	**0.77**	0.95	
	LUM		**0.96**	0.97	**0.97**	**0.98**	
Random Forest	TNBC	92.62	**0.93**	0.95	0.94	**0.99**	0.84
	HER2		**0.90**	0.49	0.63	0.94	
	LUM		0.93	**0.99**	0.96	0.97	
Logistic Regression	TNBC	93.89	0.91	0.95	0.93	**0.99**	0.87
	HER2		0.86	0.67	0.75	0.94	
	LUM		**0.96**	0.98	**0.97**	**0.98**	
LDA	TNBC	93.25	0.90	0.93	0.92	**0.99**	0.85
	HER2		0.80	0.70	0.74	**0.95**	
	LUM		**0.96**	0.97	0.96	**0.98**	
ANN	TNBC	93.17	0.92	0.94	0.93	**0.99**	0.85
	HER2		0.80	0.68	0.73	0.94	
	LUM		**0.96**	0.97	0.96	**0.98**	

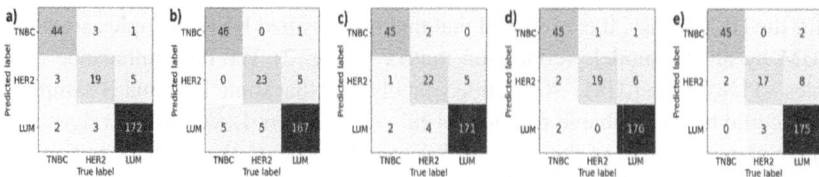

Fig. 2. Confusion matrices for SVM-RBF model for stratified 5-folds on the integrated TCGA-BRCA and GEO datasets.

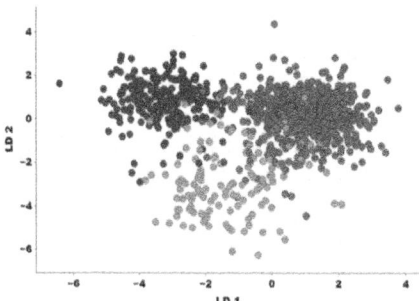

Fig. 3. LDA plot of the 1260 samples from the integrated TCGA-BRCA and GEO datasets using 35 CpG sites. Three distinct clusters corresponding to Luminal (red), HER2-enriched (green) and Triple negative (blue) subtypes are seen. (Color figure online)

4 Discussion

From the above analysis the reliability of the proposed 35 CpG sites in distinguishing between the breast cancer subtypes is seen. To understand their possible functional role in the three molecular subtypes, gene and pathway enrichment analysis of the associated genes was performed. Functional analysis of the genes using MSigDB [18] and DAVID [17] revealed 21 key genes and important pathways affected. These 21 features have non-zero feature importance scores in one or more of the five-fold SVM-RBF model, confirming their importance in correctly distinguishing between the three subtypes. A brief description of the gene and pathway enrichment analyses is given below.

4.1 Gene Set Enrichment Analysis (GSEA)

The gene set enrichment analysis of 35 differentially methylated genes was carried out using Molecular Signatures Database (MSigDB), which hosts annotated gene sets based on prior biological knowledge, such as common biochemical pathways, coexpression in experiments, coregulation, etc., including several cancer-related gene sets [19]. Top 5 gene sets enriched with the proposed 35-gene signature that are associated with hormone receptor status or molecular subtype in breast cancer are summarized in Supplementary Table 2. It is observed that the methylation status of 14 (out of 35) key genes have an overlap with these 5 gene sets. The variation in the distribution of their methylation levels for the three subtypes (TNBC, HER2, LUM) in TCGA-BRCA dataset is depicted as violin plots in Supplementary Fig. 1. A clear distinction in the mean methylation levels across the subtypes indicates their significance in distinguishing between the subtypes. A brief description of the top 5 datasets enriched with the 35 methylated gene signature is given below.

The Doane et al. dataset includes genes that are up-regulated in ER−/PR− breast tumors (do not express *ESR1* and *PGR*) [20]. Many genes of this set are shown to be either direct targets of ER, responsive to estrogen, or typically expressed in ER + BC. Eight (out of 35) genes having an overlap with this set are hypermethylated and under-expressed in the two ER− subtypes, TNBC (8) and HER2 (5) (Supplementary Fig. 1). FARMER et al. dataset includes genes that best discriminates the molecular subtypes of BC: basal *vs* luminal [21]. Of the 10 (out of 35) genes that have an overlap with this set, 8 are included in the Doane dataset, the remaining two are *CX3CL1* and *REEP1*. From the violin plots in Figure S1, we observed that 9 of these genes are hypermethylated (downregulated) in TNBC compared to LUM samples, except *CX3CL1*, which is hypomethylated in TNBC and hypermethylated in LUM subtype. The distinction in the methylation status of these 10 genes suggests their importance in distinguishing TNBC and LUM subtypes. Further, genes in Doane *et al.* and Farmer *et al.* datasets indicate androgen signalling for initiation and progression of BC as compared to estrogen signalling in Luminal group, thus suggesting the differences at the molecular level in these two subtypes. SMID et al. consists of two datasets, one in which the genes are down-regulated in basal subtype (TNBC) and the other in which genes are up-regulated in luminal B subtype. Twelve and eight genes (out of 35) have an overlap with these two datasets respectively. Of these six genes, namely, *AGR2, CCND1, CELSR1, ESR1, EVL, MLPH, SLC39A6,* and *TBC1D9*, indeed exhibit distinct methylation patterns between TNBC and LUM subtypes (Figure S1). Vantveer et al. dataset includes genes that are up regulated in ERz+ *vs* ER− tumors [22]. The overlapping of 6 genes (out of 35) with this dataset exhibit higher methylation (down regulation) in ER− subtypes (TNBC and HER2) in Figure S1. The functional role of these genes and pathways activated/repressed is briefly described below.

4.2 Pathway Analysis

Pathway enrichment analysis of the 35 signature genes using DAVID [17] identified nine genes associated with developmental (mammary gland development, nervous system development), signalling (*ERBB2* signalling, leptin signalling, signalling by non-tyrosine kinases), and cancer-related pathways. Functional analysis of these genes is summarized in Table 2 and distribution of their methylation values across the three subtypes is depicted as violin plots in Supplementary Fig. 1.

Various studies have demonstrated the role of DNA methylation in breast cancer, offering insights into subtype specific variations, and in diagnosis and prognosis. In most studies filtering of CpG sites is done based on differential methylation in tumor with respect to normal. This results in a very large number of methylated sites. To further reduce the number of methylation markers, we filtered CpG sites based on their receptor status: ER+ *vs* ER−, PR+ *vs* PR− or HER2+ *vs* HER2−, exploiting the association of hormone receptors with the molecular subtypes of BC. This additional step intrinsically captures the characteristic features responsible for difference in subtypes due to their

Table 2. Role of nine pathway enriched genes in TCGA-BRCA samples. (Methyl - methylation status, Exp - gene expression, Hypo - Hypomethylated, Hyper - Hypermethylated), ONCO - Oncogene, TSG - Tumor suppressor gene).

Gene	Subtype	Methyl	Exp	ONCO /TSG	Pathway
CCND1	TNBC	Hyper	Up	ONCO	Regulates cell cycle (G1/S phase); intermediatory role in cell cycle pathways, *viz.*, NFκB, Rac1, AMPK, *AKT* signalling pathways
ERBB2	HER2	Hypo	Up	-	Immune checkpoint signaling pathway, ErbB signaling pathway; Suppresses apoptosis leading to aggressive tumor growth and metastasis
ESR1	TNBC, HER2	Hyper	Down	-	*MAPK-Erk* pathway; ERK and Estrogen signalling pathways, cell cycle and proliferation; early prognostic marker for endocrine therapy
EVL	TNBC	Hyper	Down	-	ERK-dependent cell proliferation, cytoskeletal signaling, ER-mediated actin remodelling, inhibits BC cell motility
GRB7	HER2	Hypo	Up	ONCO	EGFR/ErbB signalling, modulates Ras signalling; RAC1/PAK1/p38/MMP2 pathway; regulation of cell migration; coexpressed/coamplified with HER2, potential prognostic marker and therapeutic target
MAML2	TNBC	Hypo	Up	ONCO	Regulates NOTCH signalling pathway; positive prognostic indicator in luminal A and HER2+

(*continued*)

Table 2. (*continued*)

Gene	Subtype	Methyl	Exp	ONCO/TSG	Pathway
RGMA	LUM, HER2+	Hyper	Down	TSG	Low expression associated with malignant growth & poor prognosis via activation of FAK/Src/PI3K/AKT signaling pathway; downregulation inhibits cell proliferation and migration in TNBC
RIPK4	LUM, HER2	Hyper	Down	ONCO/TSG	NFκB signalling, Wnt/Hedgehog /Notch; role in regulating cell proliferation, differentiation, and apoptosis, prognostic marker for bone metastasis
SOX10	LUM, HER2	Hyper	Down	-	ERK signalling, mediates proliferation through Notch4-PBP-mediated pathway, marker of TNBC/basal-like

hormone status. In numerous studies [9, 23, 24], CpG sites only from the promoter region of a gene are considered for their role in gene silencing. However, CpG sites from the gene body region are also shown to play a role in gene regulation [25, 26]. To not miss out on any relevant methylated genic regions, we considered CpG sites from both promoter and gene body regions. The differentially methylated regions (DMRs) are generally large and consist of many CpG sites within it. To reduce the number of features and consider only the most significant ones, the CpG site with its methylation level most highly correlated with the expression of the corresponding gene is considered. The integration of the two data types (methylation and gene expression) is performed based on biological relationship between the two omics data. This is unlike in any typical neural network model wherein features from each omics dataset are selected based on their relative importance in the classification task and concatenated. The number of features being small (35), these are not transformed into any new representation during the classification task, thereby providing insights into breast cancer subtype discrimination at the epigenetic level. This knowledge-based framework is quite distinct from the black-box nature of neural networks and provides an alternative to whole transcriptome analysis for molecular subtyping of breast cancer.

5 Conclusion

In this study, two omics data (DNA methylation and gene expression) are integrated to obtain a methylation-based subtype classifier for breast cancer. This has an added advantage that DNA methylation markers can provide tools for early detection and novel treatment protocols as methylation status is reversible. Twenty-one of the predicted 35 CpG sites are found to be functionally relevant. The associated genes of some of these sites are oncogenes or tumor suppressors, with their role in various cancer pathways such as ERBB2 signalling, NOTCH signalling, Wnt signalling, PI3K-AKT pathway, etc. reported in literature. These 21 sites are also observed to have high feature importance scores from SVM classifier. In six well-studied machine learning models using these 35 CpG sites as features, we observe a good classification of triple negative and luminal subtypes with a recall of 0.96 and 0.97 respectively. A slightly lower recall of 0.73 for HER2-enriched subtype is because of some Luminal B samples with HER2+ receptor status being misclassified. We expect that with more HER2 samples, the accuracy of the classifier can be improved. Since the treatment and prognosis of breast cancer patients is highly dependent on their subtype prediction, the predicted 35 CpG sites can serve as potential methylation-based prognostic markers for breast cancer. Apart from alterations in methylation, there are other factors that can influence gene expression, such as single nucleotide and copy number variations, and other epigenetic alterations induced by non-coding RNAs. Integrating other omics and variation data we hope to improve the predictive power of the classifier in future.

References

1. Haque, R., Ahmed, S.A., Inzhakova, G., Shi, J., Avila, C., Polikoff, J., et al.: Impact of breast cancer subtypes and treatment on survival: an analysis spanning two decades. Cancer Epidemiol. Biomark. Prev. **21**, 1848–1855 (2012)
2. Lubecka, K., Kurzava, L., Flower, K., Buvala, H., Zhang, H., Teegarden, D., et al.: Stilbenoids remodel the DNA methylation patterns in breast cancer cells and inhibit oncogenic NOTCH signaling through epigenetic regulation of MAML2 transcriptional activity. CARCIN. **37**, 656–668 (2016)
3. Bowler, S., Papoutsoglou, G., Karanikas, A., Tsamardinos, I., Corley, M.J., Ndhlovu, L.C.: A machine learning approach utilizing DNA methylation as an accurate classifier of COVID-19 disease severity. Sci. Rep. **12**, 17480 (2022)
4. Chen, Y., Yan, Y., Xu, M., Chen, W., Lin, J., Zhao, Y., et al.: Development of a machine learning classifier for brain tumors diagnosis based on DNA methylation profile. Front. Bioinform. **1**, 744345 (2021)
5. Hosseini, M., Lotfi-Shahreza, M., Nikpour, P.: Integrative analysis of DNA methylation and gene expression through machine learning identifies stomach cancer diagnostic and prognostic biomarkers. J. Cell Mol. Med. **27**, 714–726 (2023)
6. Levy, J.J., Titus, A.J., Petersen, C.L., Chen, Y., Salas, L.A., Christensen, B.C.: MethylNet: an automated and modular deep learning approach for DNA methylation analysis. BMC Bioinform. **21**, 108 (2020)
7. Gomes, R., Paul, N., He, N., Huber, A.F., Jansen, R.J.: Application of feature selection and deep learning for cancer prediction using DNA methylation markers. Genes **13**, 1557 (2022)
8. Gao, F., et al.: DeepCC: a novel deep learning-based framework for cancer molecular subtype classification. Oncogenesis **8**, 44 (2019)

9. Zhang, S., Wang, Y., Gu, Y., Zhu, J., Ci, C., Guo, Z., et al.: Specific breast cancer prognosis-subtype distinctions based on DNA methylation patterns. Mol. Oncol. **12**, 1047–1060 (2018)
10. Lin, Y., Zhang, W., Cao, H., Li, G., Du, W.: Classifying breast cancer subtypes using deep neural networks based on multi-omics data. Genes **11**, 888 (2020)
11. Xu, J., Wu, P., Chen, Y., Meng, Q., Dawood, H., Dawood, H.: A hierarchical integration deep flexible neural forest framework for cancer subtype classification by integrating multi-omics data. BMC Bioinform. **20**, 527 (2019)
12. Wang, T., Shao, W., Huang, Z., Tang, H., Zhang, J., Ding, Z., et al.: MOGONET integrates multi-omics data using graph convolutional networks allowing patient classification and biomarker identification. Nat. Commun. **12**, 3445 (2021)
13. Koboldt, D., Fulton, R., McLellan, M., Schmidt, H., McMichael, J., Fulton, L., et al.: Comprehensive molecular portraits of human breast tumours. Nature **490**, 61–70 (2012)
14. Edgar, R.: Gene Expression Omnibus: NCBI gene expression and hybridization array data repository. Nucleic Acids Res. **30**, 207–210 (2002)
15. Müller, F., et al.: RnBeads 2.0: comprehensive analysis of DNA methylation data. Genome Biol. **20**, 55 (2019)
16. Love, M.I., Huber, W., Anders, S.: Moderated estimation of fold change and dispersion for RNA-seq data with DESeq2. Genome Biol. **15**, 550 (2014)
17. Dennis, G., Sherman, B.T., Hosack, D.A., Yang, J., Gao, W., Lane, H.C., et al.: DAVID: database for annotation, visualization, and integrated discovery. Genome Biol. **4**, R60 (2003)
18. Liberzon, A., Subramanian, A., Pinchback, R., Thorvaldsdóttir, H., Tamayo, P., Mesirov, J.P.: Molecular signatures database (MSigDB) 3.0. Bioinformatics **27**, 1739–1740 (2011)
19. Subramanian, A., Tamayo, P., Mootha, V.K., Mukherjee, S., Ebert, B.L., Gillette, M.A., et al.: Gene set enrichment analysis: a knowledge-based approach for interpreting genome-wide expression profiles. Proc. Natl. Acad. Sci. U.S.A. **102**, 15545–15550 (2005)
20. Doane, A.S., Danso, M., Lal, P., Donaton, M., Zhang, L., Hudis, C., et al.: An estrogen receptor-negative breast cancer subset characterized by a hormonally regulated transcriptional program and response to androgen. Oncogene **25**, 3994–4008 (2006)
21. Farmer, P., Bonnefoi, H., Becette, V., Tubiana-Hulin, M., Fumoleau, P., Larsimont, D., et al.: Identification of molecular apocrine breast tumours by microarray analysis. Oncogene **24**, 4660–4671 (2005)
22. van 't Veer, L.J., et al.: Gene expression profiling predicts clinical outcome of breast cancer. Nature **415**, 530–536 (2002)
23. Wu, Z.-H., Tang, Y., Zhou, Y.: DNA methylation based molecular subtypes predict prognosis in breast cancer patients. Cancer Control **28**, 1073274820988519 (2021)
24. Bediaga, N.G., Acha-Sagredo, A., Guerra, I., Viguri, A., Albaina, C., Ruiz Diaz, I., et al.: DNA methylation epigenotypes in breast cancer molecular subtypes. Breast Cancer Res. **12**, R77 (2010)
25. Jones, P.A.: Functions of DNA methylation: islands, start sites, gene bodies and beyond. Nat. Rev. Genet. **13**, 484–492 (2012)
26. Győrffy, B., Bottai, G., Fleischer, T., Munkácsy, G., Budczies, J., Paladini, L., et al.: Aberrant DNA methylation impacts gene expression and prognosis in breast cancer subtypes: Prognostic value of methylation in breast cancer subtypes. Int. J. Cancer **138**, 87–97 (2016)

Repeated Measures Latent Dirichlet Allocation for Longitudinal Microbiome Analysis

Namitha Viona Pais[1](✉), Nalini Ravishanker[1], Sanguthevar Rajasekaran[2], and George Weinstock[3]

[1] Department of Statistics, University of Connecticut, Storrs, CT, USA
namitha.pais@uconn.edu
[2] Department of Computer Science and Engineering, University of Connecticut, Storrs, CT, USA
[3] Jackson Laboratory for Genomic Medicine, Farmington, CT, USA

Abstract. Topic modeling algorithms generally examine a set of documents, referred to as a corpus in Natural Language Processing (NLP), and analyze the words observed in a document to uncover themes that run through each document in a collection. In the microbiome framework, they are used to identify co-occurring microbial species and reveal hidden patterns or relationships within the microbial communities. Longitudinal microbiome data analysis provides a robust framework for studying microbiome compositions over time. By collecting multiple samples from the same individuals at different time points, researchers can capture the temporal variation within an individual's microbiome and evaluate its impact on the subjects' health status during each of their visits. This paper extends the Latent Dirichlet Allocation (LDA) modeling technique to a repeated measures framework. We propose Repeated Measures Latent Dirichlet Allocation (RM-LDA) where each document (subject) is assumed to be a collection of multiple sub-documents (visits associated with a given subject). In this study, we examine microbiome data on subjects making multiple visits to a medical facility to provide data on their microbiome counts. Our model allows us to analyze hidden patterns in the microbiome data over multiple visits, estimate the latent topic correlation structure within each subject, and study their association with the individual's health status over each visit.

Keywords: health status · latent topic · microbiome counts · repeated measures · topic models

1 Introduction

The human microbiome has been linked to a wide range of diseases, with the interactions among bacterial taxa, environmental factors, and genetic factors playing a crucial role. In recent years, there has been growing recognition of the significant impact the composition and function of the microbiome can have on various aspects of human health, including metabolism, immune system development, and disease susceptibility [1–3]. Therefore, it is important to investigate the dynamics of the microbiome over time to gain deeper insights into its role in health and disease progression [4–6].

Repeated measures or longitudinal microbiome data analysis provides a robust framework for studying the microbiome compositions over time. For instance, [7] identifies differences in immune, metabolic, and gut microbiome interactions in multiple sclerosis patients versus healthy controls. [1] studies the association of physical activity with gut microbiome richness, diversity, and composition, using a large, population-based cohort. By collecting multiple samples from individuals at different time points, researchers can capture the temporal variation within an individual's microbiome and evaluate its response to various factors. Longitudinal data collection enables the examination of both short-term fluctuations and long-term trends in the composition of the microbiome, offering a comprehensive understanding of its dynamics.

Traditional topic modeling techniques, such as Latent Dirichlet Allocation (LDA), have proven useful in uncovering latent themes within independent documents in a corpus. However, they do not capture how topics evolve over multiple sub-documents associated with a single document. Repeated measures LDA addresses this limitation by modeling longitudinal text data and extracting meaningful insights from its repeated measures nature. In this paper, we develop a methodology called "Repeated Measures Latent Dirichlet Allocation" (RM-LDA) that extends the Latent Dirichlet Allocation in a repeated measures framework by assuming each document to be a collection of multiple sub-documents. We introduce a document-specific correlation matrix to capture the within-document correlation of the latent topical structure.

We apply this model to analyze longitudinal microbiome data, where each subject makes multiple visits to a facility and provides data on their microbiome counts and health status. We treat each visit as a sub-document associated with the corresponding subject (document). Using the RM-LDA model, we extract the latent topic structure, which represents the underlying themes or patterns within the microbiome data across visits for each subject. We use the latent topic structure estimated in the longitudinal framework to fit a multinomial logistic regression model in order to classify visits of subjects into different health status.

2 Motivating Example

To examine how microbiomes affect human health, we analyze data provided by the Jackson Laboratory in Farmington, CT. Medical professionals collected blood from $M = 47$ subjects during their visits to the medical facility over a period of time. The number of visits n_d ranges from 5–25 on each subject for $d = 1, 2, \ldots M$. The cohort contains healthy individuals and individuals with prediabetes for host molecular omics profiling. At each visit, medical professionals collected two types of samples (stool and nasal swabs). Microbiome profiling then recorded counts on each $V = 109$ bacteria profiled using 16S gene sequencing. In addition to the bacteria counts, a battery of molecular and clinical laboratory tests were conducted and complemented by self-reported online surveys, which documented changes in medication, physical activity, diet preference, and perceived stress level. This information was observed as health status, denoted by $y_{d,r}$ for $d = 1, 2, \ldots, M$ and $r = 1, 2, \ldots, n_d$ was recorded for each subject at each visit. The health status levels were primarily self-reported by the subjects based on their health perceptions during the visits and, therefore can be considered *fuzzy* labels that provide ballpark information about an individual's health. The health status observed in the study is categorized into $C = 11$ levels: Healthy, Immunization, Infection, Post-Travel, Surgery, Antibiotics, Colonoscopy, Fiber, Stress, Weight-gain, and Weight-loss. The distribution of observed health status across a total of 511 visits of M subjects is shown in Fig. 1. The detailed description of the raw data is provided in [8].

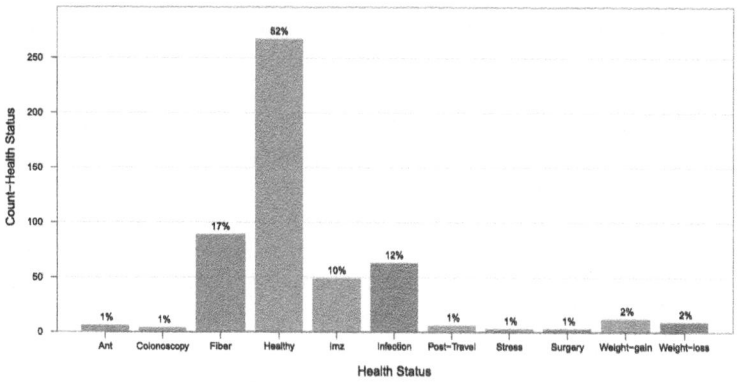

Fig. 1. Barchart of observed health status over $M = 47$ subjects recorded during each of their visits.

3 Method

In this section, we begin by reviewing Latent Dirichlet Allocation (LDA) in Sect. 3.1, originally proposed in [9]. In Sect. 3.2, we describe the Repeated Measures Latent Dirichlet Allocation (RM-LDA) model in detail.

3.1 Unsupervised Latent Dirichlet Allocation

LDA is an unsupervised, mixed membership model used mainly for document analysis, and is designed to discover latent topics in a collection of documents by analyzing the distribution of words in documents. The model assumes that each document is a mixture of a small number of topics, and each topic is a probability distribution over words. LDA enables researchers to uncover the underlying themes or topics present in a large corpus of text data. LDA has found applications in various domains, including information retrieval, recommendation systems, and text classification, making it a fundamental tool for understanding and organizing textual data. Besides, text data which includes categorical or count data, LDA can be modified to handle different types of numerical data. This can be achieved by modifying the data-generating distribution within the model.

Model Setup

Suppose we have a corpus $\mathcal{C}$ consisting of M documents $\mathcal{D}_1, \ldots, \mathcal{D}_M$. Suppose the dth document has N_d words, i.e., $\mathcal{D}_d = \{\boldsymbol{w}_{d,1}, \boldsymbol{w}_{d,2}, \ldots, \boldsymbol{w}_{d,N_d}\}$. Suppose the vocabulary from which words are drawn is indexed by $\{1, 2, \ldots, V\}$. Then, each word $\boldsymbol{w}_{d,n}$ is represented by a V-dimensional unit (basis) vector, i.e., $\boldsymbol{w}_{d,n} = (w_{d,n}^1, \ldots, w_{d,n}^V)'$ such that,

$$w_d^u = \begin{cases} 1 & \text{if } u = v, \\ 0 & u \neq v, \end{cases} \tag{1}$$

when $\boldsymbol{w}_{d,n}$ corresponds to the vth word from the vocabulary. LDA assumes the following generative process for the dth document $\mathcal{D}_d$ in the corpus $\mathcal{C}$, consisting of N_d words.

For $d = 1, 2, \ldots, M$:

- Choose $N_d \sim Pois(\lambda)$.
- Let J be the number of latent topics.
- Choose $\boldsymbol{\theta}_d \mid \boldsymbol{\alpha} \sim Dir(\boldsymbol{\alpha})$, where $\boldsymbol{\theta}_d = (\theta_1, \theta_2, \ldots, \theta_J)$ and $Dir(.)$ is a symmetric Dirichlet distribution with $\boldsymbol{\alpha} = \alpha \mathbf{1}_J$. Here, $\mathbf{1}_J$ indicates a J dimension vector of ones.
- For each of the N_d words, $\boldsymbol{w}_{d,n}$:
  • Choose a topic $\boldsymbol{Z}_{d,n} \mid \boldsymbol{\theta}_d \sim Mult(1, \boldsymbol{\theta}_d)$.
  • Choose a word $\boldsymbol{w}_{d,n} \mid \{\boldsymbol{Z}_{d,n}, \boldsymbol{\beta}\} \sim Mult(1, \boldsymbol{\beta}_{\boldsymbol{Z}_{d,n}})$, a multinomial probability distribution conditioned on the topic $\boldsymbol{Z}_{d,n}$.

LDA assumes J unobserved topics associated with a collection of documents where each document exhibits these topics in different proportions. This model uses the observed words as features to infer the hidden topic structure in each document. Once the structure of the model has been defined, the goal is to estimate the model parameters and compute the posterior distribution for inference. The LDA model estimation uses a variational expectation-maximization (VEM) algorithm to estimate the model parameters and variational inference (VI) algorithm to approximate the posterior distribution described in [9]. In the following section, we provide a novel extension of the LDA model to incorporate repeated measures data structure in the topic modeling framework.

3.2 Repeated Measures LDA

Suppose we have a corpus $\mathcal{C}$ consisting of M documents i.e., $\mathcal{C} = \{\mathcal{D}_1, \mathcal{D}_2, \ldots \mathcal{D}_M\}$. We consider the dth document to be a collection of n_d sub-documents i.e., $\mathcal{D}_d = \{\mathcal{D}_{d,1}, \mathcal{D}_{d,2}, \ldots \mathcal{D}_{d,n_d}\}$, where $\mathcal{D}_{d,r}$ corresponds to the rth sub-document of the dth document for $d = 1, 2, \ldots, M$. Each sub-document $\mathcal{D}_{d,r}$ is represented by a collection of $N_{d,r}$ words for $r = 1, 2, \ldots n_d$ and $d = 1, 2, \ldots, M$ i.e., $\mathcal{D}_{d,r} = (\boldsymbol{w}_{d,r,1}, \boldsymbol{w}_{d,r,2}, \ldots, \boldsymbol{w}_{d,r,N_{d,r}})$. Each word $\boldsymbol{w}_{d,r,n}$ for $n = 1, 2, \ldots, N_{d,r}$, $r = 1, 2, \ldots, n_d$ and $d = 1, 2, \ldots, M$ is denoted by a $V \times 1$ unit basis vector as shown in Equation (1). In the longitudinal microbiome analysis, we treat each subject as a document, and each visit is treated as a sub-document associated with that subject (document). RM-LDA assumes the following generative process for each of the rth sub-document of the dth document in corpus $\mathcal{C}$.

For $d \in \{1, 2, \ldots M\}$,

- Let J be the number of latent topics we assume tries to capture the hidden thematic structure in the corpus. In the microbiome analysis, the J topics capture the underlying distinct clusters or latent patterns within the microbiome data.
- For $r \in \{1, 2, \ldots n_d\}$,
  - Draw $\log(\boldsymbol{\alpha}_{d,r}) = (\log(\alpha_{d,r,1}), \log(\alpha_{d,r,2}), \ldots, \log(\alpha_{d,r,J})) \sim N_J(0, \boldsymbol{\Sigma}_d)$ where $\boldsymbol{\Sigma}_d$ is the $J \times J$ dimensional document-specific covariance matrix corresponding to the concentration parameter $\boldsymbol{\alpha}_{d,r}$.
  - Draw topic proportions from $\boldsymbol{\theta}_{d,r} \mid \boldsymbol{\alpha}_{d,r} \sim Dir(\boldsymbol{\alpha}_{d,r})$, where, $\boldsymbol{\alpha}_{d,r}$ is a J dimensional vector of Dirichlet concentration parameters.
  - For each word $\boldsymbol{w}_{d,r,n}$ for $n = 1, 2, \ldots N_{d,r}$. In the microbiome analysis, the words represent the observed microbiome counts.
    * Draw the per-word topic assignment $\boldsymbol{Z}_{d,r,n} \mid \boldsymbol{\theta}_{d,r} \sim Mult(1, \boldsymbol{\theta}_{d,r})$
    * Draw a word $\boldsymbol{w}_{d,r,n} \mid \{\boldsymbol{Z}_{\{d,r,n\}}, \boldsymbol{\beta}\} \sim Mult(1, \boldsymbol{\beta}_{Z_{\{d,r,n\}}})$

We describe the generative process assumed by RM-LDA as a hierarchical model written as

$$\boldsymbol{w}_{d,r,n} \mid \{\boldsymbol{Z}_{\{d,r,n\}}, \boldsymbol{\beta}\} \sim Mult(1, \boldsymbol{\beta}_{Z_{\{d,r,n\}}}), \qquad (2)$$

$$\boldsymbol{Z}_{d,r,n} \mid \boldsymbol{\theta}_{d,r} \sim Mult(1, \boldsymbol{\theta}_{d,r}), \qquad (3)$$

$$\boldsymbol{\theta}_{d,r} \mid \boldsymbol{\alpha}_{d,r} \sim Dir(\boldsymbol{\alpha}_{d,r}), \qquad (4)$$

$$\log(\boldsymbol{\alpha}_{d,r}) \sim N_J(0, \boldsymbol{\Sigma}_d). \qquad (5)$$

The dependence between the latent topic structure of sub-documents of the dth document is introduced through a document-specific $J \times J$ dimensional working correlation matrix $\boldsymbol{\Sigma}_d$ for $d = 1, 2, \ldots, M$. We write out the distribution of the latent topic structure $\boldsymbol{\theta}_{d,r}$ for the rth sub-document of the dth document below.

$$p(\boldsymbol{\theta}_{d,r} \mid \boldsymbol{\Sigma}_d) = p(\boldsymbol{\theta}_{d,r} \mid \boldsymbol{\alpha}_{d,r}) \times p(\boldsymbol{\alpha}_{d,r} \mid \boldsymbol{\Sigma}_d), \qquad (6)$$

where $p(\boldsymbol{\theta}_{d,r} \mid \boldsymbol{\alpha}_{d,r})$ and $p(\boldsymbol{\alpha}_{d,r} \mid \boldsymbol{\Sigma}_d)$ are defined in Equation (5) and (5). We see that the distribution of $\boldsymbol{\theta}_{d,r}$ depends on the working correlation matrix $\boldsymbol{\Sigma}_d$, for $r = 1, 2, \ldots, n_d$ and $d = 1, 2, \ldots, M$.

Approximate Inference

Given the rth sub-document of the dth document $\mathcal{D}_{d,r}$, the posterior distribution of the latent variables $(\boldsymbol{\theta}_{d,r,1:N_{d,r}}, \boldsymbol{Z}_{d,r,1:N_{d,r}}, \boldsymbol{\alpha}_{d,r})$ for $r = 1, 2, \ldots, n_d$ and $d = 1, 2, \ldots, M$ is given by

$$p(\boldsymbol{\theta}_{d,r}, \boldsymbol{Z}_{d,r}, \boldsymbol{\alpha}_{d,r} | \mathcal{D}_{d,r}, \boldsymbol{\beta}, \boldsymbol{\Sigma}_d) = \frac{p(\boldsymbol{\theta}_{d,r,1:N_{d,r}}, \boldsymbol{Z}_{d,r,1:N_{d,r}}, \boldsymbol{\alpha}_{d,r}, \mathcal{D}_{d,r} | \boldsymbol{\beta}, \boldsymbol{\Sigma}_d)}{p(\mathcal{D}_{d,r} | \boldsymbol{\beta}, \boldsymbol{\Sigma}_d)}, \quad (7)$$

for $r = 1, 2, \ldots, n_d$ and $d = 1, 2, \ldots, M$. The numerator term in Equation (7), $p(\boldsymbol{\theta}_{d,r}, \boldsymbol{Z}_{d,r}, \boldsymbol{\alpha}_{d,r}, \mathcal{D}_{d,r} | \boldsymbol{\beta}, \boldsymbol{\Sigma}_d)$ corresponds to the joint distribution of the latent variables $(\boldsymbol{\theta}_{d,r,1:N_{d,r}}, \boldsymbol{Z}_{d,r,1:N_{d,r}}, \boldsymbol{\alpha}_{d,r})$ and, observed document $\mathcal{D}_{d,r}$ and is given by

$$p(\boldsymbol{\theta}_{d,r}, \boldsymbol{Z}_{d,r}, \boldsymbol{\alpha}_{d,r}, \mathcal{D}_{d,r} | \boldsymbol{\beta}, \boldsymbol{\Sigma}_d) = p(\boldsymbol{\alpha}_{d,r} | \boldsymbol{\Sigma}_d) \times p(\boldsymbol{\theta}_{d,r} | \boldsymbol{\alpha}_{d,r}) \times \left(\prod_{n=1}^{N_{d,r}} p(\boldsymbol{Z}_{d,r,n} | \boldsymbol{\theta}_{d,r}) p(\boldsymbol{w}_{d,r,n} | \boldsymbol{Z}_{d,r,n}, \boldsymbol{\beta}) \right),$$

for $r = 1, 2, \ldots, n_d$ and $d = 1, 2, \ldots, M$. The denominator term in (7) i.e., $p(\mathcal{D}_{d,r} | \boldsymbol{\beta}, \boldsymbol{\Sigma}_d)$ represents the marginal distribution of the rth sub-document of the dth document and is given by,

$$p(\mathcal{D}_{d,r} | \boldsymbol{\beta}, \boldsymbol{\Sigma}_d) = \int \left(p(\boldsymbol{\alpha}_{d,r} | \boldsymbol{\Sigma}_d) \times p(\boldsymbol{\theta}_{d,r} | \boldsymbol{\alpha}_{d,r}) \times \sum_{\boldsymbol{Z}_{d,r,n}} \left(\prod_{n=1}^{N_{d,r}} p(\boldsymbol{Z}_{d,r,n} | \boldsymbol{\theta}_{d,r}) p(\boldsymbol{w}_{d,r,n} | \boldsymbol{Z}_{d,r,n}, \boldsymbol{\beta}) \right) \right) d\boldsymbol{\alpha}_{d,r} d\boldsymbol{\theta}_{d,r}$$

for $r = 1, 2, \ldots, n_d$ and $d = 1, 2, \ldots, M$. The posterior distribution in (7) is not analytically tractable. Therefore, we use variational methods that consider a simple family of distributions over the latent variables, indexed by free variational parameters. The variational parameters are estimated to minimize the Kullback-Leibler (KL) divergence between the variational distribution and the true posterior distribution. This allows us to use the variational distribution with the estimated variational parameter as an approximate to the true posterior.

We use the automatic variational inference algorithm, called Automatic Differentiation Variational Inference (ADVI) described in [10]. ADVI is a powerful algorithm supported by Stan for automatic and efficient variational inference. ADVI's flexibility enables it to handle a wide range of models, including those with non-conjugate priors, hierarchical structures, and latent variables. ADVI works by transforming the original model into an equivalent form that uses unconstrained real-valued latent variables. This ensures that all latent variables are in the same space, which makes it easier to approximate the posterior distribution. ADVI uses a factorized Gaussian variational approximation to approximate the posterior, also known as a mean-field approximation. This approximation assumes that the variational distribution can be factored into independent Gaussians, which makes it computationally efficient.

Parameter Estimation

Given a corpus $\mathcal{C}$ consisting of M documents with multiple sub-documents represented as $\mathcal{D}_{1:M}$, one is interested to find parameters Σ_d and β that maximize the log-likelihood of the data.

$$l(\beta, \Sigma_1, \ldots, \Sigma_M \mid \mathcal{D}_{1:M}) = \sum_{d=1}^{M} \sum_{r=1}^{n_d} \log p(\mathcal{D}_{d,r} \mid \Sigma_d, \beta). \tag{8}$$

Since $p(\mathcal{D}_{d,r}, y_{d,r} \mid \Sigma_d, \beta)$ is intractable, we employ ADVI to estimate the model parameters. ADVI utilizes Monte Carlo integration to approximate the ELBO and optimizes the ELBO in the real-coordinate space using stochastic gradient ascent, aiming to find the parameters that maximize the log-likelihood of the data [11]. This approach enables efficient and effective parameter estimation in complex models with large corpora. We conduct the analysis in R Stan and the code can be accessed from https://github.com/NamithaVionaPais/.

For parameter estimation, we resort to frequentist methods to estimate the unknown parameters of the RM-LDA model. An alternative approach is to employ fully Bayesian modeling, which allows us to incorporate prior knowledge about the hyperparameters and quantify uncertainty more comprehensively. To set up a Bayesian model for hyperparameter estimation in the RM-LDA model, we can set up a Dirichlet prior on β and inverse Wishart or Lewandowski-Kurowicka-Joe (LKJ) prior on the document-specific covariance matrix Σ_d for $d = 1, 2, \ldots, M$. In the next section, we apply the RM-LDA to the longitudinal microbiome data.

4 Longitudinal Microbiome Analysis Using RM-LDA

In this section, we present the results of the RM-LDA model applied to analyze the longitudinal microbiome data described in Sect. 2. We treat the observed microbiome counts over visits as the words (features) contained in the sub-documents for each subject (document) in our model. We set $M = 47$ (number of subjects), each having n_d visits (number of visits per subject) ranging from 5 to 25 for $d = 1, 2, \ldots, M$ and $V = 109$ (number of unique words/bacteria types). We set the number of latent topics $J = 11$ and estimate the topic proportions $\theta_{d,r}$ for $r = 1, 2, \ldots, n_d$ and, $d = 1, 2, \ldots, M$, and the matrix of word probabilities β. The choice of J aligns with the number of observed health status levels recorded over visits and is chosen for the interpretability of the estimated topics. Figure 2 shows the distribution of the top six bacteria types across each of the $J = 11$ latent topics. The top six bacteria types remain the same across each topic, which may be a result of the large counts associated with these bacteria types. However, their distribution varies across the different topics. We observe that *Bacteroides* is highly probable within topics $1, 3, 5, 7, 9$ and 11, *Faecalibacterium* is highly probably within topics 4 and 8 and *unclassified_Lachnospiraceae* is highly probable within topics 2 and 8.

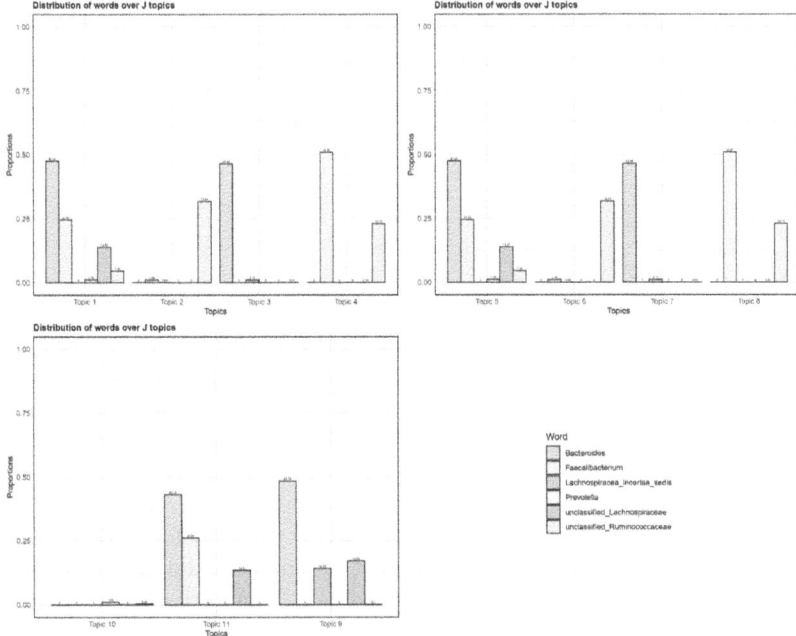

Fig. 2. Estimated distribution of the top six words over each $J = 11$ latent topics.

Within-subject Topic Correlation

For each subject, we can estimate the per-subject correlation matrix Σ_d for $d = 1, 2, \ldots, M$, which provides a measure of the strength of association among the J latent topics within each subject. For example, the estimated per-subject correlation matrix for one of the M subjects (denoted by S_d) is represented as a heat map in Fig. 3. Among the J latent topics topic 4 and topic 10 exhibit a positive correlation of 0.69. This indicates a high positive association between these two topics within the subject's profile.

Latent Topic Structure and Observed Health Status Associations

Our main goal is to estimate the underlying latent topic structure $\theta_{d,r}$ for $r = 1, 2, \ldots, n_d$ and, $d = 1, 2, \ldots, M$ for every subject, for each of their visits. The estimated latent topic structure can be used to group every visit of each subject into different health groups. While the latent topic structure with $J = 11$ topics itself is not able to capture these health groups, we can use the estimated latent topic proportions to obtain information on an individual's health status during each of their visits. We investigate the interaction between the latent topic structure and individual health statuses by fitting a multinomial logistic regression model using the estimated latent topic structure as predictors.

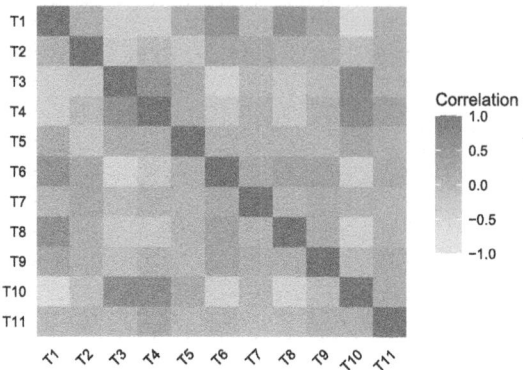

Fig. 3. Subject specific Correlation Matrix Σ_d represented as a heat map on one of the M subjects (denoted by S_d).

To analyze the relationship between the latent topic structure and observed health status recorder over visits, we employ the multinomial logistic regression model implemented using the *nnet* R package [12]. We treat the estimated J dimensional latent topic proportions $\boldsymbol{\theta}_{d,r}$ as predictors and the observed health status $y_{d,r}$ recorded for every subject during each of their visits as the response variable. The form of multinomial logistic regression model is shown below.

$$\text{logit}(P[y_{d,r} = c | \boldsymbol{\theta}_{d,r}]) = \beta_{0d} + \boldsymbol{\beta}_k^\top \boldsymbol{\theta}_{d,r},$$

for $r = 1, 2, \ldots, n_d$, $d = 1, 2, \ldots, M$ and, $c = 1, 2, \ldots C$. We include a common intercept β_{0d} for the n_d visits associated to the dth subject for $d = 1, 2, \ldots, M$. Empirical analysis of the observed health status indicates that few health status have rare occurrences as shown in Fig. 1. Therefore we set up a weighted logistic regression by giving more weight to the observations corresponding to rare classes. If $\boldsymbol{p} = (p_1, p_2, \ldots, p_C)$ is the proportions of occurrence for $C = 11$ observed health statuses, we define the weights $w_{d,r}$ as

$$w_{d,r} = \begin{cases} 4 & p_{y_{d,r}} \leq 0.05 \\ 1 & p_{y_{d,r}} > 0.05, \end{cases} \quad (9)$$

for $r = 1, 2, \ldots, n_d$ and $d = 1, 2, \ldots, M$. The choice of weights assigned to rare classes is determined through a grid search. The resulting confusion matrix on the multinomial logistic regression model is presented in Table 1. We are able to achieve an accuracy of 70.1%.

Table 1. Confusion matrix obtained from RM-LDA using data from $M = 47$ subjects.

	Ant	Colonoscopy	Fiber	Healthy	Imz	Infection	Post-Travel	Stress	Surgery	Weight-gain	Weight-loss
Ant	6	0	0	4	0	0	0	0	0	0	0
Colonoscopy	0	4	2	3	0	0	0	0	0	0	0
Fiber	0	0	78	15	4	0	0	0	0	0	0
Healthy	0	0	5	197	22	25	0	0	0	5	1
Imz	0	0	2	6	13	2	0	0	0	0	0
Infection	0	0	0	22	2	34	0	0	0	0	0
Post-Travel	0	0	0	0	1	0	6	0	0	0	0
Stress	0	0	0	0	0	1	0	3	0	0	0
Surgery	0	0	0	1	0	0	0	0	3	0	0
Weight-gain	0	0	2	12	1	0	0	0	0	7	1
Weight-loss	0	0	0	7	6	1	0	0	0	0	7

5 Comparison to Multinomial Logistic Regression Model in Repeated Measures Framework

We compare the performance of the RM-LDA to the multinomial logistic regression model for repeated measures. To demonstrate this, we train the two models on $M = 47$ subjects on the first $n_d - 1$ visits (with a total of 464 observations) while reserving the last visit on each subject as an observation in the test data for $d = 1, 2, \ldots, M$. We fit the multinomial logistic regression model for repeated measures using the R package *mclogit* [13] using $V = 109$ observed microbiome counts as predictors. We assign weights to rare classes as defined in Equation (9) while fitting the multinomial logistic regression model for repeated measures. The multinomial logistic regression model for repeated measures achieves a training accuracy of 82.3%, outperforming the RM-LDA model, which achieves a training accuracy of 72%. However, the usefulness of an approach and generalizability is best assessed by its performance on test data. Tables 2 and 3 show the confusion matrices obtained on the test data. In the test data, RM-LDA performs better with an overall accuracy of 60%, while the multinomial logistic regression model achieves a test accuracy of 48.9%.

Table 2. Confusion matrix on the test data on $M = 47$ to predict their health status during their last visit based on RM-LDA model.

	Ant	Colonoscopy	Fiber	Healthy	Imz	Infection	Post-Travel	Stress	Surgery	Weight-gain	Weight-loss
Ant	0	0	0	1	0	0	0	0	0	0	0
Colonoscopy	0	0	1	0	0	0	0	0	0	0	0
Fiber	0	0	3	4	0	0	0	0	0	0	0
Healthy	0	0	0	23	1	0	0	0	0	0	0
Imz	0	0	0	3	0	0	0	0	0	0	0
Infection	0	0	0	4	0	2	0	0	0	0	0
Post-Travel	0	0	0	1	0	0	0	0	0	0	0
Stress	0	0	0	0	0	0	0	0	0	0	0
Surgery	0	0	0	1	0	0	0	0	0	0	0
Weight-gain	0	0	0	0	0	0	0	0	0	0	0
Weight-loss	0	0	0	3	0	0	0	0	0	0	0

Table 3. Confusion matrix on the test data on $M = 47$ to predict their health status during their last visit based on the multinomial logistic model for repeated measures.

	Ant	Colonoscopy	Fiber	Healthy	Imz	Infection	Post-Travel	Stress	Surgery	Weight-gain	Weight-loss
Ant	0	0	0	1	0	0	0	0	0	0	0
Colonoscopy	0	0	0	1	0	0	0	0	0	0	0
Fiber	0	0	1	3	0	0	0	0	0	0	0
Healthy	0	0	1	19	0	0	0	0	0	0	0
Imz	0	0	0	0	1	0	0	0	0	0	0
Infection	0	0	0	5	0	2	0	0	0	0	0
Post-Travel	0	0	1	8	0	0	0	0	0	0	0
Stress	0	0	0	0	0	0	0	0	0	0	0
Surgery	0	0	0	1	0	0	0	0	0	0	0
Weight-gain	0	0	1	1	0	0	0	0	0	0	0
Weight-loss	0	0	0	1	0	0	0	0	0	0	0

6 Discussion

In this paper, we present the framework for an RM-LDA model for analyzing repeated measures data. We apply this model to a longitudinal microbiome dataset, where we have information on the observed gut microbiome counts along with the corresponding health status of different subjects across multiple visits to a medical facility. We use the RM-LDA model to uncover latent topic structures within the longitudinal microbiome data that are associated with the observed health status. While the latent topic structure alone cannot be directly used to cluster each visit for every subject into different health statuses (without incorporating the information from the observed health labels), we propose using a multinomial logistic regression model to classify each visit for every subject into different health status. The multinomial logistic regression model uses the estimated latent topic structure from the RM-LDA model as predictors to capture the relationship between the underlying latent topic structure in the microbiome data and the observed health status. The RM-LDA step can be viewed as a dimension reduction step that helps to reduce the high-dimensional microbiome data into a lower-dimensional latent topic space. Our comparative study results indicate that even with a dimension reduction in the predictor space from $V = 109$ to $J = 11$, the RM-LDA model combined with the multinomial logistic regression model performs better than the multinomial logistic regression model for repeated measures in the test data. Future development efforts will explore an alternative to our two-step model (RM-LDA followed by a multinomial logistic regression model), using a supervised Repeated Measures Latent Dirichlet Allocation (sRM-LDA) Model.

References

1. Holzhausen, E.A., et al.: Assessing the relationship between physical activity and the gut microbiome in a large, population-based sample of wisconsin adults. PLoS ONE **17**(10), e0276684 (2022)
2. Fukuda, S., Ohno, H.: Gut microbiome and metabolic diseases. Semin. Immunopathol. **36**(1), 103–114 (2013). https://doi.org/10.1007/s00281-013-0399-z
3. Dominguez-Bello, M.G., Godoy-Vitorino, F., Knight, R., Blaser, M.J.: Role of the microbiome in human development. Gut **68**(6), 1108–1114 (2019)
4. Proctor, L.M., Creasy, H.H., et al.: The integrative human microbiome project. Nature **569**(7758), 641–648 (2019)
5. Tipton, L., et al.: Measuring associations between the microbiota and repeated measures of continuous clinical variables using a lasso-penalized generalized linear mixed model. BioData Min. **11**, 1–20 (2018)
6. Brislawn, C.J., et al.: Multi-omics of the gut microbial ecosystem in inflammatory bowel diseases. Nature **569**(7758), 655–662 (2019)
7. Cantoni, C., et al.: Alterations of host-gut microbiome interactions in multiple sclerosis. EBioMedicine **76**, 1037980 (2022)
8. Zhou, W., et al.: Longitudinal multi-omics of host-microbe dynamics in prediabetes. Nature **569**(7758), 663–671 (2019)
9. Blei, D.M., Ng, A.Y., Jordan, M.I.: Latent dirichlet allocation. J. Mach. Learn. Res. **3**, 993–1022 (2003)
10. Kucukelbir, A., Ranganath, R., Gelman, A., Blei, D.: Automatic variational inference in stan. In: Advances in Neural Information Processing Systems, vol. 28 (2015)
11. Lim, K.-L., Jiang, X.: Variational posterior approximation using stochastic gradient ascent with adaptive stepsize. Pattern Recogn. **112**, 107783 (2021)
12. Ripley, B.: Feed-forward neural networks and multinomial log-linear models [R package nnet version 7.3-12] (2013)
13. Elff, M., Elff, M.M., Suggests, M.A.S.S.: R Package mclogit (2022)

Improving Disease Comorbidity Prediction with Biologically Supervised Graph Embedding

Xihan Qin(✉) and Li Liao

Department of Computer and Information Sciences, University of Delaware, Newark, DE 19716, USA
{xihan,liliao}@udel.edu

Abstract. Comorbidity is vital for disease understanding and management. In graph machine learning, it is seen as a result of mutations in disease-associated genes linked through the protein-protein interactions (PPI) of the human interactome. The incomplete human interactome however presents challenges in extracting useful features for comorbidity prediction. In this study, we introduce a new method called Biologically Supervised Graph Embedding (BSE) to select the most relevant features from the graph embedding for disease subgraphs relation representation, improving the accuracy of predicting comorbid disease pairs. Our investigation into BSE's impact on both centered and uncentered embedding methods showcases its consistent superiority over the state-of-the-art techniques and its adeptness in selecting features enriched with vital biological insights, thereby improving prediction performance significantly, up to 50% when measured by ROC AUC score. Further analysis indicates that BSE consistently and substantially improves the ratio of disease associations to gene connectivity, affirming its potential in uncovering latent biological factors affecting comorbidity. The study also reveals additional statistically significant enhancements across various metrics, further highlighting BSE's potential to introduce novel avenues for precise disease comorbidity predictions and other potential applications. The GitHub repository containing the source code can be accessed at the following link: https://github.com/xihan-qin/Biologically-Supervised-Graph-Embedding.

Keywords: Comorbidity · Human Interactome · Graph Embedding · Supervised Embedding · Isomap

1 Introduction

Comorbidity, the coexistence of multiple medical conditions within an individual [1], plays a pivotal role in disease management, treatment, and prognosis [2]. Understanding comorbidity patterns is essential for unraveling complex disease relationships and identifying shared molecular mechanisms [3]. In recent years, the study of disease comorbidity has gained considerable attention, leading to the development of network-based strategies for predicting and analyzing such relationships [2, 4–7].

In 2015, Menche et al. [8] introduced the concept of a disease module to investigate the human interactome, which comprises curated protein-protein interactions (PPI)

within the human cell. A disease module is essentially a subgraph consisting of genes known to be associated with a disease. Rather than simply relying on shared genes among different diseases, they introduced disease module separation (S_{AB}), a method measuring the distance between two disease modules. This innovative approach identifies comorbidity connections that traditional methods might overlook, as it can link disease modules using S_{AB} without requiring shared genes. Through their analysis, the authors demonstrated that shorter S_{AB} distances between disease modules in the graph correlate with a higher likelihood of comorbidity. This work sheds light on the potential of using the incomplete human interactome for studying disease relationships and beyond.

Human Interactome (HI), a large and incomplete graph, poses challenges in extracting valuable features for precise comorbidity prediction. Graph embedding, a potent technique used in various domains like PPI networks [9], social networks [10], and recommender systems [11], effectively captures significant features in a low-dimensional Euclidean space [12]. A commonly used embedding type transforms the whole graph into vectors, with each vector representing a node [13], including methods like spectral embedding [14], node2vec [15], and graph convolutional neural network [16]. In a study by Akram and Liao [17], they introduced Geometric Embedding (GE) that incorporates node geodesic distance into comorbidity prediction. Utilizing the same HI data, their approach outperformed the S_{AB}, achieving a significant enhancement ranging from 1.24 to 1.65 times. Among their experiments, the dimension reduction method Isomap [18] with a selected dimension of 20, and the SVM RBF classifier provided the best performance through 10-fold cross-validation for predicting disease pair comorbidity. This method has remained the current state-of-the-art in the field.

In this work, we introduce a novel method, Biologically Supervised Graph Embedding (BSE), designed to automatically select biologically relevant embeddings for comorbid disease pair prediction. Our extensive analysis demonstrates that BSE consistently outperforms the state-of-the-art GE method across all valid metrics when starting with the same centered embedding (Isomap). Furthermore, our findings highlight that BSE enhances disease associations and gene connectivity with distinct preferences for centered and uncentered embedding methods, thereby refining prediction performance. Notably, the ratio of disease associations to gene connectivity significantly improves—up to four times—when using BSE compared to the singular value rank method. These consistent enhancements across various performance metrics underscore BSE's potential to reveal hidden "biological meanings" and open new possibilities for accurate disease pair predictions and other potential applications.

2 Related Work

GE [17] utilize Isomap [18] to estimate the geodesic distance in a lower-dimensional space from a shortest distance matrix, D, among a chosen number of neighbors or within a specified neighborhood radius in the graph. Then, multidimensional scaling (MDS) [19] is performed to obtain the final embedding. MDS uses D to estimate the centered Gram Matrix, $\overline{G}$, using the formula below.

$$\overline{G} = -\frac{1}{2}HD^2H \qquad (1)$$

Here, H is the centering matrix:

$$H = I - \frac{1}{n}J \quad (2)$$

With n as the number of nodes in the largest connected component of the human interactome, I as the n × n identity matrix, and J as the n × n matrix of all 1 s.

The spectral decomposition of $\overline{G}$ generates the centered embedding matrix Z.

$$\overline{G} = U \Lambda U^T \quad (3)$$

$$Z = U_d \Lambda_d^{-\frac{1}{2}}, d < n \quad (4)$$

In the above equations, U is a n × n matrix, with orthonormal eigenvectors u_1, u_2, ..., u_n, and Λ is a n × n matrix diagonal matrix with eigenvalues $\lambda_1, \lambda_2, ..., \lambda_n$. The value of d is set by the user, representing the top d dimensions ranked by eigenvalues.

The state-of-the-art GE [17] redefines the distance between disease modules. While both S_{AB} and GE estimate distances using the shortest path between nodes, S_{AB} focuses on the mean shortest distance difference of nodes within the same subgraph and from different subgraphs, whereas GE calculates geodesic distances between all node pairs in the lower-dimensional space. The notable improvement with GE suggests that considering the geodesic distances between all node pairs, rather than just comparing mean shortest distances within and between disease modules, enhances the description of disease distance in the context of comorbidity. This implies that GE captures more hidden biological information by involving more nodes compared to S_{AB}. However, owing to the inherent noise in graph data, considering all node pairs in the neighborhood graph in reduced dimensions based on eigenvalues may not be the most effective way to achieve an optimally biologically meaningful embedding. In this work, we seek an improved and more biologically meaningful approach to embedding the graph for disease pair comorbidity tasks.

3 Materials

We obtained HI data, comorbid disease pairs, and relative risk (RR) dataset from [8], also used in GE [17]. The HI has 13,460 gene IDs (nodes) coding for proteins. The largest connected subgraph has 13,329 nodes, while smaller ones have only a few nodes each. We used 10,743 pairs of diseases with clinically reported RR scores for labeling. To ensure a fair comparison, we applied two thresholds: RR = 0 (RR0 dataset) and RR = 1 (RR1 dataset), as employed in GE [17]. In RR0, disease pairs with RR > 0 were labeled as positive (comorbid, "1"), while RR ≤ 0 were labeled as negative (non-comorbid, "0") and similarly in RR1. RR0 contains 82.6% positive pairs, while RR1 has 58.4% positive pairs.

4 Methods

Given the inherent incompleteness and noise in HI data, coupled with its high dimensionality, directly formulating a mathematical function to extract key "biological meanings" for the tasks is extremely challenging. Our proposed BSE method aims to leverage known knowledge which, in our case, are the HI data and disease comorbidity labels. In contrast to unsupervised embedding that relies on eigenvectors ranked by eigenvalues, we use a training dataset to identify the most discriminative eigenvectors for comorbid and non-comorbid disease pairs. We then use these selected eigenvectors to assess their performance on a separate testing dataset.

Algorithm 1: Biologically Supervised Embedding

```
Input: M - a matrix, the collection of original vectors, d - desired
number of vectors/dimensions
Output: out_vec - the selected vectors
ran_id = random permutation of M's indices
auc_avg_ori = 0                     # init the original average AUC score
out_vecs = empty column vector      # init the output vectors
for i = 0, 2, ......, d-1:
  auc_delta = []                    # init AUC score difference list
  for j = 0, 1, ......, len(rand_id)-1:
    ran_vec = M[rand_id[j]]
    cand_vecs = column_wise_append(out_vecs, ran_vec)
    feature_vecs = construct_feature_vectors(cand_vecs)
    avg_score = classifer_cross_validation (feature_vecs, labels)
    auc_delta.append((avg_auc - avg_auc_ori))
  End for
  max_idx = index_for_max(auc_delta)
  out_vecs = column_wise_append(out_vecs, M[: , max_idx])
End for
Return out_vecs
```

Algorithm 1 outlines our BSE method, which takes an input raw embedding matrix Z_0 obtained from any unsupervised graph embedding method. The user specifies the desired dimension number, d. The selection of each dimension is determined through a default 5-fold cross-validation process. The construction of the feature vector for each disease pair follows the approach described in [17], which is summarized briefly below.

The feature vector F_i for disease i, derived from embedding matrix Z is denoted as $\left[F_i^0 \ldots F_i^k \ldots F_i^{m-1}\right]$ with a length of m, corresponding to the row size (number of columns) of Z. F_i^k is calculated as the sum of the k-th column in Z for the gene set (set S) associated with disease i, using Eq. 5. For each disease pair (a and b), the feature vector is $\left[F_a^0 \ldots F_a^k \ldots F_a^{m-1}, F_b^0 \ldots F_b^k \ldots F_b^{m-1}\right]$ with a length of 2 m.

$$F_i^k = \sum_{j \in S} Z_j^k \tag{5}$$

In BSE, the output embedding matrix Z_d is built through supervised learning, optimizing the biological relevance, measured by a select metric (we use ROC AUC score).

Unlike the conventional method of selecting d columns by eigenvalues to form the embedding matrix, we choose columns based on their contributions to maximize prediction performance. For the first embedding dimension, we search through all n columns in the input Z_0 to find the best n × 1 Z_1. In case of multiple dimensions with the same top performance, we randomly choose one. When searching for the second embedding dimension, BSE explores the remaining n−1 columns to find one that, when added to Z_1, maximizes performance for the new n × 2 Z_2. We continue this process, adding one dimension at a time, until we have an n × d Z_d. This iterative approach helps us progressively identify the most informative vectors and uncover the key biological insights relevant to the task.

Table 1. Designed experiment steps to test BSE

E1. emb_select	E2. Emb_rank	E3. Vect_select	E4. Vect_rank	E5. Iso_select	E6. Iso_rank
1. Obtain the largest connected graph, G, from HI 2. Get shortest distance matrix, D, from G 3. SVD on D get $Z_0 = U_{100}\Sigma_{100}$ 4. Run BSE with Z_0, d = 20 to get Z_d	1. Same as E1 2. Same as E1 3. Same as E1 4. Select dimensions by singular values ranking, get $Z_d = U_{20}\Sigma_{20}$	1. Same as E1 2. Same as E1 3. Generate $Z_0 = U_{100}$ 4. Run BSE with Z_0, d = 20 to get Z_d	1. Same as E3 2. Same as E3 3. Same as E3 4. Trough singular values ranking, Get $Z_d = U_{20}$	1. Same as E1 2. Same as E1 3. Isomap on D to get Z_0 with 100 dimensions 4. Run BSE with Z_0, d = 20 to get Z_d	1. Same as E5 2. Same as E5 3. Isomap on D to get Z_d with 20 dimensions

Table 1 describes the 6 variations designed for testing BSE (Algorithm 1) for comorbid disease pair prediction. These variations include the Biologically Supervised Embedding (E1. emb_select), Ranked Embedding (E2. emb_rank), Biologically Supervised Vectors (E3. vect_select), Ranked Vectors (E4. vect_rank), Biologically Supervised Geometric Embedding (E5. iso_select), and Geometric Embedding (E6. iso_rank) methods. The goal is to evaluate how BSE can enhance both the centered and uncentered embedding techniques. In centered embedding approach (E5 and E6), we employ Isomap (Eqs. 1–4) to generate the embedding. For the uncentered approach (E1, E2, E3, and E4), we explain the design as below.

To implement the uncentered approach, the process initiates by generating the shortest distance matrix D directly from the graph. This stands in contrast to the GE method, where D is computed based on the k-nearest neighbor graph. Subsequently, the resulting matrix D undergoes singular value decomposition (SVD), denoted as

$$D = U\Sigma V^T \tag{6}$$

Here, U denotes the left singular vectors, which are equivalent to the eigenvectors for DD^T, which are the same as D^2 in Eq. 1.

$$DD^T = U\Sigma^2 U^T \qquad (7)$$

Likewise, V represents the right singular vectors, which are equivalent to the eigenvectors for $D^T D$, which is the same as D^2 in Eq. 1.

$$D^T D = V\Sigma^2 V^T \qquad (8)$$

As D in this uncentered approach is a symmetric square real matrix for an undirected graph, $DD^T = D^T D = D^2$. Hence, taking spectral decomposition of D^2 is equivalent to performing SVD on D [20].

For embedding approach E1 and E2, we apply SVD on D, and then take:

$$Z_d = U_d \Sigma_d, d < n \qquad (9)$$

For vectors approach E3 and E4, we apply SVD on D and then take:

$$Z_d = U_d, d < n \qquad (10)$$

We set the dimension number (d) of the final embedding matrix Z_d to 20 for all variations. This choice provides satisfactory results for a fair comparison and does not significantly increase the computation time. For the same reason, we set Z_0 to size n x 100 in variations E1, E3, and E5. When the input Z_0 has a dimension number exceeds 100, we select the initial 100 dimensions by eigenvalues to reconstruct Z_0. These numbers are subject to modification based on the desired results or computational efficiency. To assess performance, we use stratified 10-fold cross-validation on datasets RR0 and RR1. We employ the SVM classifier with the Radial Basis Function (rbf) kernel and set C to 3.5. We evaluate performance using metrics such as accuracy, precision, recall, F1 score, and ROC AUC. Our goal in these experiments is to identify the most effective approach for predicting comorbid disease pairs and to measure how BSE enhances both centered and uncentered methods.

5 Results

We examined and compared our study results using methods E1, E2, E3, E4, E5, and E6, which are also represented for short by notations "emb_s", "emb_r", "vect_s", "vect_r", "iso_s", and "iso_r" respectively. The methods with "select" or "s" used BSE, while those with "rank" or "r" utilized rank method for dimension selection. To explore the dimensions selected by BSE and to gain deeper insights, we extracted the first 5 selected dimensions from all 10-fold cross-validation folds and used their union to assess the average performance. The results for the three embedding methods are denoted as "iso_u," "embed_u," and "vect_u."

Tables 2, 3 and 4 present the average metric scores from 10-fold cross-validation for both RR0 and RR1. The rank method generally exhibits higher recall but lower ROC AUC score compared to the BSE method. A striking example emerges from the

Table 2. Average metric scores for RR0 using methods E1- E4

Metric	emb_r	emb_s	p_val	std	vect_r	vect_s	p_val	std
precision	0.8260	**0.9074**	8.63E−12	0.0059	0.8352	**0.9075**	7.02E−11	0.0066
recall	**1.0000**	0.9742	2.25E−08	0.0045	**0.9977**	0.9743	6.49E−07	0.0061
f1	0.9047	**0.9396**	2.15E−09	0.0047	0.9093	**0.9397**	1.81E−08	0.0052
Accuracy	0.8260	**0.8966**	5.59E−10	0.0081	0.8355	**0.8967**	5.35E−09	0.0091
roc_auc	0.5000	**0.7511**	5.13E−12	0.0172	0.5315	**0.7512**	5.59E−11	0.0196

Table 3. Average metric scores for RR1 using methods E1- E4

Metric	emb_r	emb_s	p_val	std	vect_r	vect_s	p_val	std
precision	0.5859	**0.7335**	4.08E−11	0.0127	0.6456	**0.7370**	4.18E−09	0.0132
recall	**0.9900**	0.8102	1.52E−10	0.0179	**0.8649**	0.8057	8.80E−07	0.0158
f1	0.7361	**0.7698**	7.13E−06	0.0116	0.7393	**0.7697**	6.87E−06	0.0104
Accuracy	0.5859	**0.7173**	3.37E−10	0.0143	0.6440	**0.7187**	5.29E−08	0.0144
roc_auc	0.5048	**0.6987**	2.12E−11	0.0155	0.5997	**0.7013**	1.01E−08	0.0162

Table 4. Average metric scores for RR0 and RR1 using methods E5- E6

Metric	RR0				RR1			
	iso_r	iso_s	p_val	std	iso_r	iso_s	p_val	std
precision	0.8644	**0.9046**	3.71E−09	0.0057	0.6975	**0.7262**	1.07E−05	0.0104
recall	**0.9846**	0.9717	5.92E−05	0.0058	**0.8250**	0.8102	1.37E−02	0.0154
f1	0.9206	**0.9369**	1.05E−06	0.0045	0.7558	**0.7658**	4.84E−03	0.0085
Accuracy	0.8596	**0.8919**	3.69E−07	0.0078	0.6889	**0.7109**	1.21E−04	0.0108
roc_auc	0.6255	**0.7424**	4.43E−09	0.0170	0.6616	**0.6910**	3.22E−05	0.0122

"emb_r" on RR0 dataset, wherein predicted labels for both real "1"s and "0"s are "1"s. Despite achieving precision and accuracy scores of 0.826 and recall of 1, the ROC AUC score languishes at 0.5, indicating random guessing and poor classification performance. Therefore, we excluded recall scores from our evaluation.

Since in BSE, we use ROC AUC improvement as our benchmark for dimension selection, a remarkable ROC AUC enhancement is achieved of up to 50.23% (emb_r vs. emb_s in Table 2). BSE consistently improves other metrics including precision, F1 score, and accuracy by up to 25.18%, 4.57%, and 22.43%, respectively, with statistically significant p-values. Importantly, the uniform high scores achieved by all BSE methods across various embedding techniques indicate that BSE has the capability to select biologically meaningful dimensions, leading to enhanced predictive performance across different embedding methodologies.

Figure 1 visually depicts the progressive improvements in precision, f1, accuracy, and ROC AUC scores achieved by BSE compared to the rank method, with dimension concatenation order for both RR0 and RR1 datasets. Notably, substantial improvements are evident in the early concatenations, typically within the first 5, followed by gradual increments. The dimensions selected by BSE significantly differ from the ranked dimensions, with minimal overlap. Moreover, the first 5 dimensions selected by BSE, consistently exhibit a conservative pattern across all 10 folds. We further examined these first 5 dimensions in the 10-fold cross-validation for all BSE methods, using their union to test and obtain the average score (Table 5). These scores closely align with those achieved using the same embedding method with BSE, underscoring the effectiveness of the union of the first 5 selected dimensions in preserving biological relevance. Interestingly, the "vect_u" f1 score even surpasses that of "vect_s" for RR1.

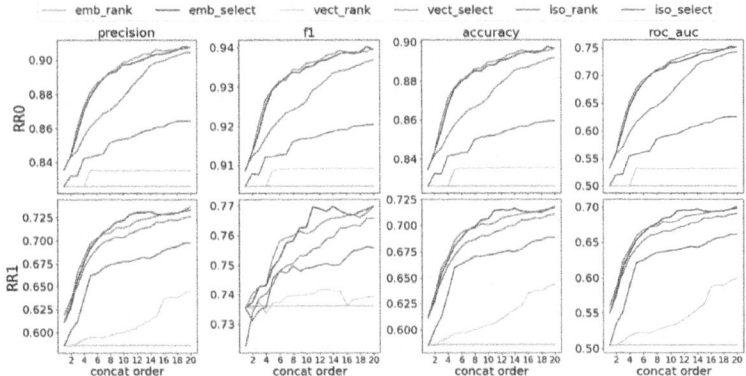

Fig. 1. Average metric scores along concatenated dimensions by BSE

Table 5. Metric scores using union dimensions.

Metric	RR0			RR1		
	iso_u	emb_u	vect_u	iso_u	emb_u	vect_u
precision	0.8863	0.8976	0.8901	0.7094	0.7377	0.7347
recall	0.9812	0.9772	0.9815	0.8264	0.8020	0.8095
f1	0.9313	0.9357	0.9335	0.7634	0.7684	**0.7702**
Accuracy	0.8805	0.8891	0.8846	0.7012	0.7180	0.7182
roc_auc	0.6918	0.7240	0.7029	0.6761	0.7011	0.6999

To further explore biological insights, we identified the top 20 genes based on embedding values in the first 5 dimensions across all methods and counted the number of associated diseases for each gene (Fig. 2). The first 5 dimensions for methods employing

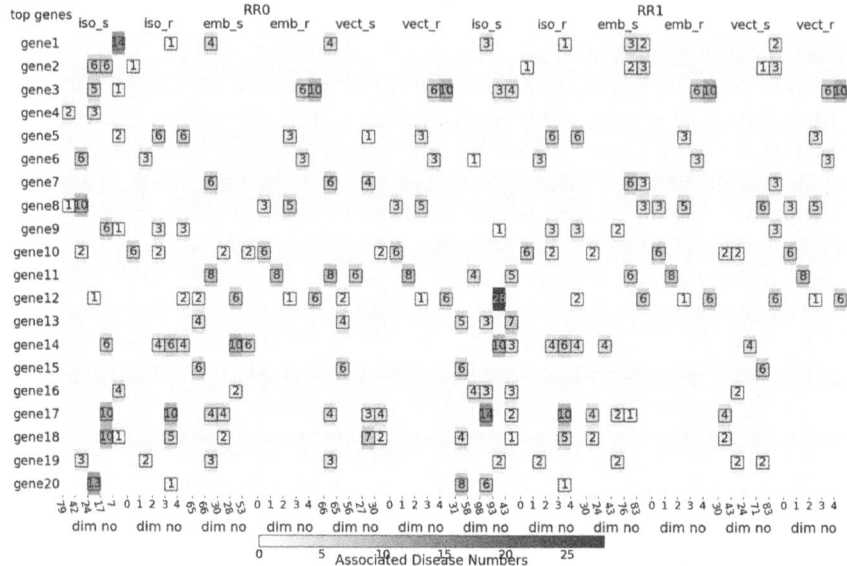

Fig. 2. Top 20 genes with associated disease numbers for first 5 dimensions.

BSE (marked as "s") were selected based on their frequency in the top 20 dimensions across all 10 folds. For comparison with the rank method, we labeled dimensions by their respective order in singular value ranking, such as "79", "42", "24", "17", "7" in "iso_s" for RR0. The top 20 genes within each dimension are displayed in corresponding columns. The color-coding signifies the number of associated diseases for each gene, with numbers indicating gene-disease associations greater than 0. Figure 2 shows that many top genes lack known disease associations from the dataset. However, for genes with known disease associations, BSE significantly enhance the tally of disease associations for the centered approach ("iso_s"). In contrast, the uncentered approaches ("emb_s" and "vect_s") display a less distinct trend.

We further depict gene connectivity, another crucial facet of "biological meaning", by plotting the node degrees for the top 20 genes (Fig. 3). Node degrees range from 1 to 870, with colors indicating degree values—darker shades represent higher degrees. Interestingly, in uncentered embedding methods, there is a noticeably lighter color pattern with BSE compared to the rank method. This suggests that BSE helps uncentered methods identify hidden and biologically significant connections, reducing connection "noise." However, this pattern is less evident in centered embedding methods.

Figure 4 provides statistics on total disease counts and total degrees for the top 20 genes within the initial 5 dimensions, across all methods and both datasets. It's evident that in centered approaches, BSE enhances disease associations, compared to the rank method. Conversely, in uncentered approaches, BSE emphasizes gene connectivity (total degrees) in contrast to the rank method. Notably, both the "emb" and "vect" embedding methods show a significant decrease (3 to 4 times) in total degrees when transitioning from the rank method to the BSE method. However, for the "iso" embedding method, the total disease count increases by up to 2 times with BSE compared to the rank

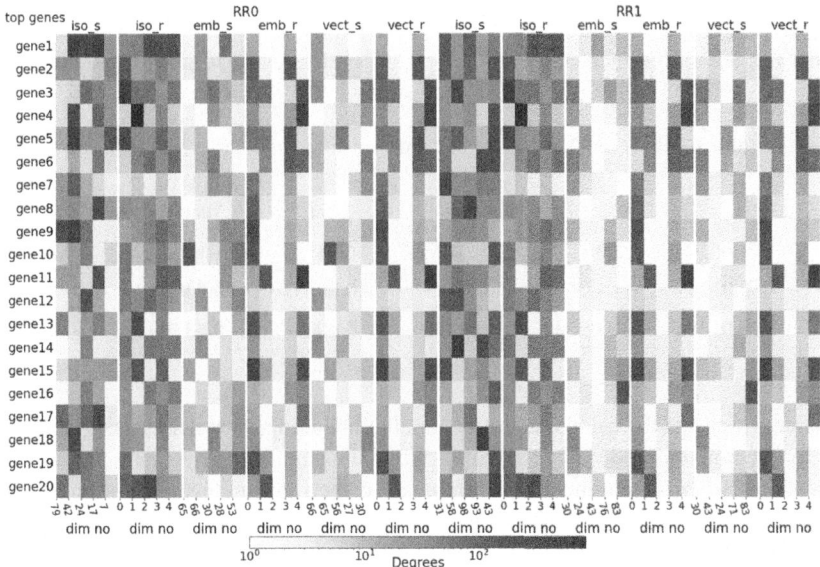

Fig. 3. Top 20 genes and their degrees for the first 5 dimensions.

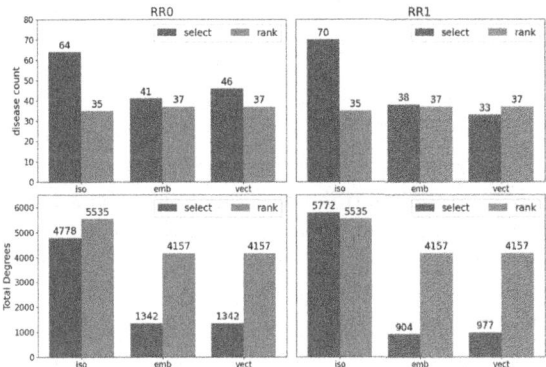

Fig. 4. Disease count and degrees for top 20 genes and 5 dimensions

method. These findings underscore the importance of both disease association and gene connectivity in this task. To quantify the constant enhancement of "biological meaning" by BSE, we define a ratio R as below.

$$R = \frac{\sum_{i \in G, j \in D} s_i^j}{\sum_{i \in G, j \in D} d_i^j} \tag{11}$$

Set D represents the chosen dimensions, wherein we selected the first 5 dimensions for all methods. Set G represents the top genes within these dimensions, wherein we selected top 20 genes for evaluation. The variable s_i^j denotes the associated disease number for gene i in dimension j, while d_i^j signifies the degree of gene i in dimension j.

Table 6. R for top 20 genes in first 5 dimensions using different methods

	Iso		Emb		Vect	
	select	rank	select	rank	select	rank
RR0	**0.0134**	0.0063	**0.0343**	0.0063	**0.0306**	0.0089
RR1	**0.0121**	0.0063	**0.0376**	0.0063	**0.0399**	0.0089

Table 6 presents the calculated ratios for all embedding approaches using BSE ("select") or the rank method. The results indicate a substantial improvement, ranging from 2 to 4 times when employing BSE compared to the rank methods. This underscores that, when embedding HI data for disease comorbidity prediction, a higher R, associated with more disease associations and lower gene connectivity in the network, signifies greater biological relevance. Across all embedding methods, BSE consistently identifies and integrates critical biological insights, significantly enhancing the ratio of disease association to gene connectivity.

6 Discussion

Our study aimed to improve disease comorbidity prediction using the Biologically Supervised Embedding (BSE) method. We explored both centered and uncentered embedding methods and assessed the impact of BSE on overall performance. Guided by ROC AUC, BSE resulted in a 50.23% increase, showcasing its ability to balance sensitivity and specificity. BSE consistently enhanced precision, f1, and accuracy metrics, contributing to performance improvements of up to 25.18%, 4.57%, and 22.43%, respectively, with statistical significance. Our analysis of top genes in HI embeddings uncovered valuable biological insights. BSE not only increased disease association counts, particularly in centered approaches, but also revealed the importance of gene connectivity in uncentered approaches. This capacity to incorporate relevant biological meanings demonstrates the versatility and potential of BSE across diverse embedding methods. Moreover, we found that the ratio of disease associations to gene connectivity matters most when embedding HI for disease comorbidity prediction. It's important to note that our algorithm is a greedy one, constructing the best embedding by iteratively selecting columns from the input matrix Z_0. There's no backtracking to change the existing embedding. While this approach yields promising results, further improvements may be achievable, albeit with increased computational costs. Additionally, our use of the RR dataset and specific thresholds is intended to benchmark against the state-of-the-art method GE. Exploring alternative comorbidity measurements to RR and considering different thresholds has the potential to enhance the performance further.

7 Conclusion

Our work introduced a novel Biologically Supervised Embedding method to enhance disease comorbidity prediction. Across benchmark datasets, we consistently observed significant performance improvements compared to traditional methods and the state-of-the-art. Our study emphasizes the importance of the ratio of disease association to gene connectivity in shaping predictive accuracy. BSE's ability to reveal hidden biological insights creates opportunities for more precise comorbid disease pair predictions. It can be adapted for more recent and larger HI data from sources like the STRING database [21]. It also has the potential to work with other popular embedding methods like Node2Vec [15] and is versatile for applications such as protein interaction analysis, drug-target interaction prediction, and gene expression pattern recognition. While initially developed for biological network analysis, this supervised embedding approach holds broader potential for analyzing graph data in diverse fields.

Acknowledgment. The authors thank Joerg Menche for sharing the comorbidity data and would also like to thank the anonymous reviewers for their valuable comments.

References

1. National Institute on Drug Abuse, Advancing Addiction Science, Research Topics: Comorbidity, U.S. Department of Health and Human Services, National Institutes of Health. https://nida.nih.gov/research-topics/comorbidity
2. Pepera, G., et al.: Epidemiology, risk factors and prognosis of cardiovascular disease in the Coronavirus Disease 2019 (COVID-19) pandemic era: a systematic review. Rev. Cardiovasc. Med. **23**(1), 28. 1 (2022)
3. Keezer, M.R., et al.: Comorbidities of epilepsy: current concepts and future perspectives. Lancet Neurol. **15**(1), 106–115 (2016)
4. Morselli Gysi, D., et al.: Network medicine framework for identifying drug-repurposing opportunities for COVID-19. Proc. Natl. Acad. Sci. **118**(19), e2025581118 (2021)
5. Charlson, M.E., et al.: A critical review of clinimetric properties. Psychother. Psychosom. **91**(1), 8–35 (2022)
6. Astore, C., et al.: LeMeDISCO is a computational method for large-scale prediction & molecular interpretation of disease comorbidity. Commun. Biol. **5**(1), 870 (2022)
7. Nam, Y., et al.: Discovering comorbid diseases using an inter-disease interactivity network based on biobank-scale PheWAS data. Bioinformatics **39**(1) (2023)
8. Menche, J., et al.: Disease networks. Uncovering disease-disease relationships through the incomplete interactome. Science **347**(6224), 1257601 (2015)
9. Yue, X., et al.: Graph embedding on biomedical networks: methods, applications and evaluations. Bioinformatics **36**(4), 1241–1251 (2020)
10. Kumar, S., et al.: Influence maximization in social networks using graph embedding and graph neural network. Inf. Sci. **607**, 1617–1636 (2022)
11. Ma, G.F., et al.: Graph neural networks for preference social recommendation. PeerJ Comput. Sci. **9**, e1393 (2023)
12. Mengjia, X.: Understanding graph embedding methods and their applications. SIAM Rev. **63**(4), 825–853 (2021)

13. Cai, H., et al.: A comprehensive survey of graph embedding: problems, techniques, and applications. IEEE Trans. Knowl. Data Eng. **30**(9), 1616–1637 (2018)
14. Rohe, K., et al.: Spectral clustering and the high-dimensional stochastic blockmodel. Ann. Stat. **39**(4), 1878–1915 (2011)
15. Grover, A., Leskovec, J.: Node2Vec: scalable feature learning for networks. KDD **2016**, 855–864 (2016)
16. Kipf, T.N., Welling, M.: Semi-supervised classification with graph convolutional networks. arXiv preprint arXiv:1609.02907 (2016)
17. Akram, P., Liao, L.: Prediction of comorbid diseases using weighted geometric embedding of human interactome. BMC Med. Genom. **12**(Suppl 7), 161 (2019)
18. Tenenbaum, J.B., et al.: A global geometric framework for nonlinear dimensionality reduction. Science **290**(5500), 2319–2323 (2000)
19. Shepard, R.N.: Multidimensional scaling, tree-fitting, and clustering. Science **210**(4468), 390–398 (1980)
20. Shores, T.S.: Applied Linear Algebra and Matrix Analysis. Undergraduate Texts in Mathematics. Springer, Cham (2007). https://doi.org/10.1007/978-3-319-74748-4
21. Szklarczyk, D., et al.: The STRING database in 2023: protein–protein association networks and functional enrichment analyses for any sequenced genome of interest. Nucleic Acids Res. **51**(D1), D638–D646 (2023)

Lightweight and Generalizable Model for COVID-19 Detection Using Chest Xray Images

Suba Suseela and Nita Parekh(✉)

International Institute of Information Technology, Hyderabad, India
suba.s@research.iiit.ac.in, nita@iiit.ac.in

Abstract. Deep learning (DL) has revolutionized the field of medical imaging, including chest radiology, by offering advanced tools for accurate and efficient detection of diseases for over a decade now. Analysis of Chest radiology images (chest X-ray - CXR and Computed tomography - CT) using DL models has widened its scope as a triaging tool since COVID-19 pandemic due to its speed, accuracy, and objectivity of disease detection, leading to better patient outcomes and more efficient healthcare delivery. CNNs are particularly well suited for image analysis tasks due to their ability to capture hierarchical features. Earlier work on Severe Acute Respiratory Syndrome (SARS) and Middle East Respiratory Syndrome (MERS) have also shown their applicability in the diagnosis of pulmonary diseases. This has led to much recent attention on the analysis of chest radiographs (CXR) using deep learning architectures for the detection of COVID-19 in a clinical setting. Applications developed for medical image analysis require high sensitivity, precision and generalizability along with reliability so as to provide radiologists and clinicians with an additional layer of information to aid in diagnoses. In this work we propose pixel-based attention mechanisms into a lightweight CNN model (Attn-CNN) trained on one of the largest publicly available COVIDx CXR-3 dataset. With much fewer training parameters, it is seen to perform better than four state-of-the-art (SOTA) deep learning models. The generalizability of the model is shown by performing analysis on external dataset. With portable chest radiography (CXR) being commonly used for early disease detection and follow up of lung abnormalities, there is a clear scope of the proposed model in assisting health experts in triaging of patients in pandemic-like situations. Data and code are available at: https://github.com/aleesuss/c19.

Keywords: Convolutional Neural Networks · Attention Mechanism · Chest X-Rays · COVID-19

1 Introduction

The need of early diagnosis for effective containment of the spread of COVID-19 disease during the pandemic has been very clear. However, shortage of RT-PCR kits, longer times required for the results, and high false negative rates during the early phase of infection have been the bottleneck faced worldwide for timely action. Further, large inflow of patients during the first year of pandemic caused heavy burden on the hospital staff worldwide, suggesting the need for automated and accurate detection of the disease in such situations. Chest radiography imaging (chest X-rays (CXRs) and computed tomography (CT) scans) were used in hospitals worldwide for faster triaging of patients and several studies have reported their advantage along with RT-PCR to confirm the diagnosis [1, 2]. Earlier work on other strains of the coronavirus family, Middle East respiratory syndrome (MERS) and severe acute respiratory syndrome (SARS) also confirm the usefulness of chest radiographs in the diagnosis of pulmonary diseases [3].

Chest X-rays are readily available in healthcare settings, including smaller clinics and hospitals, and require less time for image acquisition. This allows for a quicker diagnosis and can be particularly valuable during a pandemic when timely decisions are crucial. Use of CXRs is more practical as they use significantly less ionizing radiation than chest CT scans. This is especially important for pregnant women, children, and young adults, as excessive exposure to radiation can be harmful, particularly when used for monitoring progression of the disease and assess effectiveness of treatment. Portable chest X-ray machines can be used at the patient's bedside, minimizing the need for patient transportation within the hospital. This reduces the risk of cross-infection between patients, a concern during a highly contagious disease outbreak. However, X-ray being a 2D image, may miss out subtle lung abnormalities during early infection stages compared to CT scans which provides a 3D view of the infected region.

Major contributions of this work are:

1. Proposed a novel lightweight attention-based CNN architecture for extracting relevant features from a CXR image suitable for a clinical setting.
2. Binary and 3-class classification of CXR images using the largest openly available benchmark dataset, COVIDx CXR-3.
3. Evaluated generalizability of the model with hold out and external test sets.

The dataset used for training the models is the largest of its kind that is openly available. The model's performance is evaluated with other SOTA CNN based deep learning (DL) models such as VGG-16, ResNet-50, EfficientNet-B7 and Inception-v3. We show that though being very light-weight, the proposed model achieves comparable performance to very DL models (slightly better), with much fewer number of parameters and smaller training time, making it suitable for clinical settings.

1.1 Abnormalities in the Lungs

Chest X-Rays show characteristic abnormalities in the case of COVID-19 patients. Common findings include multiple regions of Ground Glass Opacities (GGO) sometimes also associated with Consolidations mostly in the mid and lower lobes of the lungs. The abnormalities can be bilateral and mostly seen in the peripheral areas of the lungs. Like with any pneumonia, COVID-19 also causes the lung density to increase, seen as whiteness in chest radiography images. Based on the severity of the disease the lung markings in the images are obscured. When the lung markings are partially obscured it is called a *ground glass pattern* and when the lung markings are completely obscured by the whiteness it is called *consolidation* [2, 4]. Progression of the disease is evident from Fig. 1: (a) normal, (b) showing GGOs on the first day, and (c) consolidations by the

10th day. Reading Chest X-rays requires the expertise of a radiologist which can be a bottleneck in triaging patients in pandemic situations. Computer assisted image reading may help in such situation for prioritizing the allotment of resources.

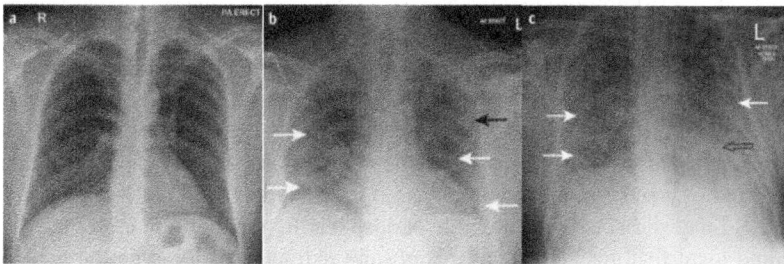

Fig. 1. Normal Posterior-Anterior (PA) chest radiograph of a patient (a) taken 12 months before COVID-19 infection, (b) when infected with COVID-19 on first day showing Ground-glass opacities (GGO) in the periphery of both lungs in the mid and lower zones, (c) Dense Consolidations in the image seen on the 10th day. (Reproduced from [2]).

2 Related Works

Recent pandemic spurred tremendous research publications and various deep learning architectures have been used for the classification of Chest X-ray images ranging from pre-defined architectures such as ResNet, VGG, Exception, UNet, etc. pretrained on ImageNet database [5] to very complex synthesized models on these deep layered architectures. For example, CovidNet [6] used a tailor-made design pattern based on ResNet architecture that comprises projection-expansion-projection-extension (PEPX) pattern of multiple Conv layers followed by a fully connected layer and the output 'softmax' layer. This highly complex design strategy was generated using a Generator-Inquisitor pair to design an optimal model for preset requirements. Another study by Khan et al. proposed CoroNet [7] on Xception architecture consisting of 71 layers, using depth-wise separable convolution layers along with residual connections. Another DL model by Sedik et al. [8] used a series of ConvLSTMs (which has the capacity to encode spatiotemporal information) and three convolutional layers for feature extraction followed by fully connected last layer for the binary classification. Brunese et al. [9] used VGG-16 model for first identifying the presence of pneumonia in CXR image and then classify Pneumonia samples into CP (COVID pneumonia) or NCP (pneumonia due to other causes) in the second step. Coro-Net model, proposed by Khobahi et al. [10], comprises two modules for 3-class classification: a Task Based Feature Extraction Network module (TFEN) and COVID-19 Identification Network module (CIN), a pre-trained ResNet-18 network. TFEN is a Feature Pyramid based AutoEncoder (FPAE) network with seven layers of convolutional encoder and decoder blocks. Rajaraman et al. [11] compared the performance of a customized CNN with a number of deep learning models pre-trained on ImageNet for extracting modality-specific features. The final prediction was made using transfer learning from various models, iteratively pruning best performing models and using ensemble of predictions by average voting. Oh et al. [12] proposed a patch-based classification technique to overcome the small dataset problem. DenseNet based model was used for segmenting CXRs which were then used for training multiple ResNet-18 models and majority voting taken to arrive at the final prediction. The study by Ozturk et al. [13] proposed a DarkNet-19 model, which is based on real-time object detection system named YOLO (You only look once). Generative Adversarial Network (GAN) was used to generate CXR images from ~300 original CXR images to overcome the limited dataset problem and improve model performances [14]. Alexnet, Googlenet and ResNet-18 were used for classification tasks in the study. Attention mechanism have been used in the analysis of CT scans in numerous studies to exploit channel features but very few studies have proposed attention based frameworks for COVID-19 classification using CXR images. In [15], an ensemble of deep learning models are used for capturing global and local features. The local features are extracted using attention mechanism and multi-instance learning by considering different parts of a single CXR image as multiple instances. However, unlike in multiple instance learning, a single CXR image has only a single label for all parts of the image even when each part is considered as a different instance. In [16], encoder built on 'self-attention' mechanism is proposed to extract global features for the classification task. The feature map extracted from a ResNet model is converted to a sequence using 'map-to-sequence' operation used in language models and fed to a 'self-attention' encoder. Many of these studies suffer from one common factor – limited image data of COVID-19 chest X-rays resulting in data imbalance. Various data augmentation methods, GANs, random under-sampling, etc. have been proposed for handling data imbalance [6–11]. A review on deep learning approaches for COVID-19 detection in chest radiograph images is given in [17].

3 Data

The Chest X-Ray dataset, COVIDx-CXR-3 [18], used in this study was downloaded from Kaggle (updated on 02/06/2022), and is collated from eight publicly available data repositories. To the best of our knowledge, this is the largest and most well-curated dataset with COVID-19 images from a multi-national patient cohort. Data is prepared for both binary (COVID-19 vs non-Covid) and three-class (COVID-19, normal and non-Covid pneumonia) classification tasks. The number of images for the three-class classification task is given in Table 1. The non-Covid class for binary classification task consists of both normal and other pneumonia images, resulting in the number of images comparable in the two classes. In Table 1, though the COVID class is $~2-3 \times$ larger compared to the other two classes, number of images (16,690) is very large compared to the patient size (2986) in COVID class, indicating multiple images from each patient, while number of images is similar to the number of patients in Normal and Pneumonia classes. Consequently, the amount of variation seen by the models in the images from the other two classes is much greater than for the COVID class and may compensate for data imbalance to some extent. For the binary classification task, 200 COVID-19 and 200 non-Covid images were randomly taken as test set and 80% of the remaining images (16490 COVID-19, 13992 non-Covid) were used for training and 20% for validation. For multi-class classification, the test set is same as the one used in binary classification with 100 images from normal and pneumonia classes and 200 images in covid class. The training set consisted of 24,104 images and 6026 images for validation across the three classes. The data split has been followed as in [18]. Another dataset from King Abdullah University hospital (KAUH), Jordan [19], was also used to evaluate the performance of the models on an unseen data. This dataset consists of CXR images from 368 confirmed COVID-19 patients with medium to severe cases requiring multiday stay in hospital. The CXR images taken after at least 5 days of admittance were included in the dataset.

Table 1. Training and test set data for three-class classification using chest X-ray radiography images. The number of patients is given in brackets.

Type	Normal	Pneumonia	COVID-19
Train	8085 (8085)	5555 (5531)	16490 (2808)
Test	100 (100)	100 (100)	200 (178)
Total	8185 (8185)	5655 (5631)	16690 (2986)

4 Methods

In this section the architecture of the proposed attention-based CNN model for binary (COVID-19/Non-COVID-19) and three-class (Covid/Normal/Pneumonia) classification of CXR images is described. The pre-processing of the CXR images and the workflow of the proposed model followed by various state-of-the-art (SOTA) DL models considered for performance comparison are given below.

Image Resizing. The input images sourced from various hospitals across the world are of different sizes because of different capturing modalities and different equipment standards. Hence all images are processed and resized to 224 × 224 pixels to provide images of the same size as input to the CNN-based models.

Normalization. The dataset is normalized within a range of 0 and 1 by multiplying each pixel in the images by a factor of 1/255. This is done to make all the images uniform in terms of pixel intensity.

Classification. The framework of the proposed Attn-CNN model is given in Fig. 2. It consists of standard Conv1 block followed by five 'depthwise' Conv blocks, an Attention Module (AM) and a standard Conv2 block followed by the output Dense layer. The input to the standard Conv block is a CXR image of size 224 × 224 and output from the AM is a set of relevant features from the image that are sent to the succeeding standard Conv2 block and Dense layer for classification. The standard Conv2 block consists of 3 'Conv' layers coupled with 'BatchNormalization' layers and a 'Maxpool' layer. The Dense layer takes the output from standard Conv2 block as input and outputs a label (Non-Covid/COVID-19) and (Normal/COVID-19/Pneumonia) for the input image for binary and 3-class classification tasks, respectively. The different blocks are color coded as shown in Fig. 2(a) and the detailed architecture of the attention module is given in Fig. 2(b). Number of kernels in the Conv blocks increases hierarchically as 16, 32, 64, 128, 256, and 512. After each Conv block, 'maxpool' and 'batchnormalization' layers are added. Standard Conv1 block consists of 'standard' convolutional kernels while blocks 2 to 6 consist of 'depthwise' convolutional kernels with kernel size 3 × 3. The 'depthwise' convolutions are chosen for reducing computational complexity. The AM inserted after this series of Conv blocks selects only the relevant pixels and discards rest of the pixels from the features of previous Conv layer. This is done by employing a 'soft attention' mechanism that adds a weight to each feature such that only the features with considerable weights are chosen and sent to the proceeding 'Dense' block. The feature maps from the previous Conv block pass through a 'Softmax' layer which assigns a value to each of the pixels in a feature map that sums up to 1. All the feature maps thus generated are added to form a single feature map which is further multiplied with the original feature map to only select the ones with higher weights. The output thus obtained is concatenated to the original feature map and sent through a 'Maxpooling' layer. Thus, the 'locally' relevant features from each feature map and the globally important features are selected before the next step. Also, as an additional residual connection, the original feature map is also sent through a 'Maxpooling' layer. The outputs of both 'Maxpool' layers are further concatenated and subjected to an activation function ('Relu'). The residual connection makes sure that not only the highly weighted pixels are selected but also any missed pixels that are still important are picked up. The final selected features are sent to the Standard Conv2 block and the 'Dense' layer for classification. The model has ~2.8M parameters which makes it lightweight and deployable in a clinical setting. The optimizer used is Adam with an initial learning rate set to 5e−5 and decay rate of first and second moments were set to the default values of 0.9 and 0.999, respectively.

Attention Module. The Attention Module assigns weights to the pixels of feature maps of the previous Conv layer using a 'Softmax' function. These weights are multiplied to the original feature maps. These features after the AM are given by:

$$F_{t+1} = \epsilon F_t(\sigma(z)_i \frac{e^{z_i}}{\sum_{j=1}^{K}(e^{z_j})}) \tag{1}$$

where, σ is the 'Softmax' function, z_i is the input pixel of the convoluted feature map, $(F_t) \in \mathbb{R}^{H \times W \times D}$ represents feature maps obtained from the previous Conv layer, and H, W and D denote the height, width and depth of the feature maps. The ϵ (<1) is a learning factor deciding the rate at which the model learns. The output, F_{t+1} is concatenated with F_t and again with the 'Maxpooled' F_t, F_t'. The final feature map sent out to Dense block can be represented as: $(F_t || F_{t+1}) || (F_t')$, where || represents the concatenation operation.

For performance evaluation of the proposed model, four state-of-the-art deep learning models, viz., VGG-16 (~138M), ResNet-50 (~23M), Inception-v3 (~25M) and EfficientNet-B7 (~60M), pre-trained on ImageNet database [5], are considered. VGG-16 model has 16 layers with weights. The fully connected layers from the model were replaced with four customized fully connected layers of 512, 128, 64 and 3 nodes in each layer respectively. ResNet-50 has Conv blocks and Identity blocks arranged alternately to form 48 layers along with 1 'Maxpool' and 1 'Avgpool' layer. An output layer with 1 node (for binary classification) and 3 nodes (for multi-class classification) were added to

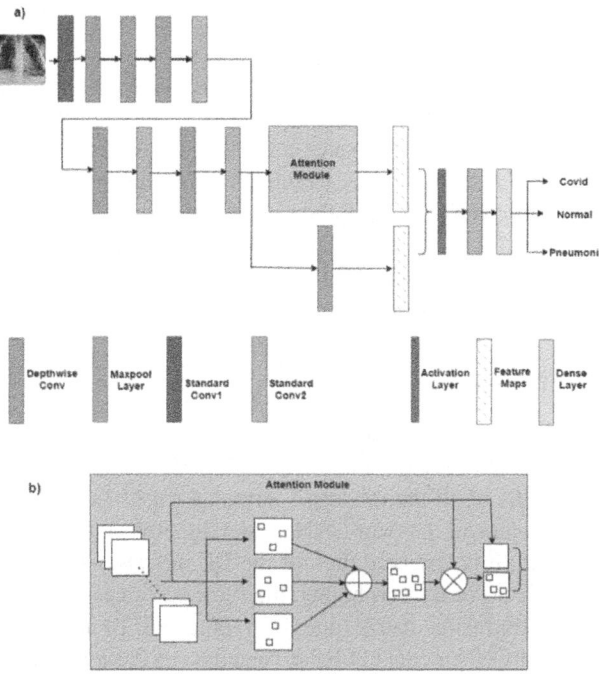

Fig. 2. A) Architecture of the Attn-CNN model. The Convolution blocks, Attention Module (AM), Max pool, and Dropout layers are colour coded as shown. b) The detailed architecture of the AM is given. The Feature Maps F_t are normalized using a 'Softmax' activation function and the aggregated output multiplied with F_t. The resultant feature maps, F_{t+1} are concatenated with F_t before sending out of AM.

the model before training. Inception-v3 has multiple symmetric and asymmetric building blocks, including convolutions, average pooling, max pooling, and fully connected layers. EfficientNet-B7 has a complex architecture with more than 800 layers and ~60M parameters. It does a compound scaling of the model in depth, width, and height dimensions with a fixed ratio to improve the performance. It has a 'stem' block followed by multiple mobile inverted bottleneck convolution ('MBConv') blocks. The 'stem' block contains convolution, 'batchnormalization' and activation layers. Each 'MBConv' block also has multiple convolutions and 'batchnormalization' layers connected among each other. The architecture was developed using automated neural architecture design strategy. The learning rate was set to 0.0001 initially for all the four models and set to reduce by 0.3 if no improvement in validation loss was observed for 2 epochs. Sigmoid activation function was used as in the proposed Attn-CNN model. Like in ResNet-50, an output layer with 1 node (for binary classification) and 3 nodes (for multi-class classification) was added to all the models for final classification.

5 Results and Discussion

Performance of the proposed Attn-CNN model is evaluated on a holdout test set of COVIDx CXR-3 dataset and on an external dataset from King Abdullah University hospital (KAUH), Jordan [19]. Performance comparison on both these datasets is carried out with four popular SOTA DL models. Results of binary and multi-class classification of the CXR images are given below. All the models were trained on dual core NVIDIA GeForce RTX 3080 Ti GPU with 32 GB DDR4 3200 MHz RAM. The operating system was Ubuntu 22.04 with CUDA version 12.0. The models were implemented using Python 3.9 and Keras package with Tensorflow version 2.4.1.

5.1 Binary Classification

For binary classification of chest X-ray images, all the models were trained on COVIDx CXR-3 training set for 20 epochs, till convergence was achieved and the time taken was ~2–3 h. The results of all the models is summarized in Table 2. It may be noted that the test accuracy value of the Attn-CNN model (96.50%) is higher than all the four DL models considered. It is observed that for the DL models, VGG-16 and Inception-v3, though the training accuracy was higher, their test accuracies were lower than with Attn-CNN. From the confusion matrix (not given) obtained, 14 incorrect classifications were made by the Attn-CNN model (11 False Negatives, 3 False Positives), much lower than with the DLs (VGG-16 (19), ResNet-50 (46), EfficentNet-B7 (47) and Inception-v3 (48)). From Table 2 we see that Attn-CNN model outperformed all the four DL models with highest values of precision (0.98), recall (0.94) and F1-score (0.96) for the COVID class. These are comparable or better to most models reported in [18] which uses the same dataset. This shows that by focusing only on the subtle patterns in the CXR images rather than the whole image it is possible to identify the discriminating features of the disease categories in an efficient way.

5.2 Multi-class Classification

Performance comparison of the Attn-CNN with the four SOTA models for the three-class classification using CXR images is summarized in Table 3. Here again, we see that Attn-CNN model outperformed with a test accuracy of 94.75%, even with a lightweight architecture compared to the four DL models with very complex architecture. From the confusion matrix (not given), the number of misclassifications for COVID-19 class are observed to be least with Attn-CNN model (5), compared to VGG-16 (19), ResNet-50 (32), Efficient-Net-B7 (41) and Inception-v3 models (38), clearly indicating its ability in capturing the characteristic features of COVID-19 in CXR images. For the normal class, performance of all the models, including Attn-CNN is comparable with 2–5 images wrongly classified as Pneumonia, while for Pneumonia class, comparatively higher number of misclassifications are observed for all models: Attn-CNN (12), VGG-16 (14), EfficentNet-B7 (16), Inception-v3 (21) and ResNet-50 (25). Most wrongly predicted images in this case went to normal class in all models. The probable reason for this could be that normal and pneumonia images are old images from legacy databases/sources and may include images with low resolution and less clarity, and thereby uniformly affected the performance of all the models. From Table 3 it is evident that the proposed model is highly efficient in distinguishing the three classes compared to the DL models. For the multi-class classification task also the training times are similar to that for binary classification for 20 epochs (~2–3 h). The model proposed in [18] reported a sensitivity of 95.5% for covid class. Another study that used an attention-based framework with 311M trainable parameters also reported similar performances; an accuracy of 95% with sensitivity (97%) and precision (98%) for COVID-19 class [16]. Both these studies used the older version of the dataset used here. From the above analysis we conclude that for building a clinical tool for detecting COVID-19, a simple CNN architecture with an attention module would suffice compared to a complex deep learning model. The number of parameters in the Attn-CNN model is ~2.8M, much less compared to VGG-16 with 138M, ResNet-50 over 23M and EfficientNet-B7 ~60M parameters.

Table 2. Performance comparison of the proposed Attn-CNN with four deep learning models for binary classification on COVIDx CXR-3 test set.

	Accuracy	Precision	Recall	F1-score
Attn-CNN				
COVID-19	**96.50**	**0.98**	**0.94**	**0.96**
Non-Covid		**0.95**	**0.98**	**0.97**
VGG-16				
COVID-19	95.25	**0.98**	0.93	0.95
Non-Covid		0.93	**0.98**	0.95
ResNet-50				
COVID-19	88.50	0.94	0.82	0.88
Non-Covid		0.84	0.95	0.89
EfficientNet-B7				
COVID-19	88.25	**0.98**	0.79	0.87
Non-Covid		0.82	**0.98**	0.89
Inception-v3				
COVID-19	87.99	0.93	0.82	0.87
Non-Covid		0.84	0.94	0.89

From the precision, recall and F1-scores on the external test set from Jordan summarized in Table 3 we observe that there is a drop in the performance of all the models in distinguishing covid images from the other classes. Even though a high recall of 0.97 is observed in the case of Attn-CNN model in the hold out test set of COVIDx CXR-3 dataset, it dropped to 0.83 on the external set. Similar behavior is observed for other DL models except EfficientNet-B7 (which exhibited improved performance), with misclassified COVID-19 images predicted as pneumonia. This is not surprising as the external set was not seen by the model during the training phase. Also, the images included in the Jordan dataset are from very severe cases of pneumonia that required multiple days of hospitalization, while COVIDx has a very wider distribution of severity. Clinically, patients in the Absorption phase or stage 4 (≥ 14 days), the ground-glass opacities and linear consolidation may be interpreted as a process of repair and reorganization, partially mediated by an organizational pneumonia [20]. That is, in late-stages COVID-19 related abnormalities transform into pneumonia, which may probably explain many COVID-19 images were misclassified as pneumonia by all the models. Further, the mean age of patients in this dataset was 63.15 years while in COVIDx dataset, majority of data is in the range 18–59 years, and higher mean age may add to complications in the patients' conditions.

Table 3. Performance evaluation of Attn-CNN model with four deep learning models shown for 3-class classification on holdout test set (COVIDx-CXR-3) and external test set (Jordan).

	Acc	COVIDx-CXR-3			External Dataset		
		Precision	Recall	F1-score	Precision	Recall	F1-score
Attn-CNN							
COVID-19	**94.75**	**0.97**	**0.97**	**0.97**	**0.98**	0.83	**0.90**
Normal		**0.90**	0.96	**0.93**	0.76	0.96	0.85
Pneumonia		**0.95**	0.88	**0.91**	0.66	0.88	**0.76**
VGG-16							
COVID-19	91.25	0.98	0.91	0.94	**0.98**	0.48	0.65
Normal		0.79	**0.98**	0.87	0.69	**0.98**	0.81
Pneumonia		0.93	0.86	0.90	0.35	0.86	0.50
ResNet-50							
COVID-19	84.0	0.94	0.84	0.89	0.91	0.28	0.43
Normal		0.67	0.94	0.78	0.60	0.94	0.73
Pneumonia		0.91	0.75	0.82	0.25	0.75	0.38
EfficientNet-B7							
COVID-19	84.0	0.95	0.80	0.87	0.97	0.81	0.88
Normal		0.71	0.94	0.81	**0.86**	0.94	**0.90**
Pneumonia		0.84	0.84	0.84	0.55	0.84	0.66
Inception-v3							
COVID-19	83.75	0.95	0.81	0.88	0.93	0.28	0.43
Normal		0.68	0.94	0.79	0.53	0.94	0.68
Pneumonia		0.87	0.79	0.83	0.28	0.79	0.42

6 Conclusion

Deep learning-based chest X-Ray analysis in the context of COVID-19 pandemic has been researched well but there is a gap in taking the research to practice and lack of understanding of the cause behind it. Many complex and very deep models with complicated architectures have been proposed for COVID-19 detection and classification using CXR images but the question of whether these models are necessary for such a task has not been discussed much. To address this here we attempt to show that a simple CNN model with attention-based architecture is able to capture the subtle and complex features of COVID-19 related abnormalities in the lungs and is able to perform better than very deep models. The typical patterns in chest radiology images associated with community acquired pneumonia are multifocal nodular opacities with patchy ground glass opacities, while that of COVID-19 are predominantly bilateral opacities (GGOs) including middle and lower zones of lungs. The manifestations of these patterns may differ from patient-to-patient and with the progression of the disease but are not highly varying as in other real-world images with respect to illumination, view-point variations, scale variations, occlusion, etc. The patterns appear mostly in very small areas within the image. This suggests that models for detecting anomalies in radiology images may not require a very deep or complex architecture. Very deep models in such cases may lead to overfitting and un-generalizable. Here we show that introducing pixel-based attention mechanisms into a lightweight CNN model helped in focusing on the local regions in the image and improving the model's performance compared to a computationally demanding model. This has an important application for deploying to small hand-held devices such as mobile phones. It also has wider applications, by applying it to detect other similar disease classification tasks with medical images.

References

1. Xie, X., et al.: Chest CT for typical 2019-nCoV pneumonia: relationship to negative RT-PCR testing. Radiology, 200343 (2020)
2. Cleverley, J., Piper, J., Jones, M.M.: The role of chest radiography in confirming COVID-19 pneumonia. BMJ **370** (2020)
3. Radiology Perspective of Coronavirus Disease 2019 (COVID-19): Lessons from severe acute respiratory syndrome and Middle East respiratory syndrome. Am. J. Roentgenol. **214**(5) (AJR)
4. Rousan, L.A., Elobeid, E., Karrar, M., Khader, Y.: Chest x-ray findings and temporal lung changes in patients with COVID-19 pneumonia. BMC Pulm. Med. **20** (2020)
5. Deng, J., et al.: ImageNet: a large-scale hierarchical image database. In: 2009 IEEE Conference on Computer Vision and Pattern Recognition, pp. 248–255 (2009)
6. Wang, L., Lin, Z.Q., Wong, A.: COVID-net: a tailored deep convolutional neural network design for detection of COVID-19 cases from chest X-ray images. Sci. Rep. **10**, 19549 (2020)
7. Khan, A.I., Shah, J.L., Bhat, M.M.: CoroNet: a deep neural network for detection and diagnosis of COVID-19 from chest x-ray images. Comput. Methods Programs Biomed. **196**, 105581 (2020)
8. Sedik, A., et al.: Deploying machine and deep learning models for efficient data-augmented detection of COVID-19 infections. Viruses **12**, 769 (2020)

9. Brunese, L., Mercaldo, F., Reginelli, A., Santone, A.: Explainable deep learning for pulmonary disease and coronavirus COVID-19 detection from X-rays. Comput. Methods Programs Biomed. **196**, 105608 (2020)
10. Khobahi, S., Agarwal, C., Soltanalian, M.: CoroNet: a deep network architecture for semi-supervised task-based identification of COVID-19 from chest X-ray images. medRxiv 2020.04.14.20065722 (2020)
11. Rajaraman, S., et al.: Iteratively pruned deep learning ensembles for COVID-19 detection in chest X-rays. IEEE Access **8**, 115041–115050 (2020)
12. Oh, Y., Park, S., Ye, J.C.: Deep learning COVID-19 features on CXR using limited training data sets. IEEE Trans. Med. Imaging **39**, 2688–2700 (2020)
13. Ozturk, T., et al.: Automated detection of COVID-19 cases using deep neural networks with X-ray images. Comput. Biol. Med. **121**, 103792 (2020)
14. Loey, M., Smarandache, F., Khalifa, N.E.M.: Within the lack of chest COVID-19 X-ray dataset: a novel detection model based on GAN and deep transfer learning. Symmetry **12**, 651 (2020)
15. Afifi, A., Hafsa, N.E., Ali, M.A.S., Alhumam, A., Alsalman, S.: An ensemble of global and local-attention based convolutional neural networks for COVID-19 diagnosis on chest X-ray images. Symmetry **13**, 113 (2021)
16. Lin, Z., et al.: AANet: adaptive attention network for COVID-19 detection from chest X-ray images. IEEE Trans. Neural Netw. Learn. Syst. **32**, 4781–4792 (2021)
17. Suba, S., Parekh, N.: Machine learning approaches in detection and diagnosis of COVID-19. In: Saxena, A., Chandra, S. (eds.) Artificial Intelligence and Machine Learning in Healthcare, pp. 113–145. Springer, Singapore (2021). https://doi.org/10.1007/978-981-16-0811-7
18. Pavlova, M., et al.: COVIDx CXR-3: a large-scale, open-source benchmark dataset of chest X-ray images for computer-aided COVID-19 diagnostics. http://arxiv.org/abs/2206.03671 (2022)
19. Fraiwan, M., Khasawneh, N., Khassawneh, B., Ibnian, A.: A dataset of COVID-19 x-ray chest images. Data Brief **47**, 109000 (2023)
20. Guarnera, A., Podda, P., Santini, E., Paolantonio, P., Laghi, A.: Differential diagnoses of COVID-19 pneumonia: the current challenge for the radiologist—a pictorial essay. Insights Imaging **12**, 34 (2021)

Decoding Heterogeneity in Quadruple-Negative Breast Cancer: A Data-driven Clustering Approach

Bikram Sahoo[1]( ), Nikita Jinna[2], Padmashree Rida[3], Zandra Pinnix[4], and Alex Zelikovsky[1]( )

[1] Department of Computer Science, Georgia State University, Atlanta, GA 30302, USA
{bsahoo1,alexz}@gsu.edu
[2] City of Hope Comprehensive Cancer Center, Duarte, CA 91010, USA
niwright@coh.org
[3] Novazoi Theranostics, Salt Lake City, UT 84105, USA
[4] Department of Biology and Marine Biology, University of North Carolina at Wilmington, Wilmington, NC 28403, USA
pinnixz@uncw.edu

Abstract. In a quest to decipher the complexities of Quadruple-Negative Breast Cancer (QNBC), this research harnesses advanced analytics applied to RNAseq gene expression data. Employing unsupervised clustering techniques, our rigorous methodology entails data preprocessing for enhanced interpretability, dimensionality reduction via autoencoders and Principal Component Analysis (PCA), and fine-tuning k-means clustering with internal validation indices. The analysis effectively discriminates two distinct QNBC subtypes, substantiated by high Silhouette (0.08) and Calinski-Harabasz (6.92) Scores. The profiles of these clusters are further unveiled through statistical analyses of top variant genes. Cluster 1 is typified by genes such as *C9ORF57* and *OR2AT4*, while Cluster 2 presents distinctive genetic features, including *KRTAP10-10* and *ADAM3A*. These data-driven clusters hold the promise of personalized assessments and interventions, contingent upon clinical validation. This study underscores the potential of integrated machine learning and statistical analysis, marking a pathway to more effective QNBC management.

Keywords: Quadruple-Negative Breast Cancer (QNBC) · QNBC subtypes · Unsupervised Clustering · Autoencoder

1 Introduction

Breast cancer, a multifaceted disease, poses significant challenges in the field of oncology. Among these challenges, triple-negative breast cancer (TNBC) stands out due to its lack of specific receptors (ER, PR) and absence of HER2 amplification. TNBC accounts for approximately 15–20% of US breast cancer cases and exhibits the lowest survival rates [16,17,19]. It is known for its high heterogeneity and diverse molecular subtypes. The first study addressing TNBC heterogeneity was conducted by Lehmann et al., who categorized TNBC into basal-like (BL1 and BL2), immunomodulatory, luminal androgen receptor (LAR), mesenchymal, and mesenchymal stem-like subtypes, each with varying responses to therapies [3,13]. Recent work by Liu et al. introduced a 4-subtype classification: immunomodulatory (IM), luminal androgen receptor (LAR), mesenchymal-like subtype (MES), and basal-like immune suppressed (BLIS) [14]. These subtypes exhibit variations in aggressiveness and clinical outcomes. While a subset of TNBC cases expresses the Androgen Receptor (AR), a substantial majority, ranging from 65% to 88%, exhibit AR-negative status, defining them as Quadruple-Negative Breast Cancer (QNBC) [8,9,18].

Extensive empirical evidence suggests that QNBC exhibits significantly increased aggressiveness when compared to AR-positive TNBC counterparts. QNBC is also epidemiologically linked to the clinically aggressive basal-like molecular phenotype, which sharply contrasts with AR-positive TNBCs that tend to display a less aggressive luminal phenotype [5,7,12,15,21]. QNBC showcases a unique molecular and genetic profile, setting it apart from other breast cancer subtypes. Although it clusters with TNBC, it has gained recognition as a distinct subtype [1,2,4,20]. This subtype demonstrates a distinct racial distribution, with a higher prevalence among the African American population [11]. Moreover, QNBC encounters challenges due to the lack of biomarkers and targeted therapies, rendering it a particularly challenging subtype to manage [2,20]. These findings underscore the complexity of QNBC, which exhibits a high degree of variability in its molecular and genetic features [1,2]. This underscores the need for further research to identify different QNBC subtypes using unsupervised clustering methods.

This study aims to address QNBC's heterogeneity using advanced analytics techniques capable of deriving meaningful insights from RNAseq gene expression data. Cluster analysis with gene expression data offers a valuable tool for identifying underlying structures in QNBC patient populations. When applied carefully, unsupervised clustering methods can unveil distinct phenotypic profiles within a heterogeneous dataset. In this paper, we present a rigorous methodology for discovering patient clusters through unsupervised machine learning on gene expression data. Our workflow encompasses several essential steps, including preprocessing data for enhanced interpretability, employing autoencoders and PCA for dimensionality reduction to distill key features, optimizing k-means clustering with internal validation indices, and performing statistical characterizations of the resulting clusters. We employ quantitative metrics and visualizations to assess cluster quality. This approach provides a robust framework for subtype identification within multidimensional gene expression data. These data-driven patient groups have the potential to inform personalized assessments and interventions tailored to each subgroup. Rigorous preprocessing and validation are essential to ensure clinically relevant clusters align with expert expectations. Our contributions include an interpretable preprocessing routine to refine features/genes, an encoding and PCA pipeline for dimensionality reduction, optimized clustering using internal metrics, statistical cluster profiling, and quantitative evaluation of cluster quality. These steps collectively constitute a robust approach for the principled discovery of QNBC patient subtypes from gene expression datasets.

2 Data

In this study, we conducted a comprehensive analysis of gene expression data sourced from breast cancer patients, utilizing publicly available RNA sequencing datasets. The RNA-seq data underwent preprocessing, including log2 median-centering, and was obtained from established repositories [6,10]. Breast cancer encompasses various subtypes, primarily characterized by hormone-receptor status determined through immunohistochemistry tests. Our investigation focused on a specific subtype known as quadruple-negative breast cancer (QNBC), a subtype of triple-negative breast cancer (TNBC). To assemble our QNBC dataset, we initially selected TNBC samples that tested negative for ER, PR, and HER2 receptors or exhibited equivocal HER2 receptor status based on FISH analysis. Furthermore, in line with the QNBC criteria, we identified samples with negative AR status by comparing their AR gene expression to the dataset's median expression. Samples with AR expression below the median were designated as AR negative. Employing these selection criteria, we curated a dataset comprising 72 QNBC samples and 20,531 genes or features.

3 Methods

3.1 Data Preprocessing

The original dataset is represented as a gene expression data matrix of QNBC samples, denoted as **Y**, comprising M samples and K features/genes:

$$Y = \begin{bmatrix} y_{11} & y_{12} & \cdots & y_{1K} \\ y_{21} & y_{22} & \cdots & y_{2K} \\ \vdots & \vdots & \ddots & \vdots \\ y_{M1} & y_{M2} & \cdots & y_{MK} \end{bmatrix} \quad (1)$$

We apply standard score normalization to each gene column, ensuring all genes have a mean of 0 and a standard deviation of 1:

$$\mathbf{s} = \frac{\mathbf{y} - \boldsymbol{\alpha}}{\boldsymbol{\beta}} \quad (2)$$

where $\boldsymbol{\alpha}$ and $\boldsymbol{\beta}$ are the mean and standard deviation vectors of **y**. Highly correlated gene pairs are detected using the Pearson correlation coefficient (p). This coefficient measures the linear association between two variables based on their covariance across the dataset:

$$p = \frac{\sum_{j=1}^{M}(y_j - \bar{y})(z_j - \bar{z})}{\sqrt{\sum_{j=1}^{M}(y_j - \bar{y})^2}\sqrt{\sum_{j=1}^{M}(z_j - \bar{z})^2}} \quad (3)$$

Gene pairs with $|p| > 0.85$ are deemed highly correlated. To reduce multicollinearity, one gene from each correlated pair is discarded, resulting in a streamlined dataset with R genes ($R < K$). This approach enhances the interpretability of the dataset while preserving the most significant variables for further analysis.

3.2 Dimensionality Reduction

To extract vital information from the raw QNBC gene expression dataset, we utilized an autoencoder neural network comprising M encoding nodes. This network transformed the input data $\mathbf{y}$ through two main stages:

1. **Encoder Layer:** The encoding layer was structured as follows:

$$\mathbf{a} = g(\mathbf{V}_1 \mathbf{y} + \mathbf{c}_1) \in \mathbb{R}^M \quad (4)$$
$$b = g(\mathbf{V}_2 \mathbf{a} + \mathbf{c}_2) \in \mathbb{R}^M \quad (5)$$

In these equations, $\mathbf{V}_1 \in \mathbb{R}^{K \times M}$ and $\mathbf{V}_2 \in \mathbb{R}^{M \times M}$ denote the weight matrices. The biases $\mathbf{c}_1$ and $\mathbf{c}_2$ facilitate the network's transformation. The function $g()$ denotes the activation function applied to introduce non-linearity into the model.

2. **Decoder Layer:** The autoencoder was trained to minimize the reconstruction error between the input data $\mathbf{y}$ and the output $\hat{\mathbf{y}}$. We utilized the mean squared error (MSE) loss function:

$$L(\mathbf{y}, \hat{\mathbf{y}}) = \|\mathbf{y} - \hat{\mathbf{y}}\|_2^2 \quad (6)$$

The code $\mathbf{b}$ represented a compressed M-dimensional representation, capturing essential genes in a compact manner.

To further improve data representation and reduce dimensionality, we applied Principal Component Analysis (PCA) to the encoded genes. The following steps were executed:

$$[\mathbf{Q}, \mathbf{R}, \mathbf{T}] = \text{Singular Value Decomposition}(\mathbf{F}) \quad (7)$$
$$\mathbf{Y} = \mathbf{F} \mathbf{T}_M \mathbf{R}_M^{-1} \mathbf{T}_M^T \in \mathbb{R}^{N \times M} \quad (8)$$

In these equations, $\mathbf{Q}$, $\mathbf{R}$, and $\mathbf{T}$ represent the outputs of the SVD, providing valuable insights into data transformation. $\mathbf{T}_M$ includes the top $M < K$ principal component loadings, facilitating dimensionality reduction while preserving essential information.

3.3 Clustering Analysis

To identify meaningful subtypes and group the samples within the QNBC dataset, we utilized the m-means clustering algorithm. This technique aimed to divide the data into m clusters, optimizing the clustering by minimizing the following objective function:

$$\min_{\mathbf{d}_1,\ldots,\mathbf{d}_m} \sum_{j=1}^{m} \sum_{\mathbf{y} \in D_j} \|\mathbf{y} - \mathbf{d}_j\|^2 \qquad (9)$$

In this equation, $\mathbf{d}_1, \ldots, \mathbf{d}_m$ denote the cluster centroids, and $\mathbf{y}$ represents the data points. The objective was to minimize the total squared distances between data points and their corresponding cluster centroids.

To determine the optimal number of clusters or QNBC subtypes, referred to as m, we assessed a range of values from 2 to 10. The selection of m was based on the Silhouette Score and the Calinski-Harabasz Index.

3.4 Cluster Characterization

To gain a deeper insight into the unique characteristics within each group D_j, we conducted a feature analysis. For every group D_j, we calculated the variance of each feature/gene to evaluate its variability within that specific group. This analysis enabled us to identify the top 100 high-variance genes, which acted as the most significant differentiators defining the properties of that particular group.

Let Var_{ik} denote the variance of feature/gene i within group k. To identify the top high-variance genes, we focused on the 100 genes with the highest values of Var_{ik} in each group.

To provide a comprehensive overview of these influential genes, we computed summary statistics for each of them. Specifically, we calculated the mean ν_{ik} and standard deviation τ_{ik} for gene i within group k. These statistics were defined as follows:

$$\nu_{ik} = \frac{1}{|D_k|} \sum_{\mathbf{y} \in D_k} y_i \qquad (10)$$

$$\tau_{ik} = \sqrt{\frac{1}{|D_k| - 1} \sum_{\mathbf{y} \in D_k} (y_i - \nu_{ik})^2} \qquad (11)$$

where: ν_{ik} represents the mean of feature/gene i in group k, τ_{ik} denotes the standard deviation of gene i in group k, y_i indicates the value of gene i for a data point $\mathbf{y}$, D_k signifies group k, and $|D_k|$ is the number of data points in group k.

These statistics provided valuable insights into the central tendencies and variations of the selected high-variance genes within each group, enhancing our understanding of their distinctive attributes.

3.5 Cluster Evaluation

To assess the effectiveness of our clustering approach, we utilized two established metrics: the silhouette score and the Calinski-Harabasz (CH) index, which provided quantitative evaluations of clustering quality and structure.

The silhouette score, denoted as Silh, quantifies the separation between clusters. It is defined as:

$$\text{Silh} = \frac{d-c}{\max(c,d)} \qquad (12)$$

Here, c represents the average intra-cluster distance, and d denotes the average nearest-cluster distance for all data points. A higher silhouette score indicates that clusters are well-separated, with data points closely grouped within their respective clusters.

The Calinski-Harabasz index, represented as CH, evaluates the compactness and separation of clusters. It is computed using the formula:

$$\text{CH} = \frac{SS_B/(m-1)}{SS_W/(p-m)} \qquad (13)$$

In this equation, SS_B indicates the between-cluster variance, SS_W signifies the within-cluster variance, m denotes the number of clusters, and p is the total number of data points. A higher CH index value suggests more distinct and well-separated clusters.

Additionally, we allocated individual data points to their corresponding clusters, facilitating a detailed examination of patients' cluster memberships. This patient-level cluster assignment enabled us to investigate QNBC subtypes and gain valuable insights into the clinical implications of our clustering outcomes.

4 Results

4.1 Identified Patient Clusters

Our analysis successfully revealed two distinct patient clusters within the QNBC dataset, achieved by optimizing the Silhouette Score (0.08) as shown in Fig. 1 and the Calinski-Harabasz Score (6.92) as displayed in Fig. 2 using k-means clustering. Cluster 1 represents one distinct QNBC subtype, while Cluster 2 signifies another. This outcome resulted from a multi-step process, beginning with the application of an autoencoder neural network and subsequent dimensionality reduction through Principal Component Analysis (PCA), as presented in Fig. 3. The insights into meaningful patient groupings were provided by k-means clustering. To enhance our understanding further, we conducted a feature variance analysis, identifying the top 100 discriminative features in each cluster. In Table 1 and 2, we present 50 features from each cluster, characterized by their mean and

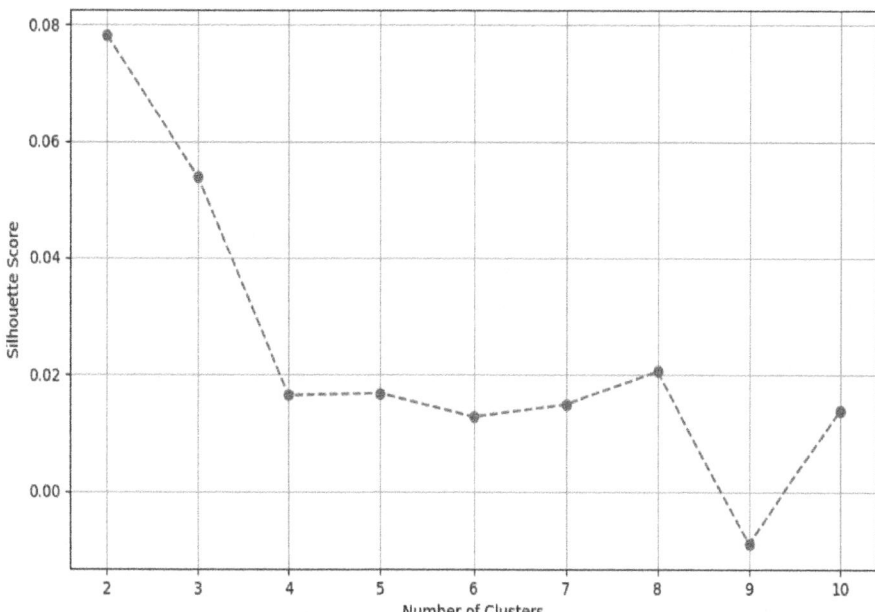

Fig. 1. This figure depicts the Silhouette Score plotted against the number of clusters. The X-axis represents the number of clusters, while the Y-axis corresponds to the Silhouette Score, offering insights into the optimal number of QNBC clusters or subtypes derived from the RNA-seq gene expression data, based on Silhouette Scores.

standard deviation, shedding light on the distinctive attributes of each subgroup. This comprehensive approach successfully elucidated clinically distinct patient subgroups within the QNBC dataset.

4.2 Evaluating Cluster Characteristics

We conducted an in-depth characterization of the identified QNBC sample clusters through a statistical analysis of the top variant genes within each subgroup. In Table 1, which contains 42 samples for Cluster 1, the top differentiated genes were found to be *C9ORF57, OR2AT4,* and *LINC00474*. These genes exhibited a mean expression of 0.08 and a high standard deviation of 1.30 within this cluster, indicating significant dysfunction in the pathways represented by these biomarkers in Cluster 1 samples.

Table 1. The table summarizes statistics for the top 50 genes identified in Cluster 1 through k-means clustering analysis of the QNBC RNA-seq gene expression dataset. It includes the gene name, cluster number, number of samples in the cluster, cluster mean, and cluster standard deviation of expression for each gene. These statistics provide information about the central tendency and dispersion of the most variant genes in Cluster 1, aiding in the characterization of the cluster's feature distribution.

Gene Name	Cluster No	No of Samples in Cluster	ClusterMean	Cluster StdDev
C9ORF57	1	42	0.08	1.30
OR2AT4	1	42	0.08	1.30
LINC00474	1	42	0.08	1.30
NLRP8	1	42	0.08	1.30
LINC00575	1	42	0.08	1.30
POTED	1	42	0.08	1.30
OR10X1	1	42	0.08	1.30
OR10V1	1	42	0.08	1.30
OR52E8	1	42	0.08	1.30
PRAMEF14	1	42	0.08	1.30
MS4A13	1	42	0.08	1.30
HNRNPCL1	1	42	0.08	1.30
TGIF2LY	1	42	0.08	1.30
OR5AP2	1	42	0.08	1.30
KRTAP9-2	1	42	0.08	1.30
SDC4P	1	42	0.08	1.30
PRAMEF9	1	42	0.08	1.30
LCN15	1	42	0.08	1.30
MYF5	1	42	0.08	1.30
OR51A4	1	42	0.08	1.30
FGF6	1	42	0.08	1.30
SNAR-G1	1	42	0.08	1.30
SNAR-B2	1	42	0.08	1.30
BHLHE23	1	42	0.08	1.30
OR9I1	1	42	0.08	1.30
SNORA66	1	42	0.08	1.30
DEFA6	1	42	0.08	1.30
CRYGB	1	42	0.08	1.30
OR5B20P	1	42	0.08	1.30
SPO11	1	42	0.08	1.30
LINC00051	1	42	0.08	1.30
LOC391628	1	42	0.08	1.30
OR5D14	1	42	0.08	1.30

(*continued*)

Table 1. (*continued*)

Gene Name	Cluster No	No of Samples in Cluster	ClusterMean	Cluster StdDev
CST9L	1	42	0.08	1.30
OCM2	1	42	0.08	1.30
LRRC30	1	42	0.08	1.30
TETM4	1	42	0.08	1.30
OR6M1	1	42	0.08	1.30
OR4D11	1	42	0.08	1.30
OR10A1	1	42	0.08	1.30
LOC286238	1	42	0.08	1.30
PRAMEF17	1	42	0.08	1.30
MRGPRG	1	42	0.08	1.30
LGALS13	1	42	0.08	1.30
TEX33	1	42	0.08	1.30
OR10G8	1	42	0.08	1.30
KRTAP10-8	1	42	0.08	1.30
KRTAP10-7	1	42	0.08	1.30
DEFB123	1	42	0.08	1.30
FAM99B	1	42	0.08	1.30

On the other hand, as shown in Table 2, Cluster 2, comprising 30 samples, displayed high variation in genes such as *KRTAP10-10, ADAM3A,* and *OR5M9.* These genes had a mean expression of 0.17 and a standard deviation of 1.53 within the cluster samples. This increased variability draws attention to abnormalities in the biological processes governed by these genes in the Cluster 2 subgroup.

The quantitative statistical characterization of the top genes in each cluster provides valuable insights into the underlying molecular differences between the QNBC sample subgroups identified through RNA-seq gene expression data. Further investigation into the implicated genes and their associated functions is warranted to fully elucidate the biological mechanisms defining each sample cluster. Nonetheless, the summary statistics offer an objective, data-driven perspective on the molecular heterogeneity between the QNBC subtypes identified through unsupervised learning.

Table 2. The table presents summary statistics for the top 50 genes differentiated in Cluster 2 identified through k-means clustering analysis of the QNBC RNA-seq gene expression dataset. For each gene, the table lists the gene name, cluster number, number of samples in the cluster, cluster mean, and cluster standard deviation of expression. This data quantifies the central tendency and dispersion of the most variant genes in Cluster 2, assisting in characterization of the cluster's feature distribution.

Gene Name	Cluster No	No of Samples in Cluster	Cluster Mean	Cluster StdDev
KRTAP10-10	2	30	0.17	1.53
ADAM3A	2	30	0.17	1.53
OR5M9	2	30	0.17	1.53
LOC286106	2	30	0.17	1.53
CELA3A	2	30	0.17	1.53
TAAR2	2	30	0.17	1.53
OR4D5	2	30	0.17	1.53
LCE6A	2	30	0.17	1.53
OR8B8	2	30	0.17	1.53
LOC644145	2	30	0.17	1.53
SLC17A2	2	30	0.17	1.53
LCE1D	2	30	0.17	1.53
IL9	2	30	0.17	1.53
OR6P1	2	30	0.17	1.53
KRTAP4-2	2	30	0.17	1.53
WFDC9	2	30	0.17	1.53
OR11-134	2	30	0.17	1.53
NCRNA00185	2	30	0.17	1.53
OR4X2	2	30	0.17	1.53
RESP18	2	30	0.17	1.53
DEFB109P1	2	30	0.17	1.53
PRAMEF23	2	30	0.17	1.53
LOC115735	2	30	0.17	1.53
RBMXL3	2	30	0.17	1.53
PRM3	2	30	0.17	1.53
OR5BF1	2	30	0.17	1.53
SSX9	2	30	0.17	1.53
MRGPRX1	2	30	0.17	1.53
TBL1Y	2	30	0.17	1.53
TMEM247	2	30	0.17	1.53
OR10A7	2	30	0.17	1.53
CT47A2	2	30	0.17	1.53
OR9K2	2	30	0.17	1.53
KRTAP12-2	2	30	0.17	1.53
TSPY4	2	30	0.17	1.53
OR6Y1	2	30	0.17	1.53
OR6C75	2	30	0.17	1.53
PISRT1	2	30	0.17	1.53

(continued)

Table 2. (*continued*)

Gene Name	Cluster No	No of Samples in Cluster	Cluster Mean	Cluster StdDev
BCYRN1	2	30	0.17	1.53
FAM183CP	2	30	0.17	1.53
OR10S1	2	30	0.17	1.53
OR5L1	2	30	0.17	1.53
RNASE11	2	30	0.17	1.53
CST8	2	30	0.17	1.53
OR5T3	2	30	0.17	1.53
KRTAP26-1	2	30	0.17	1.53
LOC100507851	2	30	0.17	1.53
BRDTP1	2	30	0.17	1.53
OR6X1	2	30	0.17	1.53
CTSL3P	2	30	0.17	1.53

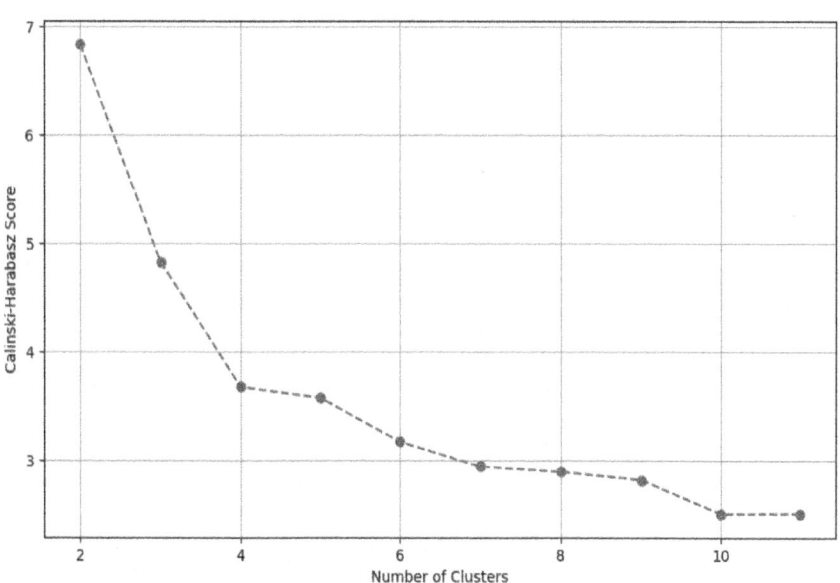

Fig. 2. This figure illustrates the relationship between the Calinski-Harabasz Score and the number of clusters. The X-axis represents the number of clusters, while the Y-axis corresponds to the Calinski-Harabasz Score. It provides insights into determining the optimal number of QNBC clusters or subtypes derived from RNA-seq gene expression data based on Calinski-Harabasz Scores.

4.3 Preliminary Cluster Evaluation

The 2D principal component projection of QNBC sample clusters, as shown in Fig. 3, clearly illustrates separation among the subgroups, signifying differentiation between QNBC subtypes. Each cluster occupies a distinct region, affirming the effectiveness of our unsupervised learning approach in identifying clinically

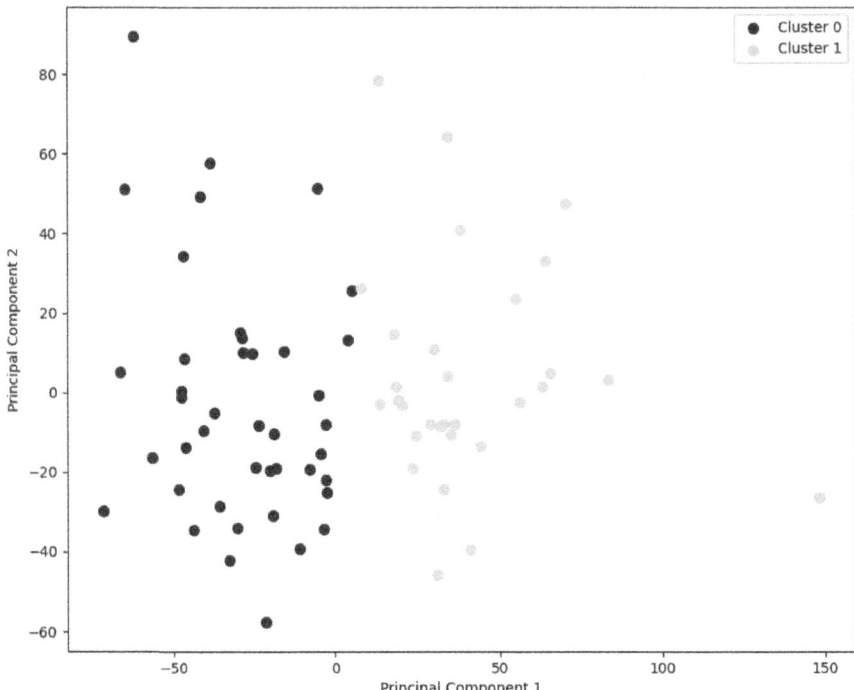

Fig. 3. This figure illustrates the clustered data in the principal component space. Each point on the plot corresponds to a data point and is color-coded to indicate its cluster membership. The legend distinguishes the number of QNBC clusters or subtypes with various colors. This representation offers insights into how data points are grouped in the PCA-transformed feature space.

distinct sample clusters. However, to ensure accurate sample assignments to these data-driven clusters, further validation methods, such as aligning individual cases with clinical knowledge, are indispensable. In summary, while 2D visualization serves as a crucial initial step in evaluating sample clusters, complete confirmation of cluster assignments necessitates additional validation methods to ensure clinical relevance and alignment with expectations.

5 Conclusion

This study underscores the potential of an integrated machine learning workflow to uncover valuable insights within heterogeneous patient data. By focusing on Quadruple-Negative Breast Cancer (QNBC), our analysis effectively identified two distinct QNBC subtypes through differentiated clusters.

A statistical characterization of the top variant genes within each cluster provided data-driven insights into their molecular profiles. Cluster 1 exhibited dysregulation in pathways marked by genes like *C9ORF57* and *OR2AT4*, while Cluster 2 displayed abnormalities in biomarkers, including *KRTAP10-10* and *ADAM3A*.

Preliminary visualization indicated separation between the clusters, affirming the utility of our unsupervised learning approach. Nevertheless, the confirmation of individual sample assignments through clinical validation remains vital to ensure accuracy. These data-driven subgroups align with the known heterogeneity of QNBC, offering promising opportunities for tailoring prognostic assessments and therapeutic interventions to the specific biological characteristics of each QNBC subtype. The implicated genes within these subtypes serve as critical targets for further research into the underlying mechanisms of QNBC heterogeneity.

In summary, this study underscores the potential of integrating machine learning with rigorous statistical analysis to reveal data-driven patient subgroups within RNA-seq gene expression datasets. After thorough validation, these clusters and biomarker insights hold promise for improving personalized management of this clinically challenging breast cancer subtype. Our approach emphasizes the importance of addressing QNBC's heterogeneity through advanced analytics techniques capable of deriving meaningful insights from RNAseq gene expression data. This methodology establishes a robust framework for the systematic discovery of QNBC patient subtypes, paving the way for enhanced patient care tailored to the diverse subgroups within the QNBC population.

References

1. Angajala, A., et al.: Quadruple negative breast cancers (QNBC) demonstrate subtype consistency among primary and recurrent or metastatic breast cancer. Transl. Oncol. **12**, 493–501 (2019). https://doi.org/10.1016/j.tranon.2018.11.008
2. Bhattarai, S., Saini, G., Gogineni, K., Aneja, R.: Quadruple-negative breast cancer: novel implications for a new disease. Breast Cancer Res. **22** (2020). https://doi.org/10.1186/s13058-020-01369-5
3. Burstein, M.D., et al.: Comprehensive genomic analysis identifies novel subtypes and targets of triple-negative breast cancer. Clin. Cancer Res. **21**, 1688–1698 (2014). https://doi.org/10.1158/1078-0432.ccr-14-0432
4. Davis, M., et al.: AR negative triple negative or "quadruple negative" breast cancers in African American women have an enriched basal and immune signature. PLoS ONE **13**, e0196909 (2018)

5. Doane, A.S., et al.: An estrogen receptor-negative breast cancer subset characterized by a hormonally regulated transcriptional program and response to androgen. Oncogene **25**, 3994–4008 (2006). https://doi.org/10.1038/sj.onc.1209415
6. geo: Home - geo - NCBI (2019). https://www.ncbi.nlm.nih.gov/geo/
7. Gucalp, A., Traina, T.A.: Triple-negative breast cancer. Cancer J. **16**, 62–65 (2010). https://doi.org/10.1097/ppo.0b013e3181ce4ae1
8. Hon, J.: Breast cancer molecular subtypes: from TNBC to QNBC (2016)
9. Huang, M., Wu, J., Ling, R., Li, N.: Quadruple negative breast cancer. Breast Cancer **27**(4), 527–533 (2020). https://doi.org/10.1007/s12282-020-01047-6
10. Institute, N.C.: The cancer genome atlas program (TCGA) - NCI (2022). https://www.cancer.gov/ccg/research/genome-sequencing/tcga
11. Jinna, N., et al.: Racial disparity in quadruple negative breast cancer: aggressive biology and potential therapeutic targeting and prevention. Cancers **14**, 4484 (2022). https://doi.org/10.3390/cancers14184484, https://www.mdpi.com/2072-6694/14/18/4484
12. Jovanović, B., et al.: A randomized phase ii neoadjuvant study of cisplatin, paclitaxel with or without everolimus in patients with stage II/III triple-negative breast cancer (TNBC): Responses and long-term outcome correlated with increased frequency of DNA damage response gene mutations, tnbc subtype, AR status, and ki67. Clin. Cancer Res. **23**, 4035–4045 (2017) An Official Journal of the American Association for Cancer Research. https://doi.org/10.1158/1078-0432.CCR-16-3055, https://pubmed.ncbi.nlm.nih.gov/28270498/
13. Lehmann, B.D., et al.: Identification of human triple-negative breast cancer subtypes and preclinical models for selection of targeted therapies. J. Clin. Invest. **121**, 2750–2767 (2011). https://doi.org/10.1172/jci45014
14. Liu, Y.R., et al.: Comprehensive transcriptome analysis identifies novel molecular subtypes and subtype-specific RNAS of triple-negative breast cancer. Breast Cancer Res. **18** (2016). https://doi.org/10.1186/s13058-016-0690-8
15. Luo, X., Shi, Y.X., Li, Z.M., Jiang, W.Q.: Expression and clinical significance of androgen receptor in triple negative breast cancer. Chin. J. Cancer **29**, 585–590 (2010). https://doi.org/10.5732/cjc.009.10673
16. Sahoo, B., Adeyeha, T., Pinnix, Z., Zelikovsky, A.: Exploring racial disparities in triple-negative breast cancer: insights from feature selection algorithms. ISBRA 2023, LNCS, pp. 487–497 (2023). https://doi.org/10.1007/978-981-99-7074-2_39
17. Sahoo, B., Pinnix, Z., Sims, S., Zelikovsky, A.: Identifying biomarkers using support vector machine to understand the racial disparity in triple-negative breast cancer. J. Comput. Biol. (2023). https://doi.org/10.1089/cmb.2022.0422
18. Sahoo, B., Pinnix, Z., Zelikovsky, A.: Deep learning reveals biological basis of racial disparities in quadruple-negative breast cancer. ISBRA 2023, LNCS, pp. 498–508 (2023). https://doi.org/10.1007/978-981-99-7074-2_40
19. Sahoo, B., Sims, S., Zelikovsky, A.: An SVM based approach to study the racial disparity in triple-negative breast cancer. ICCABS 2021. LNB, vol. 13254 (2022). https://doi.org/10.1007/978-3-031-17531-2.13

20. Saini, G., Bhattarai, S., Gogineni, K., Aneja, R.: Quadruple-negative breast cancer: an uneven playing field. JCO Global Oncol., 233–237 (2020). https://doi.org/10.1200/jgo.19.00366
21. Yu, Q., et al.: Expression of androgen receptor in breast cancer and its significance as a prognostic factor. Ann. Oncol. **22**, 1288–1294 (2011). https://doi.org/10.1093/annonc/mdq586, https://www.annalsofoncology.org/article/S0923-7534(19)38469-8/fulltext

Determining Temporal Linkages in Dynamic Epidemiological Networks Using the Earth Mover's Distance

Rahul Singh[✉] and Jiadong Yu

Department of Computer Science, University of Iowa, MacLean Hall, Iowa City, IA 52242, USA
rahul-singh@uiowa.edu

Abstract. The Dynamic Epidemiological Networks (DEN) is a multiscale and extensible data representation framework for describing the spread of a rapidly evolving infectious diseases like COVID-19. A DEN can represent information at various granularities starting from the molecular to epidemiological. It can also represent non-conservative real-world environments characterized by impersistent as well as novel data. One prototypical application of the DEN framework has been to represent molecular data collected from the Covid-19 pandemic. The rapid evolution of the SARS-CoV-2 virus leads to a continually changing molecular landscape of the etiological agent. In such settings, assignment of samples to variants and even variant definitions themselves can change as existing variants evolve, novel variants appear and older variants die-out. In this paper we present preliminary results extending the DEN framework using the transportation formulation to determine correspondences between viral lineages as they evolve. An advantage of our approach lies in the fact that correspondences between related sample clusters across time can be determined even when these clusters do not share common elements. The applicability of our approach is demonstrated by constructing and analyzing temporal molecular networks of SARS-CoV-2 genomes sequenced as part of COVID-19 tracking efforts.

Keywords: SARS-CoV-2 · Network theory · Clusters · Optimal Transport · Phylogenetics · Linkages

1 Introduction

Severe acute respiratory syndrome coronavirus 2 (SARS-CoV-2), the etiological agent of Coronavirus disease 2019 (Covid-19), has had a significant impact on both challenging and advancing the state-of-the-art in molecular epidemiology. Since its first recorded outbreak in December 2019, the genome of SARS-CoV-2 has been sequenced and made publicly available [1]. According to the Centers for Disease Control and Prevention, as of March 30, 2024, over 5 million sequences collected in the US have been made available in GISAID and over 3.5 million sequences made available through NCBI. These numbers represent data collected since January 2020 (covid.cdc.gov). A measure of the worldwide effort in tracking Covid-19 can be gleaned from the fact that almost

13 million genomes had been sequenced globally by 2021 [2]. A number of tools that were developed relatively recently, including Pango [3] and Nextstrain [4] have utilized this data to reconstruct the evolution of the virus through phylogenetic analysis and have provided a nomenclature for emerging, genetically distinct iterations of the virus, referred to by terms such as "lineages", "clades" or "variants". The World Health Organization's (WHO) working definitions for variants of interest (VOIs) and concern (VOCs) [5] are based on similar analysis frameworks that employ phylogenetic analysis. However, phylogenetics-based approaches, including the aforementioned, may be insufficient for representing the evolutionary landscape of a rapidly evolving disease where information about variants may be incomplete (*e.g.* some variants may have evaded capture in the sampling process), non-stationary (*i.e.* the information about the variants may have changed over time), and non-conservative (*i.e.* specific samples may not have persisted – either due to the variant dying out or due to data subsampling), and new samples get incorporated over time. These challenges are reflected in the fact that factors used to define variants such as evidence of persistence in a host population, transmission, and phenotypic changes have all been subject to change as the pandemic has progressed [5, 6].

In molecular epidemiology based on phylogenetic analysis, genomes sampled from infected individuals are related to one another based on sequence similarity and inferences are made about the spread and prevalence of a virus using the molecular relatedness of sampled genomes [7–9]. Relatedly, phylodynamic analysis, which studies the interaction and influence of epidemiological, immunological, and evolutionary processes on viral evolution and genetic variation, infers viral population levels over time based on phylogenetic reconstruction [10]. Regarding SARS-CoV-2, phylogenetic and phylodynamic studies have been used to estimate the source and date of origin of infection, the temporal reproductive number, geographical spread, and the role of super spreaders [8, 11, 12]. However, methods based on phylogenetic reconstruction suffer from limitations when applied to real-world data pertaining to Covid-19 due to the incompleteness, non-stationary, and non-conservative nature of the data. Additionally, the constraint of a tree structure limits the application of network science to the data under investigation [13] which may not conform to the constraints of a single source and predefined branching tree structure [13, 14]. By contrast, network (graph)-based models provide a different perspective. A number of such network representations of biological data are known and include representation of gene and protein functions, human neural networks, and contact tracing for epidemiology [13, 15–18]. The results presented in this paper utilize a network representation called Dynamic Epidemiological Networks (abbreviated as DEN hereafter) which is described in the following section.

2 Dynamic Epidemiological Networks: A Summary

A DEN is a multiscale network that is defined in a data-driven manner from the molecular constitution of the pathogen samples collected over time [14]. Thus, for a given longitudinal data set, the DEN is explicitly parameterized by time. At a particular time-point in a DEN, vertices represent sampled genomes and edges are defined between samples that are deemed to be genetically close. Furthermore, vertices are grouped into clusters (communities), with each cluster corresponding to one or more related variant(s). As one proceeds from a particular time point to the next, the DEN represents correspondences between clusters occurring at the successive time points as determined by their genomic contents. If one or more samples at a particular time point are found to significantly differ from existing samples in the DEN, then these new samples are represented by a cluster of vertices that have no prior correspondences. Thus, a DEN can be used for: (1) identification of epidemiologically relevant variants through analysis of the clusters in the network, (2) tracking the evolution of variants, including the emergence of new variants and cessation of old ones by considering the changes in the number and constitution of clusters across time, and (3) obtaining insights about the evolving viral landscape by analyzing the cluster correspondences. Furthermore, by describing within- and between-cluster relationships, DENs support a multiscale data representation. The key steps in the construction of a DEN include:

1. *Determination of sample interrelationships*: This is done by aligning the N sample sequences and defining a distance between all sequence pairs. In [14], the Hamming distance between sequences was used and an $N \times N$ all-pair distance matrix D was computed. Other genomic distances are equally applicable at this step.
2. *Embedding of the data in a low-dimensional metric space*: Classical multidimensional scaling (CMDS) was used to embed D in a low dimensional space while minimally perturbing the inter-sample distance distribution [19]. The dimension of the low-dimensional representation space was chosen to be 10 based on analyzing the eigenvalue distribution underlying the CMDS using a Scree plot.
3. *Data subsampling at individual time points*: To mitigate the impact of sample numerosity at a given timepoint t, the data was sub-sampled. For this purpose, the mean-shift algorithm was used to identify modes of the data distribution. Mean shift is a mode finding algorithm used for empirically identifying the maxima of the data density function [20]. The only parameter in this method is the bandwidth h, of a weight determining kernel function, making the algorithm particularly useful in problems where the number of clusters is not known *a priori*. Given $X = \{x_1, \ldots, x_n\}$, where $x_i \in R^{10}$ denote the positional representation of genome samples after CMDS, the sampled sequences $S = \{s_1, \ldots, s_n\}$, selected in DEN constituted sequences whose representation in X were the shortest Euclidean distance to their respective cluster modes. If multiple sequences were found to be closest to the mode, then the sample collected earliest in time was chosen. This step ensured that the selected samples were either the modes or the closest data point(s) to the mode(s). For a given value of the bandwidth h, the mean shift identified modes in both dense and low-density regions (particularly, for samples whose distance to all other samples was greater than h). Since low-density regions of the molecular landscape typically contained samples with random low frequency mutations, clusters that represented less than a fraction, f, of the dataset at a specific time t were removed from the sampled set. CMDS and mean shift were performed at each time point $t = \{t_1, \ldots, t_n\}$, with the sampled sequences at each t represented by $S(t)$. The appropriate values for h and f ($h = 2 \times 10^{-5}$, and $f = 0.0001$) were arrived at empirically by testing over a range of numbers.

4. *Network formation (at time t)*: Consider a timepoint t with n samples $\{s_1, s_2, \ldots s_n\}$. Let φ_k denote the k_{th} nearest-neighbor graph (NNG) for scale parameter k. A formal definition of φ_k is provided in Eq. (1), where for each sample s_i, $d_m(s_i)$ denotes the m_{th} closest sample in terms of their genetic distance. That is, $d_1(s_i)$ denotes the closest phenotype to s_i, $d_2(s_i)$ denotes the second closest phenotype to s_i and so on. For a set of vertices, φ_1 would be obtained by connecting each vertex to its nearest neighbor, φ_2 would be obtained by connecting each vertex to its second nearest neighbor, and φ_k would connect the kth nearest neighbors.

$$\varphi_k = (V, E) : V = \{s, \ldots, s_n\} \wedge (s_i, s_j) \in E,$$
$$if\ s = d_k(s_i) \qquad (1)$$

The complete connectivity structure at a time point t was obtained by considering the union of k-nearest neighbor graphs for successively increasing values of k (Eq. (4)) till all the vertices in the graph were Voronoi connected following [21].

$$\varphi = \varphi_1 \cup \varphi_2 \cup \cdots \cup \varphi_n \qquad (2)$$

5. *Grouping related sequences by graph partitioning*: Laplacian spectral partitioning was iteratively used to cluster the network φ. The normalized cut value was employed to determine if a specific partition required further sub partitioning.
6. *Establishing correspondences between clusters over time:* Two groups (clusters) of samples at consecutive time points were connected by an edge if they shared samples. This simple criterion represented a minimalist information theoretic perspective on relatedness of two sets of samples across time. It should be noted that edges relating clusters had a different semantic representation in a DEN as compared to edges in φ which connected similar samples. Thus, the DEN could be perceived as a "network of networks" since samples at each time point t were modeled as a nearest neighbor network and partitioned into interconnected clusters. In turn, directed edges between vertices (clusters) across consecutive time-points characterized the dynamics of the disease in terms of the appearance, disappearance, and merging/splitting of viral genotypes over time.

3 Modeling Viral Evolution as a Transportation Problem

3.1 Cluster Refinement

While the clusters defined in the DEN mostly contained related lineages, in certain situations as illustrated in Table 1, it was possible to further subcluster the DEN clusters to obtain groupings with improved lineage homogeneity. This empirical observation motivated the design of a cluster refinement step in our method. The cluster refinement was driven by the hypothesis that variations (if any) within DEN clusters were due to local low dimensional structure in the data. To identify such structures, each DEN cluster was independently embedded in a low dimensional space obtained using multi-dimensional scaling based on stress minimization using majorization (SMACOF). The target dimension for each cluster was identified independently by determining the smallest dimensionality starting from which the difference in the consecutive stress values

Table 1. Example of lineage heterogeneity in clusters obtained using Laplacian spectral partitioning in the DEN. Note that Cluster 1 of week 92 contains 35 different Pango lineages with a total of 333 sequences, with a majority of B.1, B.1.1.50, and B.1.362. However, it also contains other lineages, such as A.2.2, B.1.2, B.6, B.1.1.529.2.12, *etc.* Further sub-clustering of clusters with diverse lineage composition leads to better grouping of variants as shown in the last column.

Cluster Week	Cluster	Lineage	Number of Sequences	Sub cluster refinement
Feb 21, 2022	1	A.2	2	1.1
		A.2.2	1	1.1
		B	2	1.1
		B.1	35, 1	1.1, 1.4
		B.1.1	9	1.1
		B.1.1.1	1	1.1
		B.1.1.1.36.3	1, 1	1.3, 1.4
		B.1.1.141	1	1.1
		B.1.1.189	1	1.4
		B.1.1.216	1	1.1
		B.1.1.28.1	1	1.5
		B.1.1.294	6, 1	1.1, 1.5
		B.1.1.294.1	4	1.1
		B.1.1.315	1	1.1
		B.1.1.397	3	1.1
		B.1.1.50	161, 1	1.1, 1.3
		B.1.1.529.2.12	1	1.2
		B.1.1.70	5	1.1
		B.1.160	3	1.1
		B.1.177	6	1.1
		B.1.177.35	8	1.1
		B.1.177.73	2	1.1
		B.1.2	4, 1	1.1, 1.5
		B.1.221.2	1	1.1
		B.1.258	6, 1	1.1, 1.5
		B.1.351	11, 1	1.1, 1.4
		B.1.362	35, 1	1.1, 1.4
		B.1.504	5	1.1
		B.1.505	1	1.1

(continued)

Table 1. (*continued*)

Cluster Week	Cluster	Lineage	Number of Sequences	Sub cluster refinement
		B.1.526	1	1.5
		B.1.575	1	1.1
		B.36	1	1.1
		B.39	1	1.1
		B.40	1	1.1
		B.6	1	1.1

became small compared to prior differences. In Fig. 1, we show how the stress values for two clusters from our data (including the cluster in Table 1) change as the target dimension increases. Typically, the target dimensionality ranged between three and five for clusters in our data. In the interest of space, this paper includes results obtained by selecting the target dimension of 5 for all DEN clusters. Subsequently, the mean-shift algorithms was used to identify subclusters.

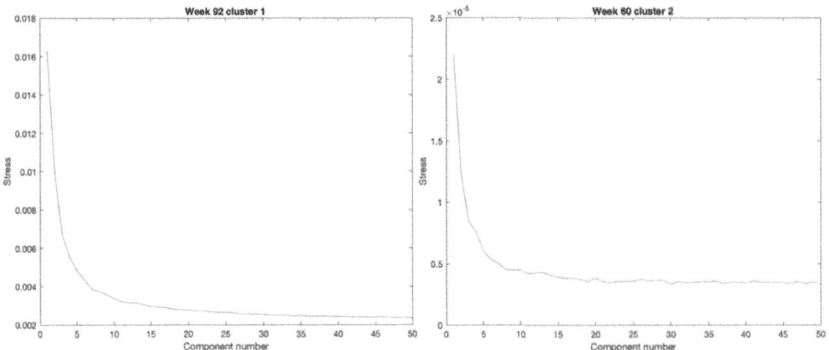

Fig. 1. Plots of stress and component number using SMACOF MDS. Left: Cluster 1 of week 92 with a total of 333 sequences. Right: Cluster 2 of week 60 with a total of 35 sequences.

3.2 The Transportation Formulation for Edge Determination

The problem of establishing correspondences between DEN clusters at two consecutive time points can be interpreted as that of matching two sets of quantized data descriptions (or histograms). Several measures have been proposed to address this problem and they can be broadly classified into two categories. Most fall into the first category of bin-by-bin dissimilarity measures. This includes the χ^2 statistics, histogram intersection, L_p distances, Kullback-Leibler divergence, Jeffry divergence, and Jensen-Shannon divergence. A fundamental assumption underlying these techniques is that the domain of the histograms can be aligned. However, this assumption may often not hold due to noise, sub-optimal quantization, different number of clusters (bins), or as in our case, the inherent nature of the data encompassing evolving variants and different numbers of samples at each time point. The second category of measures is called cross-bin measures. Cross-bin measures utilize the ground distance between representative features in different bins to compare both aligned and non-aligned bins. The Earth Mover's distance (EMD) [22] is an example of such a measure and is used by us.

The EMD is also known as the Wasserstein distance or the optimal transport distance and measures the similarity between two probability distributions in a given metric space. Consider the metric space R^n with the marginal distributions $s_1(x)$ and $s_2(y)$ (s_1 and s_2 being the marginal distributions of some joint distribution $G(s_1, s_2)$ defined on $R^n \times R^n$). Then the Wasserstein distance between s_1 and s_2 is defined as:

$$W_p(s_1(x), s_2(y)) = \left(\inf_{\gamma(x,y) \in G(s_1, s_2)} \iint d(x,y)^p \gamma(x,y) dx dy \right)^{1/p} \quad (3)$$

Linear programming can be used to compute the Wasserstein distance by finding the minimum distance in the joint distribution G. In the following, we briefly present a description of the process used to obtain correspondences between a set of clusters in the DEN which employs a notation common to EMD formulations.

Given two sets of clusters in a DEN at consecutive time points (for brevity we henceforth refer to these sets as P and Q), one of them can be interpreted as a mass distribution spread on the underlying space and the other as a collection of holes in that same space. If a unit of work corresponds to transporting a unit of mass by a unit of ground distance, then the correspondence problem can be defined as determining the least amount of work required to fill the holes. This precisely corresponds to the EMD between the two distributions underlying the two sets. We formalize the method as follows: Let the first set of clusters be represented by a set of tuples $P = \{<p_1, w_{p1}>, <p_2, w_{p2}>, \ldots, <p_m, w_{pm}>\}$, where the ith cluster is represented by the tuple $<p_i, w_{pi}>$ with p_i denoting an appropriately chosen cluster representative (such as the consensus sequence) and w_i the weight of the ith bin, given by the fraction of sequences at that point in time which belong to this cluster. Similarly, let Q = $\{<q_1, w_{q1}>, <q_2, w_{q2}>, \ldots, <q_n, w_{qn}>\}$ be the tuple set representing the second set and d_{ij} denote the ground distance between the clusters p_i and q_j (we used the Hamming distance as the ground distance for all results in this paper). Establishing the correspondence between the clusters by computing the EMD requires solving the following minimization problem, where f_{ij} denotes the flow between clusters p_i and q_j:

$$\arg\min_{f_{ij}} \left(\sum_{i=1}^{m} \sum_{j=1}^{n} d_{ij} f_{ij} \right) \quad (4)$$

The minimization is subject to the constraints (5)–(8) below, where the constraint (5) ensures that the mass is moved in only one direction, constraint (6) and (7) ensure that the mass sent by clusters in P and the mass received by clusters in Q is limited to their weights, and constraint (8) requires that the maximum possible amount of mass is moved.

$$f_{ij} \geq 0, i = 1, 2, \ldots, m; j = 1, 2, \ldots, n \quad (5)$$

$$\sum_{j=1}^{n} f_{ij} \leq w_{p_i}, i = 1, 2, ..., m \tag{6}$$

$$\sum_{i=1}^{m} f_{ij} \leq w_{q_j}, j = 1, 2, ..., n \tag{7}$$

$$\sum_{i=1}^{m}\sum_{j=1}^{n} f_{ij} = min\left(\sum_{i=1}^{m} w_{p_i}, \sum_{j=1}^{n} w_{q_j}\right) \tag{8}$$

Given the optimal flows f_{ij} obtained from solving the transportation problem as described above, the EMD between the two residue contexts is defined as in Eq. (8), where the numerator denotes the resulting work and the denominator describes the total flow.

$$EMD(P, Q) = \frac{\sum_{i=1}^{m}\sum_{j=1}^{n} d_{ij} f_{ij}}{\sum_{i=1}^{m}\sum_{j=1}^{n} f_{ij}} \tag{9}$$

4 Experiments and Results

4.1 Data

Analysis of SARS-CoV-2 molecular evolution within a population was performed on data collected in Israel, henceforth referred to by us as the Israel Data Set (IDS) [11]. In this dataset, the collection date was known for each sequence. The IDS was downloaded from the GISAID database (https://www.gisaid.org) [2]. The sequences were filtered for completeness, high coverage, and known dates of collection till Apr 5, 2022. This resulted a total of 16,929 genomes. A reference sequence (originating from the first recorded outbreak in Wuhan, China) was downloaded from GenBank (https://www.ncbi.nlm.nih.gov/genbank/, accession number MN908947). The genomes for each dataset were aligned to the reference sequence with MAFFT [23]. Non-coding regions were removed from all genomes according to the reference sequence annotation. Additionally, samples were removed from further downstream analysis if the coding region contained more than 1% ambiguous nucleotides or gaps. Insertions and deletions (indels) were ignored due to lack of clarity between indels and ambiguous nucleotides. This results a final size consisted of 16526 samples for IDS. After removal of non-coding regions, sequences had a nucleotide length of 29,703.

The majority (52.75%, 8717/16,526) of IDS consisted of B.1.617.2 and descendant lineages (Delta), with descendant lineages comprising 95.98% (8367/8717) of the B.1.617.2 group. Lineage B.1.1.7 (Alpha) was next after B.1.617.2 by prevalence (31.54%, 5213/16,526). B.1.1.7 descendant lineages were <1% (49/5213) of the B.1.1.7 group. The WHO classified variants, Beta (lineage B.1.351 and descendants) and Omicron (B.1.1.529 and descendants), were respectively found to be approximately 0.8% (131/16,526) and 0.8% (132/16,526) of the data. Other lineages of notable frequency (and their respective descendants) were B.1.1.50 (5.97%, 987/16,526), B.1.362 (2.27%, 375/16,526), and B.1.1.294 (0.5%, 90/16,526).

Of the remaining lineages, B.1.1 represented 0.5% (86/16,256), B.1 represented 1.8% (293/16,526), and B represented less than 0.01% (6/16,526) of the data. Finally,

0.08% (13/16,526) of samples were within the A lineage group; the remainder (77 B sub-lineages) are not outlined here (3.5%, 574/16,526). As many lineages were low frequency, sub-lineages were grouped with their ancestral lineages, unless otherwise noted. In the initial year of the pandemic, B.1 happened to be the most frequent lineage, followed by B.1.1, B, and A. By December 2020, lineage B.1.1.7, B.1.1.50, and B.1.362 began to emerge. These lineages grew until approximately September 2021 after which point B.1.617.2 became dominant.

4.2 EMD Can Be Used to Determine Cluster Correspondences Across Time Even in the Absence of Common Sequences

DEN builds correspondence based on common sequences shared between consecutive time frames. However, if clusters do not share common sequences the approach in DEN cannot relate them even if the respective consensus sequences are highly similar. Since EMD creates correspondences based on the ground distance between clusters, it can identify similarities even when the clusters do not intersect. In Fig. 2 we show a section of the DEN representing data from weeks 62 and 63 of the IDS. For this data, we manually removed all the common sequences between two consecutive time frames to simulate data that does not persist. In DEN this would lead to clusters that are not temporally connected *i.e.* no intercluster edges would exist, even if the genomic constitutions of the clusters were similar. With EMD however, a number of edges between clusters can be retained based on similarity of the cluster consensus sequences. Figure 2(a) shows the intercluster connections in DEN. Figure 2(b) depicts the edges that were retained because of the EMD.

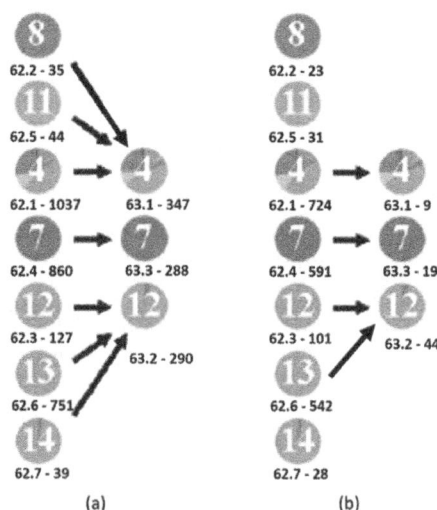

Fig. 2. Data from weeks 62 and 63 of the IDS. 62.x indicates cluster x of week 62 and the number following the hyphen indicates the total number of sequences in that cluster. (a) shows the intercluster edges in the DEN and (b) shows the edges which are retained due to EMD once common samples between clusters were removed.

5 Conclusions

In this paper, we have presented preliminary work demonstrating how variant evolution can be represented across using the transportation formulation as defined by the EMD. While the EMD has been used to analyze contact networks underlying HCV outbreaks [24], to the best of our knowledge, this is the first application of the transportation framework to analyze pandemic-scale data from SARS-CoV-2. In particular, our aim was to extend the DEN framework in two ways: (1) through the refinement of existing clusters by local embedding and sub-clustering and (2) determining cluster linkages across time using the EMD. The latter extension allows extending the framework of DEN by allowing definition of linkages even when the samples in the clusters are distinct.

Acknowledgement. The authors would like to acknowledge Fiona Senchyna who worked with RS on developing the original DEN formulation. The data used in our paper was collected as part of that research.

Funding. This research was funded in part by the National Science Foundation (NSF) grant IIS-1817239.

References

1. Wu, F., Zhao, S., Yu, B., et al.: A new coronavirus associated with human respiratory disease in China. Nature **579**, 265–269 (2020). https://doi.org/10.1038/s41586-020-2008-3
2. Khare, S., Gurry, C., Freitas, L., et al.: GISAID's role in pandemic response. China CDC Weekly **3**, 1049–1051 (2021). https://doi.org/10.46234/ccdcw2021.255
3. Rambaut, A., Holmes, E.C., O'Toole, Á., et al.: A dynamic nomenclature proposal for SARS-CoV-2 lineages to assist genomic epidemiology. Nat. Microbiol. **5**, 1403–1407 (2020). https://doi.org/10.1038/s41564-020-0770-5
4. Hadfield, J., Megill, C., Bell, S.M., et al.: Nextstrain: real-time tracking of pathogen evolution. Bioinformatics **34**, 4121–4123 (2018). https://doi.org/10.1093/bioinformatics/bty407
5. Konings, F., Perkins, M.D., Kuhn, J.H., et al.: SARS-CoV-2 variants of interest and concern naming scheme conducive for global discourse. Nat. Microbiol. **6**, 821–823 (2021). https://doi.org/10.1038/s41564-021-00932-w
6. SARS-CoV-2 clade naming strategy for 2022. https://nextstrain.org//blog/2022-04-29-SARS-CoV-2-clade-naming-2022. Accessed 15 Apr 2024
7. Wymant, C., Hall, M., Ratmann, O., et al.: PHYLOSCANNER: inferring transmission from within- and between-host pathogen genetic diversity. Mol. Biol. Evol. **35**, 719–733 (2018). https://doi.org/10.1093/molbev/msx304
8. Sledzieski, S., Zhang, C., Mandoiu, I., Bansal, M.S.: TreeFix-TP: phylogenetic error-correction for infectious disease transmission network inference. In: Pacific Symposium on Biocomputing, vol. 26, pp. 119–130 (2021)
9. Didelot, X., Kendall, M., Xu, Y., et al.: Genomic epidemiology analysis of infectious disease outbreaks using TransPhylo. Curr. Protoc. **1**, e60 (2021). https://doi.org/10.1002/cpz1.60
10. Volz, E.M., Koelle, K., Bedford, T.: Viral phylodynamics. PLoS Comput. Biol. **9**, e1002947 (2013). https://doi.org/10.1371/journal.pcbi.1002947
11. Miller, D., Martin, M.A., Harel, N., et al.: Full genome viral sequences inform patterns of SARS-CoV-2 spread into and within Israel. Nat. Commun. **11**, 5518 (2020). https://doi.org/10.1038/s41467-020-19248-0

12. Danesh, G., Elie, B., Michalakis, Y., et al.: Early phylodynamics analysis of the COVID-19 epidemic in France. Epidemiology (2020)
13. Zarrabi, N., Prosperi, M., Belleman, R.G., et al.: Combining epidemiological and genetic networks signifies the importance of early treatment in HIV-1 transmission. PLoS ONE 7, e46156 (2012). https://doi.org/10.1371/journal.pone.0046156
14. Senchyna, F., Singh, R.: Dynamic epidemiological networks: a data representation framework for modeling and tracking of SARS-CoV-2 variants. J. Comput. Biol. **30**, 446–468 (2023). https://doi.org/10.1089/cmb.2022.0469
15. Kuchaiev, O., Stevanović, A., Hayes, W., Pržulj, N.: GraphCrunch 2: software tool for network modeling, alignment and clustering. BMC Bioinform. **12**, 24 (2011). https://doi.org/10.1186/1471-2105-12-24
16. Hayes, W., Sun, K., Pržulj, N.: Graphlet-based measures are suitable for biological network comparison. Bioinformatics **29**, 483–491 (2013). https://doi.org/10.1093/bioinformatics/bts729
17. Skums, P., Zelikovsky, A., Singh, R., et al.: QUENTIN: reconstruction of disease transmissions from viral quasispecies genomic data. Bioinformatics **34**, 163–170 (2018). https://doi.org/10.1093/bioinformatics/btx402
18. Vecchio, F., Miraglia, F., Maria Rossini, P.: Connectome: graph theory application in functional brain network architecture. Clin. Neurophysiol. Pract. **2**, 206–213 (2017). https://doi.org/10.1016/j.cnp.2017.09.003
19. Torgerson, W.S.: Multidimensional scaling: I. Theory and method. Psychometrika **17**, 401–419 (1952). https://doi.org/10.1007/BF02288916
20. Fukunaga, K., Hostetler, L.: The estimation of the gradient of a density function, with applications in pattern recognition. IEEE Trans. Inform. Theory **21**, 32–40 (1975). https://doi.org/10.1109/TIT.1975.1055330
21. Singh, R., Beasley, R., Long, T., Caffrey, C.R.: Algorithmic mapping and characterization of the drug-induced phenotypic-response space of parasites causing schistosomiasis. IEEE/ACM Trans. Comput. Biol. Bioinform. **15**, 469–481 (2018). https://doi.org/10.1109/TCBB.2016.2550444
22. Rubner, Y., Tomasi, C., Guibas, L.J.: The earth mover's distance as a metric for image retrieval. Int. J. Comput. Vision **40**, 99–121 (2000). https://doi.org/10.1023/A:1026543900054
23. Katoh, K., Rozewicki, J., Yamada, K.D.: MAFFT online service: multiple sequence alignment, interactive sequence choice and visualization. Brief. Bioinform. **20**, 1160–1166 (2019). https://doi.org/10.1093/bib/bbx108
24. Melnyk, A., Knyazev, S., Vannberg, F., et al.: Using earth mover's distance for viral outbreak investigations. BMC Genom. **21**, 582 (2020). https://doi.org/10.1186/s12864-020-06982-4

Functional Connectivity Disruptions in Alzheimer's Disease: A Maximum Flow Perspective

Emma T. Stubby[1], Seyed Majid Razavi[2], and Sina Khanmohammadi[1,2(✉)]

[1] School of Computer Science, University of Oklahoma, Norman 73019, USA
sinakhan@ou.edu
[2] Data Science and Analytics Institute, University of Oklahoma, Norman 73019, USA

Abstract. Alzheimer's disease is a neurological disorder characterized by functional and structural atrophy, leading to symptoms like memory loss and cognitive decline. This study seeks to analyze the disruptions of functional connectivity pathways within the brain caused by Alzheimer's disease from the maximum flow perspective. More specifically, we computed the maximum flow pathways within the functional brain networks, and compared it between healthy controls and Alzheimer's patients. Our results suggest that the Alzheimer's patients utilize pathways related to the default mode network (DMN) more frequently and display significant alterations in the usage of paths connected to the striate cortex (SC). The increased usage of DMN pathways might point to a compensation mechanism that facilitates interregional communications in Alzheimer's patients. Understanding the nature of such a compensation mechanism could help develop new treatment options for Alzheimer's patients.

Keywords: Brain Connectivity · Alzheimer's Disease · Functional Connectivity · Maximum Flow · Default Mode Network · Neurological Disorder

1 Introduction

Alzheimer's disease (AD) is the primary cause of dementia characterized by synaptic loss and senile plaques (SP) in the brain [1]. Common symptoms of AD include the progressive loss of episodic memory, cognitive decline, language impairment, and visuospatial deficiencies [2]. Currently there are an estimated 6.7 million people living with Alzheimer's disease across the United States, and this estimate is expected to increase to 13.8 million by 2060 [3]. The diagnosis of AD presents a particular challenge due to its intricate pathophysiology and multifaceted clinical presentations [4]. Furthermore, the most predominant biomarkers of AD are the cerebrospinal fluid (CSF) levels of beta-amyloid and Tau proteins, which involve invasive sampling methods [4]. Therefore, with the rapid progression of computational methodologies and neuroimaging modalities, many researchers have begun to investigate alternate methods of early AD detection and diagnosis, including methods based on disruptions of brain network connectivity [5].

The use of modern network analyses on fMRI measurements has been pivotal not only in identifying unique connectivity patterns, but also in understanding intrinsic dynamics of neural circuitry associated with Alzheimer's disease [6,7]. For example, Dai et al. [8] used resting-state fMRI measurements of Alzheimer's patients to construct and analyze graph representations of both the functional connectivity and structural connectivity networks of the brain, finding significant disruptions within the AD networks in the form of increased characteristic path length and segregation. Another study by Wang et al. [9] found connectivity disruptions in AD patients by analyzing networks created using low-frequency fluctuations (LFF) of blood oxygen level-dependent signals, finding decreased connection between anterior and posterior lobes. Taking a different approach, Brier et al. [10] investigated functional connectivity from the perspective of resting-state subnetworks, which consistently exhibited overall loss of correlation within and across networks as AD progresses. A full review of advanced computational techniques used for mapping brain connectivity in Alzheimer's patients is available in Yu et al. [11] as well as van den Heuvel and Hulshoff Pol [12].

Here, instead of focusing on the topological properties of functional networks, we took a different approach and analyzed the disruptions of functional connectivity in Alzheimer's patients from a maximum flow perspective. Using the maximum flow formulation provides a quantitative framework to assess the efficiency of information flow between different brain regions while taking into account indirect connections and bottlenecks in the network. We constructed the functional flow networks using cross-regional correlations as a measure of edge capacities, where the larger correlation values indicate higher capacity for neural co-activation between two regions of the brain. After performing a maximum flow analysis on the networks for each of the fMRI scans, we then determined which connections are utilized more in the maximum flow path. Our results suggest that AD patients use alternative connections in the maximum flow path compared to the common neural pathways used in control subjects.

2 Methods

2.1 Data Set

This study utilized an fMRI dataset from the Alzheimer's Disease Neuroimaging Initiative (ADNI) which contains a large collection of imaging, biological, and clinical datasets recorded from multiple subjects over several years [13]. The data used in this study was randomly selected from the most recent cohort, ADNI-3. We selected 8 subjects diagnosed with Alzheimer's Disease (AD) and 6 Healthy Control (HC) subjects. Six of the subjects were male and eight were female. The average age of the subjects was 73.46 years old with a standard deviation of 6.98 years. The collected data included T-1 weighted anatomical MRI scans, resting state fMRI scans, clinical data, and demographic data. For each subject, on average, three fMRI scans were recorded over multiple sessions.

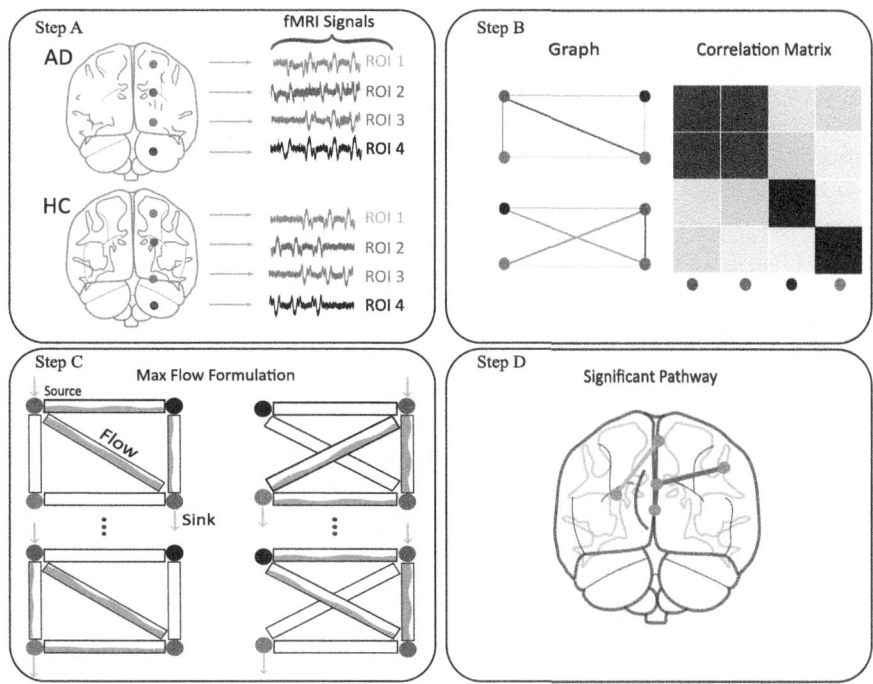

Fig. 1. The maximum flow analysis framework for identifying information flow pathways in functional brain networks. In Step A, we preprocessed the fMRI signals followed by constructing the functional connectivity networks from each scan in step B. In Step C, we conducted the maximum flow analyses on the graphs, followed by performing statistical analyses on the maximum flow pathways in Step D.

2.2 Framework

In this section, we present our proposed framework to study the changes of maximum flow pathways in Alzheimer's Disease. The proposed framework includes four main steps of 1) Data Preprocessing: to refine and prepare the datasets, 2) Graph Construction: to identify the functional connections, 3) Maximum Flow Analysis: to identify the maximum flow pathways, and 4) Statistical Analysis: to identify the significantly different pathways between Alzheimer's patients and control subjects (Fig. 1).

2.2.1 Data Preprocessing

For preprocessing, we first performed brain extraction and tissue segmentation, followed by normalizing the data into MNI152-NLin2009cAsym template [14]. Next, we performed head motion correction using rigid-body transformation followed by co-registration of the functional images to the normalized anatomical image [14]. Once the time series data were extracted, we detrended the data and applied a band-pass filter with a range of 0.006 to 0.0533 Hz. The time series were then standardized to zero mean and unit variance, and the first 10 frames of each scan were dropped for signal stabilization. Finally, we applied the MSDL (Multi-Subject Dictionary Learning) atlas to parcellate our data into 39 distinct brain regions.

2.2.2 Graph Construction

We constructed a unique flow network graph $G(V, E)$ for each fMRI scan, where V represents the set of nodes (brain regions) and E the set of edges (flow connections). Flow networks are directed graphs with each edge e characterized by a flow $f(e)$ and a capacity $c(e)$. First, for every scan, we measured the functional connectivity between each of the 39 regions of interest by computing a 39×39 correlation matrix using Pearson's correlation coefficient. The correlation coefficient r between two brain regions was then mapped to an edge capacity c such that $c(e) = |r|$, representing the capacity of co-activation between brain regions. As correlation values are not inherently directed, we treat each edge of the graph as bidirectional, implying that both (u, v) and (v, u) possess the same capacity $c(e)$. This graph construction was consistently applied across all scans.

2.2.3 Maximum Flow Analysis

For each of the flow network graphs, we performed a maximum flow analysis in order to determine which functional connections are most frequently used in the maximum flow paths. In a maximum flow analysis, there is a source node, denoted by s, which supplies the network with flow, and a sink node, denoted by t, which dispossesses the flow in the network. The overarching goal becomes the optimization problem:

$$\text{Maximize} \sum_{(s,v) \in E} f(s, v)$$
$$\text{subject to } 0 \leq f(u, v) \leq c(u, v) \quad \forall (u, v) \in E \quad (1)$$
$$\sum_{(u,v) \in E} f(u, v) = \sum_{(v,w) \in E} f(v, w) \quad \forall v \in V - \{s, t\},$$

where the objective is to maximize the flow f from the source to the sink, while ensuring that flow does not exceed the edge capacities denoted by c, and for any node other than the source and sink, the total inflow equals the total outflow. We assigned the source and sink nodes to test all possible combinations, and performed a total of 1,482 maximum flow analyses for each functional network using the Dinic's algorithm [15]. Lastly, we calculated the total amount of flow that traveled through each of the 741 edges for all optimal paths for each scan.

2.2.4 Statistical Analysis

After obtaining the maximum flow values for each edge in each fMRI scan, we conduct a statistical t-test on each of our 741 edges to determine the significantly different flow paths between the AD and HC networks. We filtered out all edges with a p-value of 0.05 or greater and then calculated the average flow values in AD and HC groups for each of the remaining significant edges.

3 Results

Figure 2 shows the significantly different max flow pathways between the control subjects and Alzheimer's patients. The θ in Fig. 2 represents the difference in average flow where the red color signifies an increase in the average flow value for the AD group compared to healthy controls. The blue signifies a decrease in average flow value for the AD group compared to the healthy controls. Table 1 shows the top ten most significantly different pathways ranked based on the p-value. The striate cortex (SC), also known as the primary visual cortex, and the default mode network (DMN) both had multiple edges with significant differences between AD and control subjects. The SC had significant increases in flow for AD subjects in its pathways connected to the right DMN, medial DMN, and right posterior temporal lobe. In addition to its connections to the SC, the DMN saw increased flow in its connections to the left intraparietal sulcus, left lateral occipital complex, and the right superior temporal sulcus.

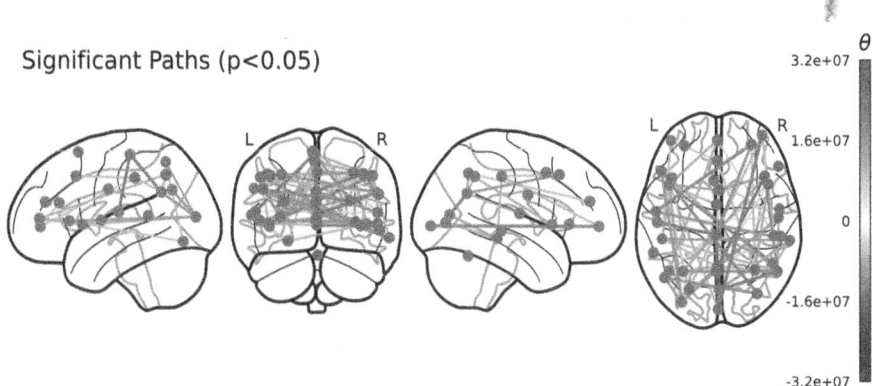

Fig. 2. Brain network representation of the max flow pathway differences between Alzheimer's patients and control subjects. The red edges signify an increased flow value in the maximum flow paths of Alzheimer's patients, whereas the blue edges signify decreased flow values. Opacity of each edge represents the significance levels, where the more transparent edges are less significant.

Table 1. Top ten edges with the most significant difference in flow values between AD and HC.

	Connection	Difference in Average Flow Values	p value
1	Motor - L STS	−1575.2417	0.0002
2	Basal - L Ins	1090.6047	0.0017
3	Med DMN - L IPS	1140.6867	0.0027
4	Striate - R DMN	2137.8761	0.0046
5	R Post Temp - R Ant IPS	990.5943	0.0047
6	Striate - Med DMN	3214.9617	0.0048
7	Striate - R Post Temp	1394.0277	0.0055
8	L Ins - Cing	−1049.3722	0.0058
9	Med DMN - L LOC	1614.3369	0.0061
10	R DMN - R STS	1147.5153	0.0063

4 Discussion

This study complements the current research in identifying functional connectivity disruptions in Alzheimer's patients. Unlike most studies that consider the microscopic and macroscopic properties of the functional networks, here we formulated our analysis as maximum flow problem. More specifically, we looked at the utilization of different functional connections as maximum flow pathways in the brain. We found several disruptions in the maximum flow pathways of Alzheimer's patients, specifically in the default mode network and striate cortex.

The maximum flow formulation has previously been used on structural brain networks as a measurement of brain resiliency [16], whereas here we have considered functional connectivity and looked at the problem from optimal pathway perspective. Additionally, Kumar et al. [17] have implemented the maximum flow technique on networks defined by transfer entropy as an alternative measure of functional connectivity between brain regions, but they have not considered the direct application of maximum flow formulation in analyzing functional brain networks. To the best of our knowledge, this is the first study of its kind where the functional connectivity of Alzheimer's patients has been investigated from a maximum flow perspective for identifying the disruptions in optimal pathways.

We observed that the striate cortex as well as the default mode network displayed significant differences in the optimum pathways of AD patients. Previous studies have consistently found connections between the onset of AD and the alteration of brain activity in the DMN [18]. It has even been proposed that the neural activity in DMN could serve as a potential biomarker for the diagnosis of AD [19]. A possible explanation of this altered activity is the compensatory mechanisms of DMN to reallocate processing resources [18]. Our results coincide with these conclusions, as the alteration in optimal path usage could be interpreted as the reallocation of resources.

Although there has been less research investigating the link between Alzheimer's disease and the striate cortex, several studies have found pathological effects of AD within the primary visual cortex [20,21]. One such study found increased numbers of senile plaques (SP) and neurofibrillary tangles (NFT) as well as decreased numbers of neurons in the SC [22]. Another study found increased numbers of SP and NFT, in the cuneal and lingual gyri areas of the visual cortex [23]. Furthermore, visual impairment is commonly reported as an early clinical symptom of AD [24]. The findings of our study support these observations of physical and functional atrophy of the SC, although it is unclear whether these changes are because of a compensatory mechanism or other factors.

In our current methodology, the network construction relies on Pearson's correlation to determine the weights or capacities of the network edges, emphasizing linear relationships between brain regions. Recognizing that neural interactions might manifest in nonlinear dynamics, it is important for our future research to consider nonlinear interactions between brain regions. Furthermore, in our maximum flow formulation of the functional connectivity analysis, we only considered the usage of optimal pathways whereas other properties such as minimum cuts and edge residual capacities could also provide useful insights about the functional connectivity disruptions in Alzheimer's patients.

5 Conclusion

In conclusion, we considered the irregularities within the maximum flow pathways of functional brain networks in Alzheimer's patients. Utilizing functional connectivity networks from fMRI data, we formulated a maximum flow problem and considered the usage of functional connections in the optimal pathways. Our results show that the default mode network and the striate cortex in AD patients were significantly different in terms of their utilization for optimal pathways in maximum flow formulation. Understanding the underlying neural mechanisms that cause such changes could help developing new diagnostic and treatment methods for Alzheimer's disease.

Acknowledgement. The fMRI data used in this study was obtained from Alzheimer's Disease Neuroimaging Initiative (ADNI), which was funded by the National Institutes of Health (Award U01 AG024904) and Department of Defense (Award W81XWH-12-2-0012). The authors would also like to thank Connor Behnen for the preparation and preprocessing of the fMRI data.

References

1. Breijyeh, Z., Karaman, R.: Comprehensive review on alzheimer's disease: causes and treatment. Molecules **25**(24), 5789 (2020). https://doi.org/10.3390/molecules25245789

2. Silva, M.V.F., Loures, C.d.M.G., Alves, L.C.V. et al.: Alzheimer's disease: risk factors and potentially protective measures. J. Biomed. Sci. **26**, 33 (2019). https://doi.org/10.1186/s12929-019-0524-y
3. 2023 Alzheimer's disease facts and figures. Alzheimer's Dement. **19**, 1598–1695 (2023). https://doi.org/10.1002/alz.13016
4. García-Morales, V., et al.: Current understanding of the physiopathology, diagnosis and therapeutic approach to Alzheimer's disease. Biomedicines **9**(12), 1910 (2021). https://doi.org/10.3390/biomedicines9121910
5. Bullmore, E., Sporns, O.: Complex brain networks: graph theoretical analysis of structural and functional systems. Nat. Rev. Neurosci. **10**, 186–198 (2009). https://doi.org/10.1038/nrn2575
6. Sperling, R.: Potential of functional MRI as a biomarker in early Alzheimer's disease. Neurobiol. Aging **32** Suppl 1(Suppl 1), S37–43 (2011). https://doi.org/10.1016/j.neurobiolaging.2011.09.009
7. Supekar, K. et al.: Network analysis of intrinsic functional brain connectivity in Alzheimer's disease. PLoS Comput. Biol. **4**(6), e1000100 (2008). https://doi.org/10.1371/journal.pcbi.1000100
8. Dai, Z., Lin, Q., Li, T. et al.: Disrupted structural and functional brain networks in Alzheimer's disease. Neurobiol. Aging **75**, 71–82 (2019). https://doi.org/10.1016/j.neurobiolaging.2018.11.005
9. Wang, K. et al.: Altered functional connectivity in early Alzheimer's disease: a resting-state fMRI study. Hum. Brain Mapp. **28**(10), 967–978 (2006). https://doi.org/10.1002/hbm.20324
10. Brier, M.R. et al.: Loss of intranetwork and internetwork resting state functional connections with Alzheimer's disease progression. J. Neurosci. **32**(26), 8890–8899 (2012). https://doi.org/10.1523/JNEUROSCI.5698-11.2012
11. Yu, M., Sporns, O., Saykin, A.J.: The human connectome in Alzheimer disease - relationship to biomarkers and genetics. Nat. Rev. Neurol. **17**, 545–563 (2021). https://doi.org/10.1038/s41582-021-00529-1
12. van den Heuvel, M.P., Hulshoff Pol, H.E.: Exploring the brain network: a review on resting-state fMRI functional connectivity. Eur. Neuropsychopharmacol. **20**(8), 519–534 (2010). https://doi.org/10.1016/j.euroneuro.2010.03.008
13. Petersen, R.C., et al.: Alzheimer's disease neuroimaging initiative (ADNI): clinical characterization. Neurology **74**(3), 201–209 (2010). https://doi.org/10.1212/WNL.0b013e3181cb3e25
14. Esteban, O., et al.: fMRIPrep: a robust preprocessing pipeline for functional MRI. Nat. Methods **16**, 111–116 (2019). https://doi.org/10.1038/s41592-018-0235-4
15. Dinitz, Y.: Algorithm for solution of a problem of maximum flow in networks with power estimation. Soviet Math. Dokl. **11**, 1277–1280 (1970)
16. Wook Yoo, S., Han, C., Shin, J. et al.: A network flow-based analysis of cognitive reserve in normal ageing and Alzheimer's disease. Sci. Rep. **5**, 10057 (2015). https://doi.org/10.1038/srep10057
17. Kumar, S. et al.: An information network flow approach for measuring functional connectivity and predicting behavior. Brain Behav. **9**(8), e01346 (2019). https://doi.org/10.1002/brb3.1346
18. Mevel, K., Chételat, G., Eustache, F., Desgranges, B.: The default mode network in healthy aging and Alzheimer's disease. Int. J. Alzheimers Dis. **2011**, 535816 (2011). https://doi.org/10.4061/2011/535816
19. Koch, W. et al.: Diagnostic power of default mode network resting state fMRI in the detection of Alzheimer's disease. Neurobiol. Aging **33**(3), 466–478 (2012). https://doi.org/10.1016/j.neurobiolaging.2010.04.013

20. Kusne, Y., Wolf, A.B., Townley, K., Conway, M., Peyman, G.A.: Visual system manifestations of Alzheimer's disease. Acta Ophthalmol. **95**(8), e668–e676 (2017). https://doi.org/10.1111/aos.13319
21. Brewer, A.A., Barton, B.: Changes in visual cortex in healthy aging and dementia. Update on Dementia. InTech (2016). https://doi.org/10.5772/64562
22. Leuba, G., Kraftsik, R.: Visual cortex in Alzheimer's disease: occurencee of neuronal death and glial proliferation, and correlation with pathological hallmarks. Neurobiol. Aging. **15**(1), 29–43 (1994). https://doi.org/10.1016/0197-4580(94)90142-2
23. Armstrong, R.A.: Visual field defects in Alzheimer's disease patients may reflect differential pathology in the primary visual cortex. Optom. Vis. Sci. **73**(11), 677–682 (1996). https://doi.org/10.1097/00006324-199611000-00001
24. Brewer, A.A., Barton, B.: Visual cortex in aging and Alzheimer's disease: changes in visual field maps and population receptive fields. Front. Psychol. **5**, 74 (2014). https://doi.org/10.3389/fpsyg.2014.00074

On Multi-phase Metagenomics Reads Binning

Francesco Tomasella and Cinzia Pizzi(✉)

Department of Information Engineering, University of Padova, Padua, Italy
`cinzia.pizzi@dei.unipd.it`

Abstract. Metagenomics is the study of heterogeneous microbial samples extracted directly from their natural environment, e.g., from soil, water, or the human body. The detection and quantification of species that populate microbial communities have been the subject of many recent studies based on classification and clustering, motivated by being the first step in more complex pipelines (e.g. for functional analysis, de-novo assembly or comparison of metagenomes).

In this paper we explore the idea of improving the overall quality of metagenomics binning at reads-level by proposing a framework that sequentially combine two complementary read binning approaches: one based on species abundances determination and another one relying on reads overlap in order to cluster reads together.

Our preliminary results show that the combination of the two tools can lead to the improvement of the clustering quality in realistic conditions where the number of species is not known beforehand.

Keywords: sequence analysis · metagenomics · read binning · alignment-free algorithms

1 Introduction

Metagenomics is the study of heterogeneous microbial communities sampled from their natural living environment (e.g., soil, water, or the human gut or saliva, etc.) with the primary goal of determining the taxonomical identity of the microorganisms residing in the samples. The advent of metagenomics has revolutionized the field of Microbiology by shifting the focus from the individual microbe study to that of a complex microbial community.

The benefits of metagenomics are manyfold. First, traditional genomic-based approaches require prior clone and lab culturing for further investigation [1]. However, not all bacteria can be cultured. Metagenomics overcome this problem by directly sampling an environment and sequencing the entire microbial community it contains [2]. Other advantages include the possibility of study interactions among microbes living in the same environment [3] and comparing samples taken from similar environments or at different points in time for environmental monitoring or health screenings [4,5].

Alongside opening new research perspectives, metagenomics brings along both experimental challenges for a correct environmental sampling, and computational challenges for quality control, assembly, taxonomic and functional classification of large-scale complex communities [6]. In particular, the detection and quantification of the species in a metagenomic sample is of paramount interest both as a challenging computational problem per se, and as the first step in complex pipelines [7]. Despite extensive studies, accurate identification at read level remains challenging [8,9]. Supervised methods can obtain high precision levels, but they rely on reference database completeness. Moreover, the construction of k-mers DB is usually very demanding in terms of RAM and disk space. For these reasons, when using supervised methods, the number of unassigned reads can be very high [10,11]. On the other hand, unsupervised classification tools, also known as genome binning algorithms, are based on the observation that the k-mer distributions of the DNA fragments from the same genome are more similar than those from different genomes. Thus, without using any reference genome, one can determine if two fragments are from genomes of similar species based on their k-mer distributions. In this study, we will focus on the unsupervised detection of species in a sample without the use of reference genomes and considering as fragments the short reads provided by the sequencing process.

One of the major problem when processing metagenomic data is the fact that the proportion of species in a sample, i.e. abundance rate, can vary greatly. Some tools, e.g. AbundanceBin [12] explicitly exploits this variability, clustering reads based on their abundance ratio. While this approach works well if all the species in the sample have a different abundance, in presence of species with the same abundance the approach struggle to distinghuish among them. Approaches based on exploiting reads overlap, and subsequently clustering them, are capable of better distinguish among single species, even if their abundance is similar. In recent years several such approaches have been proposed, mainly differing in the techniques used for feature extraction and the distance measure they use to define similarity [13–17].

In this paper we explore the idea of improving the overall quality of metagenomics binning at reads-level by proposing a framework that sequentially combine two complementary read binning approaches. Indeed also the authors of AbundanceBin tried out a similar approach, combining their tool with MetaCluster. However, in their experiments they did assume that the exact number of species in the sample was known beforehand. Motivated by the curiosity of testing this idea on a more realistic framework, we paired AbundanceBin with MetaProb, which has the capability of estimating the number of species and proved in separate experiments [15] to outperform MetaCluster. Our preliminary experiments suggests that our intuition is correct and that the combination of complementary tools can indeed be beneficial for metagenomic binning at read-level in more realistic unsupervised settings.

2 Methods

In this section we will describe the combined framework we used for our analysis, starting with the methodological details of the the two tools we used: AbundanceBin [12] and MetaProb [15].

2.1 AbundanceBin

AbundanceBin is a tool for metagenomic binning based on abundance estimation. The working hypothesis of AbundanceBin is that the distribution of reads follows the Lander-Waterman model, whereby the coverage of the various nucleotide positions is modeled via a Poisson distribution. The metagenomic sequencing procedure can be viewed as a set of Poisson distributions, with each of these representing a different species. In the presence of m different species, therefore, it is possible to identify m Poisson distributions. The mean of each of these distributions represents the abundance of the species and is therefore the element that must be calculated to obtain an estimate of their abundance. AbundanceBin thus solves an optimization problem through the use of an Expectation-maximization (EM) algorithm. Once the EM algorithm has converged, it is possible to calculate the probability of assigning a read to a bin, even if there is the possibility that the read remains unassigned. The EM algorithm requires the number of bins as input. To solve this problem, AbundanceBin adopts a recursive approach that is based on dividing the dataset into two bins, subsequently iterating the process until bins with very different abundances are obtained. AbundanceBin perform well in situations in which the abundance of species is different, although not less than a 1:2 ratio. In cases where there is less variability and the species have a comparable abundance, AbundanceBin is no longer an optimal choice and has very high error rates, since it will most likely group different species with a similar abundance in the same bin.

2.2 MetaProb

Metaprob is a two-step approach for metagenomic read binning. The reads are first grouped together based on their overlap, measured in terms of the number of shared q-mers, with $q = 31$ by default. The output of this phase is a relatively large number of small groups of very connected reads that are therefore likely to belong to the same species. Next, within each group, a set of representative, not overlapping (to avoid redundancy) reads is chosen and from it an l-mer profile ($l = 5$ by default) is extracted and normalized to obtain a group signature. Such signatures are finally given in input to the k-means clustering algorithm that will group together signatures (and their corresponding groups of reads) so to obtain the final clusters that represent the different species in the sample.

Similarly to EM, also the k-means algorithm requires previous knowledge of the number k of clusters to obtain. Metaprob can both accept this parameter as input, or estimate the value of k exploiting the Kolmogorov-Smirnov test.

2.3 Combined Framework

The idea of our framework is a two-phases approach. First, in Phase 1, we partition the reads so that all the reads of species with the same abundance are clustered together with the abundance-based algorithm AbundanceBin.

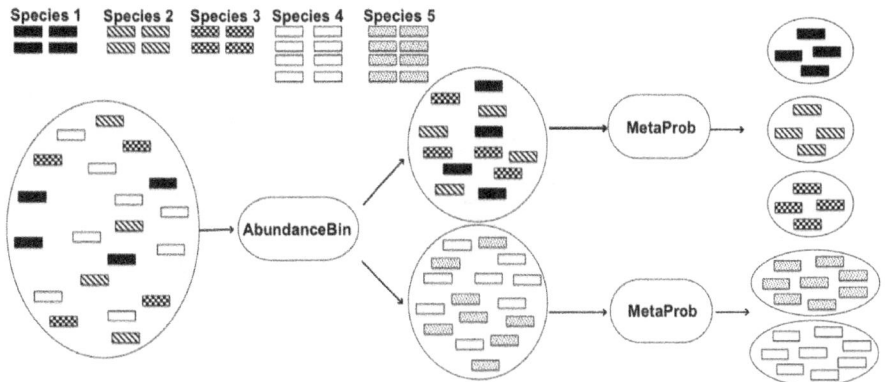

Fig. 1. Ideal pipeline of the combined framework.

Next, in Phase 2, the overlap-based approach MetaProb is applied to each of the obtained clusters in order to separate the species within it. Figure 1 shows the ideal pipeline of our approach. In reality, especially if the final number of expected species is not known and given, the combinations of the two tools is not as smooth as in the ideal pipeline. Besides the fact that none available read binning algorithms is capable of perfect clustering, AbundanceBin, unlike most read binners (including MetaProb), has not been designed to take into account of pair-end reads. This means that is possible that reads that are paired (and thus belong to the same species) are assigned by AbundanceBin to different clusters. This raises the problem on how to deal with paierd-end reads that have been wrongly separated by AbundanceBin because MetaProb needs in input sets of paired-end reads. Possible options include deleting all unpaired reads or design a re-assignment strategy. However, determining the destination cluster for each read is complex, requiring a case-by-case evaluation based on the results obtained and the overall composition of the clusters. In our experiments we used two possible approaches: i) the reassignment of reads only from clusters with a very high percentage of unpaired reads; ii) reassignment of unpaired reads starting from the cluster with the highest percentage of unpaired reads, and iteration of the reassignment until no more unpaired reads remain. In Fig. 2 we can see a more realistic pipeline, with added intermediate processing.

3 Results and Discussion

In this section we will describe the experimental settings, and we will discuss the results obtained with our analysis. The two tools were all run with default parameters, but *without* giving the final number of expected clusters in input. All the experiments were performed on a machine with Intel(R) Xeon(R) Gold 5220 CPUs @ 2.20/3.90GHz and 2TB of RAM.

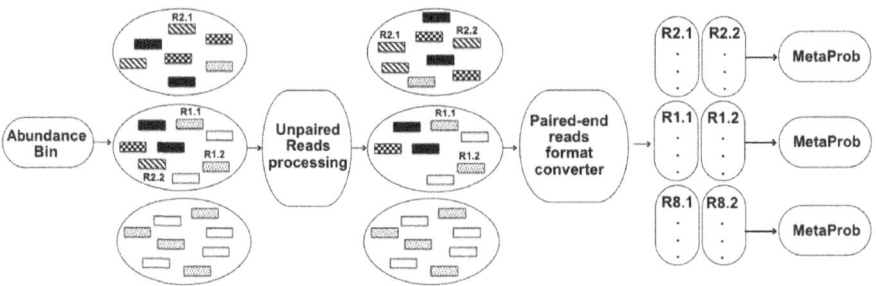

Fig. 2. A more realistic pipeline (for space limits we omit the input to AbundanceBin and the output from Metaprob).

3.1 Dataset

We used one of the datasets used in several previous papers on metagenomic read binning [11,14–16]. In particular, we chose the dataset S7 which is suitable for verifying our hypothesis. The S7 dataset is composed of short paired-end reads, and includes 5 species with abundance ratios 1:1:1:4:4 and with phylogenetic distance at order and genre to which they belong. The dataset was simulated using MetaSim, a tool for the generation of metagenomic reads, using the Illumina error profile with an error rate of 1%. Details of the dataset composition are given in Table 1.

Table 1. Details of the dataset S7.

Species	Nome	Coverage
Species 1	Actinobacillus pleuropneumoniae serovar 5b str. L20	10
Species 2	Aliivibrio salmonicida LFI1238	10
Species 3	Haemophilus somnus 129PT	10
Species 4	Pasteurella multocida 36950	40
Species 5	Vibrio cholerae M66-2	40

3.2 Evaluation Metrics

To evaluate the results we used three popular performance evaluation metrics (precision, recall and f-measure) as defined in other read-binning papers [11,14–18]. Let n be the number of species in the simulated dataset, and let C be the number of clusters returned by the algorithm, A_{ij} is the number of reads from species j assigned to cluster i:

$$\text{Precision} = \frac{\sum_{i=1}^{C} \max_j A_{ij}}{\sum_{i=1}^{C} \sum_{j=1}^{n} A_{ij}} \quad (1)$$

$$\text{Recall} = \frac{\sum_{j=1}^{n} \max_i A_{ij}}{\sum_{i=1}^{C} \sum_{j=1}^{n} A_{ij} + \#\text{unassigned reads}} \quad (2)$$

$$\text{F-measure} = \frac{2 * \text{Precision} * \text{Recall}}{\text{Precision} + \text{Recall}} \quad (3)$$

3.3 Experimental Analysis

Phase 1. We obtained from AbundanceBin three different clusters, one more than expected. From Table 2 we can see the actual partition, and in particular we can notice the presence of paired reads that have been assigned to different clusters, and their abundance in the cluster. The three resulting clusters highlight an effective grouping of the species with the highest abundance, i.e. species 4 and 5 (as illustrated in Table 3), which are mainly located within Cluster 1. The three species with lower abundance are instead distributed between Cluster 2 and Cluster 3, the latter mainly characterized by the presence of unpaired reads.

Table 2. Cluster compositions generated by AbundanceBin

Cluster	Total Reads	Paired Reads	Unpaired reads	% Unpaired reads
Cluster 1	2 405 443	2 152 948	252 495	10.5 %
Cluster 2	786 613	453 664	332 949	42.33 %
Cluster 3	115 044	20 506	94 538	82.18 %

Although the obtained number of clusters differs from the actual one, when looking in more details at the composition of the clusters, we can state that the partitioning of the species within the clusters is congruent with the expected abundance, demonstrating the effectiveness of AbundanceBin in identifying and grouping together species with the same abundance in dataset S7.

Table 3. Number of reads per species in the clusters generated by AbundanceBin.

Cluster	Species 1	Species 2	Species 3	Species 4	Species 5
Cluster 1	10 000	18 723	8 097	1 065 139	1 303 484
Cluster 2	225 827	108 303	203 762	109 648	139 073
Cluster 3	48 029	23 340	40 507	1 389	1 779

Moreover, Tables 4, 5, and 6 highlight, for each species within each cluster, how many of the associated reads are actually paired or unpaired. The % Read column also allows one to see at a glance the structure of the cluster and the percentage with which the species appear within it.

Table 4. Cluster 1 detailed composition.

Species	Reads	% Reads	Of which paired	Of which unpaired
Species 1	10 000	0.42%	30.04 %	69.96 %
Species 2	18 723	0.78%	27.98 %	72.02 %
Species 3	8 097	0.34%	30.04 %	69.96 %
Species 4	1 065 139	44.30%	90.62 %	9.38 %
Species 5	1 303 484	54.16%	90.30 %	9.70 %

Table 5. Cluster 2 detailed composition.

Species	Reads	% Reads	Of which paired	Of which unpaired
Species 1	225 827	28.70%	80.51 %	19.49 %
Species 2	108 303	13.77%	74.09 %	25.91 %
Species 3	203 762	25.90%	81.82 %	18.18 %
Species 4	109 648	13.94%	9.90 %	90.10 %
Species 5	139 073	17.69%	10.10 %	89.90 %

Table 6. Cluster 3 detailed composition.

Species	Reads	% Reads	Of which paired	Of which unpaired
Species 1	48 029	41.75%	18.46 %	81.54 %
Species 2	23 340	20.29%	17.93 %	82.07 %
Species 3	40 507	35.21%	18.40 %	81.60 %
Species 4	1 389	1.21%	0 %	100 %
Species 5	1 779	1.55%	0 %	100 %

As it can be seen, more than 90% of the reads of the species with the highest abundance is included in Cluster 1. The remaining 10%, mainly composed of unpaired reads, are distributed between Clusters 2 and Cluster 3. The reads of the species with lower abundance are mainly distributed between between Cluster 2 and Cluster 3, with some smaller quantities erroneously assigned to Cluster 1. For each of these three species, it is noteworthy that the vast majority of reads are assigned to Cluster 2 as a pair of paired-end reads, while the opposite trend is observed within Cluster 3, in which the prevalence of reads for each species consists of unpaired reads.

The computed values of precision, recall and F-measure for AbundanceBin are 0.47, 0.87 ad 0.61, respectively, in line with those obtained in [15]. The low value of precision is indeed expected, since the AbundanceBin principle is to clusters together reads with the same abundance that, in our case study, can belong to different species. Similarly, we do expect a high value of recall, since most read of the same species will be included in the same cluster.

Phase 2. Before applying Metaprob to each of the clusters obtained with AbundanceBin a processing was needed to avoid the presence of unpaired reads. We tried two approaches:

1. reassign all unpaired reads of clusters with a composition of unpaired reads above a threshold T (we chose $T = 80\%$) to the cluster of their paired-read;
2. reassign all unpaired reads of the cluster with the highest percentage of unpaired reads; iterate the process until no more unpaired reads are left.

In practice, in our case study, with Approach 1) we just reassigned the unpaired reads of Cluster 3 to their counterpart in Cluster 1 and Cluster 2, and then remove any other unpaired read that was left. Instead, with Approach 2) we did not remove any read, and ended up with more than 250.000 reads re-paired with respect to Approach 1). For space limits we report here in details only the results of the iterative Approach 2) since from our experience it appears to be the most effective one, as we will discuss later. The details of this processing are shown in Tables 7, 8 and 9.

Cluster 1. In ideal conditions, Cluster 1 generated by AbundanceBin should have contained only reads from the two most abundances species. However, as previously shown in Table 3, some noise from other species was also introduced. Nonetheless, the results confirm that MetaProb is able to effectively identify two subclusters covering about 90% of the reads of Cluster 1 and containing the majority of the reads from these two species: 92% of Cluster 1.A contains reads that belong to Species 4, while more than 99% of Cluster 1.C contains reads from Species 5. The remaining Cluster 1.B covers just 11% of the total reads of Cluster 1. Of these, 70% belongs to Species 5 and 24% reads belong to Species 4, with the remaining 6% consisting of reads from the minority species. The size and mixture of this cluster are therefore those of a "spurious" cluster possibly produced by the noise introduced by AbundanceBin or by a wrong estimate

Table 7. Partitioning of Cluster 1 after applying MetaProb.

Cluster	Species	Pairs of Reads
Cluster 1.A	Specie 1	4 370
	Specie 2	15 285
	Specie 3	3 211
	Specie 4	**541 144**
	Specie 5	22 428
Cluster 1.B	Specie 1	4 077
	Specie 2	812
	Specie 3	3 649
	Specie 4	37 410
	Specie 5	**110 605**
Cluster 1.C	Specie 1	51
	Specie 2	7
	Specie 3	21
	Specie 4	3 967
	Specie 5	**581 932**

number of clusters from Metaprob or by a combination of both.

Furthermore, it is important to highlight that MetaProb is naturally better suited to work with clusters that contain species with similar abundances. This is evident in the correct classification of species 4 and 5. However, MetaProb encounters difficulties when the variation in species abundance is more marked, as in the case of Cluster 2. This aspect motivates the approach adopted in this experiment, which aimed to improve the performance of MetaProb by removing one of its weak points, a limitation similar to that found in other software based on DNA composition.

Cluster 2. Giving the reads of Cluster 2 as input to MetaProb caused the generation of 10 different clusters. Considering that 95% of the reads in Clusters 2 belongs to Species 1, Species 2 and Species 3 this result was somehow unexpected. However, by carefully looking into each cluster composition we can observe that among these clusters, four of them (2.A, 2.C, 2.F and 2.I) contain a number of paired-reads greater than 30000. Specifically, reads from Species 1 and Species 3 are mainly assigned, respectively, to Cluster 2.I and 2.C, and in both cases they represent about 98% of the total composition of these clusters. Species 2 instead has been mainly split between Clusters 2.A and 2.F. that together contain more than 90% of the reads of this species.

The remaining six clusters have been assigned a smaller number of reads and can either been seen as a relatively small erroneous splitting of a species (e.g. Cluster 2.J that is almost entirely composed by reads of Species 1), a correct identification of a species that should have not been in Cluster 2 (Cluster 2.J that mainly contains reads from Species 5), or a mixture generated by the intrinsic similarity in terms of k-mers that some species may share.

Overall, this in-depth analysis allows us to conclude that the output of MetaProb consists of four main clusters characterized by reads of the three species expected from this cluster, plus some noise. We will discuss later ideas on how to further improve this result.

Table 8. Partitioning of Cluster 2 after applying MetaProb.

Cluster	Species	Paired Reads	Cluster	Species	Paired Reads
Cluster 2.A	Species 1	4 698	Cluster 2.F	Species 1	554
	Species 2	**20 952**		**Species 2**	**30 002**
	Species 3	12 111		Species 3	0
	Species 4	3 449		Species 4	1
	Species 5	1 853		Species 5	0
Cluster 2.B	Species 1	6 508	Cluster 2.G	Species 1	133
	Species 2	64		Species 2	585
	Specie 3	**6 628**		Species 3	5
	Species 4	757		Species 4	441
	Species 5	72		**Species 5**	**2 087**
Cluster 2.C	Species 1	2 381	Cluster 2.H	Species 1	1 608
	Species 2	258		**Species 2**	**3 111**
	Species 3	**91 188**		Specie 3	299
	Species 4	53		Species 4	345
	Species 5	1		Species 5	3 039
Cluster 2.D	**Species 1**	**3 178**	Cluster 2.I	**Species 1**	**98 564**
	Species 2	1 246		Species 2	0
	Species 3	2 235		Species 3	1 079
	Species 4	347		Species 4	0
	Species 5	86		Species 5	0
Cluster 2.E	Species 1	268	Cluster 2.J	**Species 1**	**11 104**
	Species 2	766		Species 2	2
	Species 3	**1 257**		Species 3	774
	Species 4	114		Species 4	60
	Species 5	24		Species 5	41

Cluster 3. Once the unpaired reads had been moved, Cluster 3 contained only 10,253 pairs of paired-end reads, which represents less than 1% of the total number of reads in the dataset, confirming the fact that it is a "spurious" cluster generated by AbundanceBin. Although it exclusively contained reads from

the three species with lower abundance, given their not representative nature, MetaProb could not fully exploit its distinguish power based on overlap and composition signature. Nonetheless, as it can be seen in Table 9, it was able to clearly distinguish reads of Species 1 (80% of Cluster 3.A is composed of reads from this species), while the distinction between Species 2 and Species 3 was more difficult. It is worth noting that within Cluster 3, Species 1 has basically the double of the reads of the other two species. We speculate this could have helped Metaprob in finding the overlaps it needs to build its initial clusters.

Table 9. Partitioning of Cluster 3 after applying MetaProb.

Cluster	Species	Paired Reads
Cluster 3.A	**Species 1**	**3 281**
	Species 2	303
	Species 3	564
	Species 4	0
	Species 5	0
Cluster 3.B	Specie 1	708
	Species 2	519
	Specie 3	**1 502**
	Species 4	0
	Species 5	0
Cluster 3.C	Specie 1	445
	Species 2	1 271
	Species 3	**1 660**
	Species 4	0
	Species 5	0

3.4 Comparison with Standalone Tools

To summarize our results we computed quality measures of the final binning we obtained with those obtained by the single use of AbundanceBin and MetaProb. Moreover we included the results of the current available version of MetaCluster (5.0) that includes a low/high abundance partitioning phase before the final clustering.

Results shown in Table 10 support our intuition: among the datasets analyzed in previous studies, S7 was one of the most challenging as AbundanceBin struggles to obtain a good precision while MetaProb shows lower performances in both precision and recall with respect to other datasets. MetaCluster 5.0, which already includes an high/low abundance partitioning, showed the highest precision, but the worst recall. Our combined approach slightly improves over

Table 10. Standalone tools vs combined framework.

	MetaCluster	MetaProb	AbundanceBin	AB+MP
Precision	**0.925**	0.818	0.477	0.912
Recall	0.671	0.745	**0.879**	0.812
F-measure	0.778	0.780	0.618	**0.859**

MetaProb in terms of both precision (showing results comparable to those of MetaCluster) and recall, while with respect to AbundanceBin it can keep the recall high while doubling the precision. The values of F-measure confirm the overall better performances of the proposed combined complementary approach.

4 Conclusions and Future Work

This study explored the combined use of complementary tools for metagenomics read binning in order to improve the overall quality of the binning process, when several species have the same abundance ratio and no knowledge of the actual number of species is given, as in a realistic context. Our results showed that a combined framework can exploit the strengths of different binning approaches to obtain better values in terms of clustering quality metrics with respect to the usage of a single tool. As a future work we plan to expand the experimental assessment including other datasets with different complex species distributions. Moreover, a fair analysis of our results needs to consider that the total number of clusters produced by the current pipeline is higher than the exact number of clusters. While some over-estimation is expected since the exact estimation of the number of clusters is a challenge itself, it would be interesting to see if and to which extent this estimation can be improved. Two directions we plan to investigate aims at: i) the reduction of the noise introduced by AbundanceBin not taking into consideration paired-end reads by developing a strategy that cluster paired-end reads together or by adopting other strategies for reads re-assignment; and ii) the further merge of "minor" clusters by adding a post-processing step.

Acknowledgements. Cinzia Pizzi is supported by the National Recovery and Resilience Plan (NRRP), National Biodiversity Future Center - NBFC, NextGenerationEU.

References

1. Felczykowska, A., Bloch, S.K., Nejman-Faleczyk, B., Baraska, S.: Metagenomic approach in the investigation of new bioactive compounds in the marine environment. Acta Biochim. Pol. **59**(4), 501–5 (2012)
2. Kang, D.D., Froula, J., Egan, R., Wang, Z.: MetaBAT, an efficient tool for accurately reconstructing single genomes from complex microbial communities. PeerJ **3**, e1165 (2015)

3. Shreiner, A.B., Kao, J.Y., Young, V.B.: The gut microbiome in health and in disease. Curr. Opin. Gastroenterol. **31**(1), 69–75 (2015)
4. Ondov, B.D., et al.: Mash: fast genome and metagenome distance estimation using MinHash. Genome Biol. **17**(1), 132 (2016)
5. Pellegrina, L., Pizzi, C., Vandin, F.: Fast approximation of frequent k-Mers and applications to metagenomics. J. Comput. Biol. **27**(4), 534–549 (2020)
6. Bharti, R., Grimm, D.G.: Current challenges and best-practice protocols for microbiome analysis. Brief Bioinform. **22**(1), 178–193 (2021)
7. Mande, S.S., Mohammed, M.H., Ghosh, T.S.: Classification of metagenomic sequences: methods and challenges. Brief. Bioinform. **13**(6), 669–81 (2012)
8. Sczyrba, A., et al.: Critical assessment of metagenome interpretation - A benchmark of metagenomics software. Nat. Methods **14**(11), 1063–1071 (2017)
9. Comin, M., Di Camillo, B., Pizzi, C., Vandin, F.: Comparison of microbiome samples: methods and computational challenges. Brief. Bioinform. **22**(1), 88–95 (2021)
10. Lindgreen, S., Adair, K.L., Gardner, P.P.: An evaluation of the accuracy and speed of metagenome analysis tools. Sci Rep. **6**, 19233 (2016)
11. Girotto, S., Comin, M., Pizzi, C.: Higher recall in metagenomic sequence classification exploiting overlapping reads. BMC Genomics **18**(Suppl 10) (2017). https://doi.org/10.1186/s12864-017-4273-6
12. Wu, Y.W., Ye, Y.: A novel abundance-based algorithm for binning metagenomic sequences using l-tuples. J. Comput. Biol. **18**(3), 523–34 (2011)
13. Wang, Y., Leung, H.C.M., Yiu, S.M., Chin, F.Y.L.: MetaCluster 5.0: a two-round binning approach for metagenomic data for low-abundance species in a noisy sample, Bioinformatics **28**(18), i356-i362 (2012)
14. Vinh, L.V., Lang T.V., Binh, L.T., Hoai, T.V.: A two-phase binning algorithm using l-mer frequency on groups of non-overlapping reads. Algorithms Mol. Biol. **10**(1), 2 (2015). https://doi.org/10.1186/s13015-014-0030-4
15. Girotto, S., Pizzi, C., Comin, M.: MetaProb: accurate metagenomic reads binning based on probabilistic sequence signatures. Bioinformatics **32**(17), i567–i575 (2016)
16. Andreace, F., Pizzi, C., Comin, M.: MetaProb 2: metagenomic reads binning based on assembly using minimizers and K-Mers statistics. J. Comput. Biol. **28**(11), 1052–1062 (2021)
17. Balvert, M., Luo, X., Hauptfeld, E., Schönhuth, A., Dutilh, B.E.: OGRE: overlap graph-based metagenomic Read clustEring. Bioinformatics **37**(7), 905–912 (2021)
18. Girotto, S., Comin, M., Pizzi, C.: Metagenomic reads binning with spaced seeds. Theoret. Comput. Sci. **698**, 88–99 (2017)

A Unified Machine Learning Framework for Multi-subtype Tumour Classification Across Diverse Datasets

Ankur Yadav[✉] and Ovidiu Daescu

Department of Computer Science, The University of Texas at Dallas, 800 W Campbell Rd., Richardson 75080, TX, USA
ankur.yadav@utdallas.edu, daescu@utdallas.edu

Abstract. Machine learning models are increasingly employed in the classification of digitized medical tissue images, including for identifying cancer types and subtypes. Most models focus on a target tumor type, using datasets from a single source, which limits generalization. To overcome this, we formulated unified frameworks and applied them to bone, colon, and prostate cancer datasets from varied origins. Rigorous testing concluded that framework 1 achieved an overall accuracy of 88.48%, while framework 2, with classification corrections, achieved an enhanced overall accuracy of 90.28%. The area under the curve (AUC) values exceeded 0.97, with average specificity and sensitivity surpassing 0.98 and 0.99 across frameworks for individual classes. Additionally, when Framework 2 was applied to an unseen breast cancer dataset, it demonstrated a notable accuracy of 88.8% for normal vs. tumor tile classification, indicating its efficacy on unseen datasets. The results reflect the potential of engineered feature extraction in predictive models.

Keywords: cancer · subtype · classification · model

1 Introduction

Cancer treatment has transitioned to more focused methodologies targeting individual molecular subtypes, accentuating the necessity for precise tumor subtype classification [16]. Machine learning (ML) models present a promising solution. However, the current approach involves developing separate models for each tumor type and subtype. A unified model for tumor types and subtypes could have significant advantages and facilitate clinical adoption. Recent studies, such as multi-task learning models, have demonstrated improved accuracy in histology subtype classification, thereby reducing the workload for trained pathologists and fostering a more streamlined and integrated analysis [8,17].

Image and feature similarities notably hinder the task of tumor subtype differentiation in histopathological slides across various cancer types [11,13,25,26]. Manual grading of digital slides is laborious and time-consuming, requiring expertise from pathologists and clinicians. The emergence of automated classification presents a viable alternative for improving speed and accuracy in clinical diagnoses [21]. However, multi-source data variance introduces complexity in feature extraction and classification tasks, on top of the computational demand associated with high-resolution whole slide images (WSI) [6].

We present two frameworks for classifying tumor subtypes across various datasets, focusing on Bone, Colon, and Prostate Cancer. Our methodology hinges on extracting topological and textural features from RGB images. Additionally, we extended our frameworks to an unseen Breast Cancer dataset from the BACH Dataset [3]. We achieved encouraging results of 88.8% accuracy in normal vs. tumor image classification, demonstrating the usability of our frameworks even on datasets not seen during the training phase.

Framework 1 and 2 exhibited accuracy of 88.48% and 90.28% respectively, with remarkable average AUC values (greater than 0.98 for classes). The average specificity and sensitivity exceeded 0.98 and 0.99, respectively, underscoring the potential of our top performing framework.

2 Materials and Methods

2.1 Datasets

In this study, we leveraged three distinct datasets corresponding to different types of cancer. For Prostate Cancer, the SICAPv2 dataset [19] was employed, encompassing 9959 prostate histology whole slide image patches annotated with global Gleason scores and patch-level Gleason grades, categorized into four classes - NC, G3, G4, G5.

For Colon Cancer, the CRC100K dataset [15] was utilized, comprising 100,000 histological images of human colorectal cancer and healthy tissue, focusing on Normal, Tumor, and Stroma classes for cancer detection and grading.

In the case of Bone Cancer, the UT-Osteosarcoma dataset [4] was used, containing 1144 H&E (Hematoxylin and Eosin) stained osteosarcoma histology images categorized as Non-Tumor, Viable Tumor, and Necrosis.

Additionally, an extension of our frameworks was tested on a subset of the BACH Dataset [3] from the ICIAR 2018 *Grand Challenge on Breast Cancer Histology Images*. We extracted 500 tiles from each of the four subtypes in the breast cancer dataset, totaling 2000 tiles, to examine the adaptability of Framework 2 on unseen data.

While there are differences in image size across these datasets, we extracted tiles of size 224×224 to ensure consistency. A balanced dataset was curated by utilizing 1500 samples randomly from each subtype. We divided this dataset into training, validation, and test subsets using a 70:10:20 split.

2.2 Feature Extraction

Persistent Homology. Persistent Homology (PH) is a technique from Topological Data Analysis (TDA) that helps in understanding the shape or structure of the data by tracking changes over a range of values [7,20]. For image analysis, a series of binary images are generated, and PH tracks the appearance and disappearance of features (connected components, holes/loops) in this image series, based on the color values of the pixels. For our frameworks, for each image tile we used eight channels (Red, Blue, Green, Gray from RGB, and Hue, Saturation, Value, and Average from HSV) and extracted 100 Betti-0 and 100 Betti-1 vectors, adding up to 1600 features per image (Fig. 1).

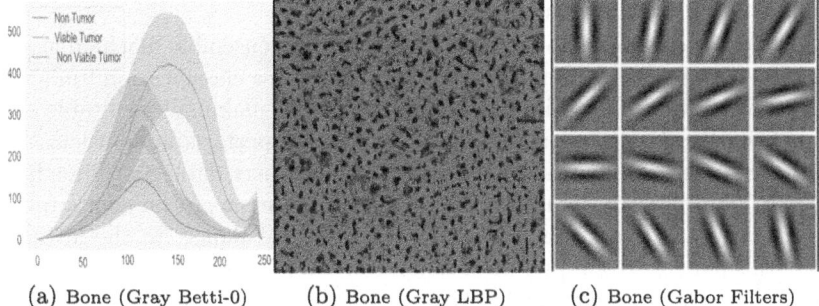

(a) Bone (Gray Betti-0)   (b) Bone (Gray LBP)   (c) Bone (Gabor Filters)

Fig. 1. (a) Betti 0 curve derived from a tile of the Bone Cancer dataset, showcasing the topological features captured. (b) Local Binary Pattern (LBP) representation of a tile from the Colon Cancer dataset, highlighting the textural features extracted. (c) Various Gabor filters applied, exhibiting the range of frequencies and orientations utilized for feature extraction in our framework

Local Binary Patterns (LBP). Local Binary Patterns (LBP) is a method used for classification that creates a binary code for each pixel by comparing it to its neighboring pixels [18,24]. Using LBP, for each image, we extracted 100 features for each channel, totaling 800 features.

Gabor Filters. Gabor Filters are used for edge detection, texture analysis, and feature extraction, and are known for their ability to pick up frequency content in images along certain directions [2,5,14]. We used Gabor Filters to generate 50 features for each channel per image, totalling 400 features.

In summary, a total of 2800 features were extracted for every input image using the three methods outlined earlier. This feature extraction step provides a rich set of features for the classification task.

3   Classification Framework

In digital histopathology, both machine learning and deep learning approaches have shown significant promise [10]. Deep learning, in particular, has shown excellent results in various histopathology tasks including tumor segmentation, grading, and classification [1]. The high accuracy levels achieved by deep learning techniques, especially Convolution Neural Networks (CNNs), have led to their wide appeal [12].

However, deep learning frameworks tend to be computationally intensive, requiring substantial computational resources for training and inference. Moreover, deep learning models often require large, balanced datasets to train effectively and achieve high performance. In practice, obtaining such balanced datasets can be difficult, if not impossible, and the models might perform suboptimally on unbalanced datasets.

Given these challenges, our approach employs an ensemble machine learning model, specifically the XGBoost model [9], to develop a classification framework. XGBoost is known for its efficiency and performance, making it a suitable choice for our task. We design two separate frameworks for performing multiclass classification of different tumor subtypes from various cancer datasets. These frameworks are structured to handle the data efficiently, manage the computational costs effectively, and address the issue of dataset imbalance. In the latter sections of this paper, we will present a comparative analysis of the performances of these frameworks, showcasing the advantages of our ensemble machine learning approach over deep learning frameworks in the context of histopathology image classification (Fig. 2).

3.1 Framework 1: Hierarchical Classification Framework

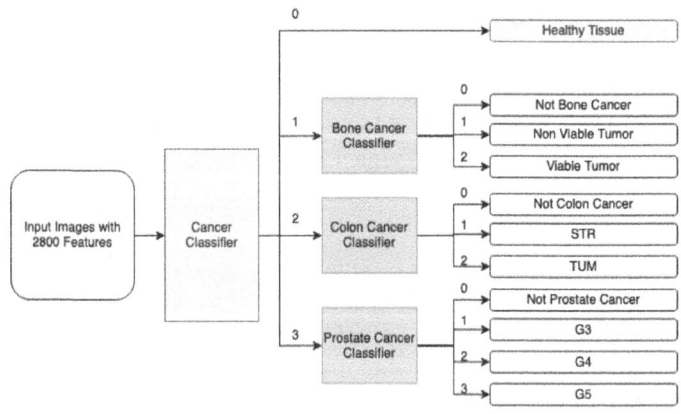

Fig. 2. Inference using framework 1.

In our first framework, a hierarchical classification approach is adopted to manage the complexity of multiclass classification and to potentially reduce the error rate. This framework consists of four XGBoost models organized in a two-level hierarchy.

The first level is handled by a primary model, which is trained to classify the input tissue images into four classes: Healthy Tissue, Bone Cancer Tissue, Colon Cancer Tissue, and Prostate Cancer Tissue. This classification serves as a broad categorization, channeling the data into more specialized models at the second level.

At the second level, three distinct models are deployed, each specialized in classifying subtypes of one of the cancer types: Bone Cancer, Colon Cancer, and Prostate Cancer. In addition to the subtypes, each of these models has a class labeled "Does not belong to this cancer type" to accommodate for possible misclassifications by the primary model at the first level. This structure aims to provide a safety net, ensuring that even if the primary model misclassifies the cancer type, the secondary models can provide valuable information regarding the incongruence in classification, prompting manual intervention by pathologists.

This hierarchical structure facilitates a more nuanced classification. The results generated from the framework can be categorized into three types:

- Correctly Classified into Subtypes: The primary model correctly identifies the cancer type, and the secondary model accurately classifies the cancer subtype.
- Incorrectly Classified into Subtypes: The primary model correctly identifies the cancer type, but the secondary model misclassifies the cancer subtype.
- Incorrectly Classified by Cancer Type Model but Correctly Classified by Submodels: The primary model misclassifies the cancer type, but the secondary models correctly identify that the tissue does not belong to the purported cancer type, triggering a review by pathologists.

Through this hierarchical approach, our framework aims to leverage the strengths of machine learning to provide accurate classifications while incorporating safeguards to identify and rectify potential misclassifications, ensuring a robust and reliable classification system (Fig. 3).

3.2 Framework 2: Probabilistic Correction Framework

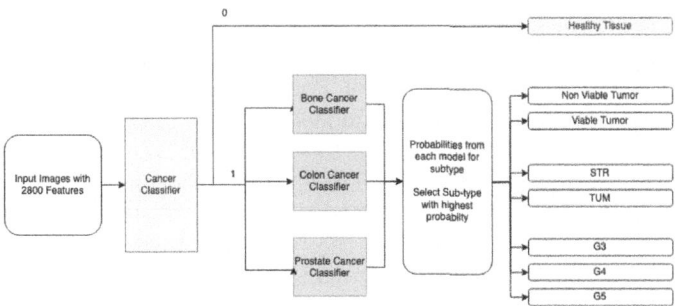

Fig. 3. Inference using framework 2.

In our second framework, we adopt a probabilistic approach to rectify potential misclassifications from the initial model, enhancing the robustness and reliability of the classification system. This framework also consists of four XGBoost models but operates in a slightly different manner compared to the first framework.

Initially, the primary model is trained to classify tissue samples into two broad categories: Healthy Tissue or Cancer Tissue. Post this broad categorization, the framework channels the samples classified as Cancer Tissue to three submodels, each specialized in identifying subtypes of Bone Cancer, Colon Cancer, or Prostate Cancer. Additionally, each of these submodels has a class labeled "Does not belong to this cancer type" to account for the possibility of incorrect categorization by the primary model.

During inference, each cancer tissue sample is passed through all three submodels to obtain the probability scores for each subtype. The framework then selects the subtype with the highest probability score across all submodels. If the highest probability corresponds to the "Does not belong to this cancer type" class in a submodel, the framework considers the subtype with the second-highest probability from that submodel. If even the second-highest probability corresponds to the "Does not belong to this cancer type" class, the framework then considers the subtype with the third-highest probability.

A unique feature of this framework is its corrective mechanism: if all three submodels yield the highest probability for the "Does not belong to this cancer type" class, the framework overrides the primary model's classification and labels the sample as Healthy Tissue. This probabilistic approach aims to correct potential misclassifications from the primary model, ensuring a more accurate and reliable classification, especially in scenarios where the primary model might be prone to errors.

Through this mechanism, the second framework capitalizes on the "collective intelligence" of the submodels to correct potential misclassifications. This method not only provides an approach to handle misclassifications but also leverages the probabilistic nature of the models to achieve a more accurate and reliable classification, potentially reducing the need for manual intervention and speeding up the diagnostic process.

3.3 Framework 2 Validation on Unseen Breast Cancer Dataset

To validate the power to generalize of Framework 2, we considered a breast cancer dataset that was not used for model training. Initially, the dataset was fed into the cancer classification model to segregate it into Normal and Tumor classes. If a tile was predicted as Tumor, it was further processed through the individual tumor classification sub-models. The aim was to observe if the submodels would categorize the tiles into the "Does not Belong" class, testing the discriminative ability of the individual models.

Table 1. Performance Metrics for both frameworks.

Model	Subtype	Spec.	Sens.	F1	Acc.	AUC
Framework1 Model	Non Tumor	0.97	1.00	0.95	85.56	0.96
	Bone Cancer	0.99	1.00	0.99	95.60	0.99
	Colon Cancer	0.99	1.00	0.99	98.10	1.00
	Prostate Cancer	0.96	1.00	0.97	96.80	0.99
Framework2 Model	Non Tumor	0.96	1.00	0.95	84.20	0.95
	Tumor	0.84	1.00	0.87	96.30	0.99
Bone Cancer	Not This Class	0.94	1.00	0.95	98.90	1.00
	Non Viable Tumor	0.99	1.00	0.99	93.10	0.98
	Viable	0.99	1.00	0.99	90.10	0.97
Colon Cancer	Not This Class	0.97	1.00	0.98	99.60	1.00
	STR	1.00	1.00	1.00	98.00	1.00
	TUM	1.00	1.00	1.00	94.20	0.99
Prostate Cancer	Not This Class	0.94	1.00	0.94	97.74	1.00
	Benign	0.98	1.00	0.98	85.40	0.91
	InSitu	0.99	1.00	0.98	81.10	0.92
	Invasive	0.99	1.00	0.99	90.40	0.97

4 Results

Our study aimed to develop a robust framework capable of classifying multiple tumor subtypes across diverse cancer categories using a machine-learning framework. We utilized three distinct datasets representing Prostate, Bone, and Colon Cancer. A balanced dataset was curated by extracting 224×224 tiles from each dataset and utilizing 1,500 samples from each subtype. Our feature extraction process encompassed eight channels from both RGB and HSV color spaces, yielding 2800 features per image. These features were derived using Persistent Homology, Local Binary Patterns, and Gabor Filters, as discussed in the Materials and Methods Sect. 2.

We devised two frameworks to perform the classification task. The first framework employed a primary model to segregate the tissues into four broad categories, followed by three submodels for subtype classification. The second framework initiated with a binary classification of healthy vs cancer tissue by the main model, followed by probabilistic inference through three submodels to ascertain the tumor subtype.

The performance of both frameworks was evaluated using standard metrics, including Accuracy, Specificity, Sensitivity, F1 Score, and AUC (Area Under the ROC Curve). These metrics provide a comprehensive insight into the models' capabilities and are pivotal for medical diagnostics applications.

The evaluation results are given in Table 1. Framework 1 demonstrated an overall accuracy of **88.48%**. The primary model within this framework exhibited a class-wise accuracy of 93.45%, while the sub-models tailored for Bone, Colon, and Prostate cancer classification achieved accuracies of 97.45%, 98.93%, and 94.11%, respectively, as detailed in Table 1. On the other hand, Framework 2 yielded a notable improvement in the main model accuracy, reaching 92.71% in distinguishing tumor from healthy tissues. The sub-models in Framework 2 mirrored the accuracies of those in Framework 1. However, a unique facet of Framework 2 was the implementation of a classification correction mechanism for the main model utilizing the three sub-models. Initially, both Framework 1 and Framework 2 manifested accuracies of 85.56% and 84.20% in healthy tissue classification. Post-correction, a total of 4.05% of samples initially misclassified were accurately re-classified as healthy tissue by the sub-models, elevating the healthy tissue classification accuracy to 88.25%. Consequently, the overall accuracy of Framework 2 escalated to **90.28%** post-correction, reflecting a substantial enhancement. Both frameworks managed to achieve commendable average Area Under the Curve (AUC) values exceeding 0.97. Furthermore, the specificity and sensitivity averages across all sub-models were highly promising, surpassing 0.98 and 0.99, respectively. These metrics underscore the robustness and reliability of the proposed frameworks in classifying multiple tumor types effectively, thereby holding significant promise for practical deployment in histopathological image analysis.

To further test Framework 2, we applied it to an unseen Breast Cancer dataset. The main model was utilized to discern between Tumor and Non-tumor tiles, achieving an accuracy of **88.8%** despite not being trained on this dataset. Subsequently, the three sub-models were employed to ascertain whether they categorize these tiles into the "Does not belong to this class" category, validating the robustness of our framework. Specifically, the Bone and Colon Cancer models classified 96% of the tumor tiles accurately into the "Does not belong to this class" category, while the Prostate Cancer model, due to the stark resemblance between Prostate and Breast Cancer tiles, only categorized 2.8% of tiles into the "Does not belong" class, showcasing the intricacies of tissue similarities across different cancer types. These results not only elucidate the versatility and applicability of our framework on unseen data but also highlight the efficacy of manual feature extraction in predictive modelling, underpinning a solid foundation for future explorations in multi-tumor classification across heterogeneous datasets.

5 Discussion

The comparison of our frameworks with existing methods demonstrates a competitive, if not superior, performance while offering a unique approach to handling misclassifications, a common challenge in histopathology image analysis. Our frameworks' ability to correct potential misclassifications embodies a significant contribution to the field, providing a robust and reliable method for tumor subtype classification across diverse cancer categories.

Our findings underscore the potential of employing ensemble machine learning approaches, like XGBoost, in conjunction with well-designed frameworks to improve the accuracy and reliability of histopathology image classification, especially in a multi-cancer, multi-subtype scenario.

In recent studies, innovative approaches for multi-tumor classification have shown promising results. One study [22] employed Heterogeneous Transfer Learning (HTL) to alleviate data bottlenecks, achieving an AUC of 0.949 in sentinel lymph node patches and 0.962 in breast patches, showcasing the potential of HTL in utilizing a large dataset of a certain cancer type to build accurate models for similar cancers. Another study [23] presented a comprehensive detection system integrating a deep learning neural network with noncoding RNA biomarkers, achieving an impressive AUC of 96.3% for binary cancer vs. healthy classification and 99 to 100% AUC for individual cancer type detection. Furthermore, their system demonstrated a notable multi-classification accuracy of 82.15% across common cancers. These findings highlight the advancements in employing machine learning techniques to improve the accuracy and reliability of multi-tumor classification, contributing significantly towards practical cancer screening at the population level.

The performance of Framework 2 on the unseen Breast Cancer dataset shows the adaptability and potential of our framework. The accuracy of 88.8% in distinguishing Normal vs Tumor tiles, alongside the correct categorization of unseen data by the sub-models, especially for tiles in the "do not belong to this class," reflects the strength of the manual feature extraction techniques employed. The resemblance in tile classification between Prostate and Breast Cancer underscores the importance of incorporating more diverse training data in future work to refine the framework's discerning ability between closely related cancer types.

The results from our frameworks show a marked improvement in multi-tumor classification, with accuracies that are competitive with, or superior to, previously reported methods. The correction mechanism in Framework 2 is particularly notable, providing a robust solution to common misclassification issues in histopathology image analysis. This contribution is pivotal in advancing cancer classification, aiding early detection and personalized treatment plans, thereby enhancing healthcare outcomes. Future work could leverage larger datasets and extend the frameworks to encompass a broader spectrum of cancer types.

References

1. Abdeltawab, H.A., Khalifa, F.A., Ghazal, M.A., Cheng, L., El-Baz, A.S., Gondim, D.D.: A deep learning framework for automated classification of histopathological kidney whole-slide images. J. Pathol. Inf. **13**, 100093 (2022)
2. Alekseev, A., Bobe, A.: GaborNet: gabor filters with learnable parameters in deep convolutional neural network. In: 2019 International Conference on Engineering and Telecommunication (EnT), pp. 1–4. IEEE (2019)
3. Aresta, G., et al.: BACH: grand challenge on breast cancer histology images. Med. Image Anal. **56**, 122–139 (2019)
4. Arunachalam, H.B., Mishra, R., Daescu, O., Cederberg, K., Rakheja, D., Sengupta, A., Leonard, D., Hallac, R., Leavey, P.: Viable and necrotic tumor assessment from WSI of osteosarcoma using ML and DL models. PLoS ONE **14**(4), e0210706 (2019)

5. Bourkache, N., Laghrouch, M., Sidhom, S.: Gabor filter algorithm for medical image processing: evolution in big data context. In: 2020 International Multi-Conference on:"Organization of Knowledge and Advanced Technologies" (OCTA), pp. 1–4. IEEE (2020)
6. Breen, J., Allen, K., Zucker, K., Hall, G., Orsi, N.M., Ravikumar, N.: Efficient subtyping of ovarian cancer histopathology whole slide images using active sampling in multiple instance learning. In: Proceedings of SPIE 12471, vol. 12471. SPIE (2023)
7. Chazal, F., Michel, B.: An introduction to topological data analysis: fundamental and practical aspects for data scientists. Front. Artif. Intell. **4**, 108 (2021)
8. Chen, K., Wang, M., Song, Z.: Multi-task learning-based histologic subtype classification of non-small cell lung cancer. Radiol. Med. (Torino) **128**(5), 537–543 (2023)
9. Chen, T., Guestrin, C.: XGBoost: a scalable tree boosting system. In: Proceedings of the 22nd ACM SIGKDD International Conference on Knowledge Discovery and Data Mining, pp. 785–794. KDD 2016, ACM, New York, NY, USA (2016). https://doi.org/10.1145/2939672.2939785, http://doi.acm.org/10.1145/2939672.2939785
10. Cooper, M., Ji, Z., Krishnan, R.G.: Machine learning in computational histopathology: challenges and opportunities. Genes Chromosom. Cancer **62**, 540–556 (2023)
11. ktefaie, Y., et al.: Integrative multiomics-histopathology analysis for breast cancer classification. NPJ Breast Cancer **7**(1), 147 (2021)
12. Hamed, E.A.R., Salem, M.A.M., Badr, N.L., Tolba, M.F.: A deep learning-based classification framework for annotated histopathology lung cancer images. In: Hassanien, A., Rizk, R.Y., Pamucar, D., Darwish, A., Chang, KC. (eds.) International Conference on Advanced Intelligent Systems and Informatics. pp. 86–94. Springer, Cham (2023). https://doi.org/10.1007/978-3-031-43247-7_8
13. Hu, B., El Hajj, N., Sittler, S., Lammert, N., Barnes, R., Meloni-Ehrig, A.: Gastric cancer: classification, histology and application of molecular pathology. J. Gastrointest. Oncol. **3**(3), 251 (2012)
14. Jusman, Y., Nur'aini, M.A., Puspitasari, S.: Gabor filter-based caries image feature analysis using machine learning. In: 2022 5th International Seminar on Research of Information Technology and Intelligent Systems (ISRITI), pp. 514–519. IEEE (2022)
15. Kather, J.N., Halama, N., Marx, A.: 100,000 histological images of human colorectal cancer and healthy tissue. Zenodo10 **5281** (2018)
16. Khader, A., et al.: Importance of tumor subtypes in cancer imaging. Eur. J. Radiol. Open **9**, 100433 (2022)
17. Li, B., Wang, T., Nabavi, S.: Cancer molecular subtype classification by graph convolutional networks on multi-omics data. In: Proceedings of the 12th ACM Conference on Bioinformatics, Computational Biology, and Health Informatics, pp. 1–9 (2021)
18. Ojala, T., Pietikainen, M., Harwood, D.: Performance evaluation of texture measures with classification based on kullback discrimination of distributions. In: Proceedings of 12th International Conference on Pattern Recognition, vol. 1, pp. 582–585. IEEE (1994)
19. Silva-Rodríguez, J., Colomer, A., Sales, M.A., Molina, R., Naranjo, V.: Going deeper through the gleason scoring scale: an automatic end-to-end system for histology prostate grading and cribriform pattern detection. Comput. Methods Programs Biomed. **195**, 105637 (2020)
20. Skraba, P., Ovsjanikov, M., Chazal, F., Guibas, L.: Persistence-based segmentation of deformable shapes. In: 2010 IEEE Computer Society Conference on Computer Vision and Pattern Recognition-Workshops, pp. 45–52. IEEE (2010)

21. Srikantamurthy, M.M., Rallabandi, V., Dudekula, D.B., Natarajan, S., Park, J.: Classification of benign and malignant subtypes of breast cancer histopathology imaging using hybrid CNN-LSTM based transfer learning. BMC Med. Imaging **23**(1), 1–15 (2023)
22. Sun, K., Chen, Y., Bai, B., Gao, Y., Xiao, J., Yu, G.: Automatic classification of histopathology images across multiple cancers based on heterogeneous transfer learning. Diagnostics **13**(7), 1277 (2023)
23. Wang, A., Hai, R., Rider, P.J., He, Q.: Noncoding RNAs and deep learning neural network discriminate multi-cancer types. Cancers **14**(2), 352 (2022)
24. Wang, X., Han, T.X., Yan, S.: An HOG-LBP human detector with partial occlusion handling. In: 2009 IEEE 12th International Conference on Computer Vision, pp. 32–39. IEEE (2009)
25. Yadav, A., Ahmed, F., Daescu, O., Gedik, R., Coskunuzer, B.: Histopathological cancer detection with topological signatures. In: 2023 IEEE International Conference on Bioinformatics and Biomedicine (BIBM), pp. 1610–1619. IEEE (2023)
26. Yadav, A., Daescu, O., Leavey, P., Rudzinski, E.: Machine learning for rhabdomyosarcoma whole slide images sub-type classification. In: Proceedings of the 16th International Conference on PErvasive Technologies Related to Assistive Environments, pp. 192–196 (2023)

AFA: Abstract Functional Analysis Identifies New Microglial Subtypes at Single Cell Level in Alzheimer's Disease

Chenyu Zhang[1]([✉]), Honglin Wang[1], Seung-Hyun Hong[1], Riqiang Yan[2], and Dong-Guk Shin[1]

[1] Computer Science and Engineering Department, University of Connecticut, Storrs, CT 06269, USA
chenyu.zhang@uconn.edu
[2] Department of Neuroscience, University of Connecticut Health Center, Farmington, CT 06030, USA

Abstract. With recent advancements in single cell sequencing technologies, routine data analysis activities include identifying cell subtypes in tissues and understanding their relationships at the single cell level. While existing algorithms excel in distinguishing different cell types based on known markers, subtyping cells based on their functions remains to be a challenge. To address this limitation, we propose a new single cell subtyping method, called Abstract Functional Analysis (AFA), which incorporates a priori known context-specific biological processes into the analysis. The key premise of AFA is that interjecting "some form of prior knowledge" (GO Biological Processes in our case) into the otherwise unbiased analysis is amenable to deriving, namely, biological function centric subtyping. We assessed our method on eight publicly available Alzheimer's Disease related single-cell mRNA datasets and demonstrated that AFA can subgroup cells based on their functional roles into subtypes such as disease associated microglia (DAM) and late-stage homeostatic microglia exhibiting DAM signature. Advantages of AFA include labeling subtypes based on functions and discovering additional biological processes enriched within each identified subtype. AFA offers a new way of subgrouping and naming cells thereby enhancing our understanding of cellular heterogeneity in a more intuitive and useful way.

Keywords: Single Cell Gene Expression Data Analysis · Biological Process · GO Analysis · Alzheimer's Disease · Data Visualization

1 Introduction

Single cell gene expression technology has enabled biomedical scientists to study the molecular landscape of a biological systems at the single-cell level. The current gold standard of analyzing single cell gene expression data is to perform an "unbiased" analysis using a data analysis pipeline such as Scanpy [1] and Seurat [2]. These methods have been very successful in subgrouping cells when distinct markers of specific cell types in different tissues are well established. For example, Villani et al 2017 report six human blood dendritic cells and four monocyte subtypes from their scRNA-seq study performed on human blood [3]. Although this unbiased way of subgrouping cells has been successful, common problems are: (i) what is the appropriate resolution for subtyping (i.e., how many subtypes of cells should be produced), and (ii) how to interpret and label each identified subgroup of cells. Problems arise particularly when the subgrouping process is artificially "forced" to produce a higher resolution of sub-grouped cells. In this case labelling each identified subgroup with proper nomenclature becomes very challenging. It is also unclear if each subgroup indeed represents any meaningful biological function or physiological condition of the cell/tissue.

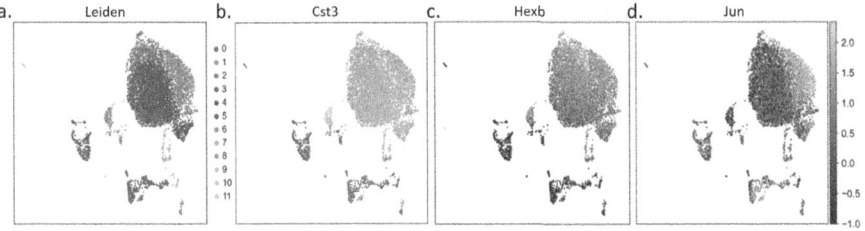

Fig. 1. UMAP visualization results from the SCANPY package, focusing on three genes: Cst3, Hexb and Jun for Singh's data set. (Color figure online)

We illustrate these issues using the example given in Fig. 1. Figure 1a is a two-dimensional display obtained by using SCANPY's leiden/UMAP processing on one snRNA-seq gene expression study sample (GSM5962713) from GSE199027 which was published by Singh et al [4]. This study aimed to assess if targeted deletion of Bace-1 in adult 5xFAD mice microglia enhances phagocytic capabilities of some beneficial subpopulation of microglia and thus could be responsible for reducing accumulation of amyloid plaques, the known hallmark pathology of Alzheimer's disease (AD).

While Fig. 1a shows an unbiased way of subdividing cells into 12 subgroups, it is debatable whether the spade shaped clump of cells at the top right of the figure should be further divided into four subgroups (colored by green, blue, orange, and purple). One may attempt an explanation by combining three other figures of Fig. 1b–d which include single gene distribution patterns of various genes. Fig. 1b shows an elevated level of Cst3 over that entire cell population, suggesting that the large clump of the cells made up of cluster IDs, 0 (blue), 1 (orange), 2 (green) and 4 (purple) in Fig. 1a most likely constitutes homeostatic microglia as Cst3 is a well-known marker for this type of microglia. Furthermore, when the elevated levels of Jun (limited to cluster ID 1 – orange color; Fig. 1d) and Hexb (most likely limited to cluster ID 2 – green color; Fig. 1c) are factored in, this combination may explain why clusters 0, 1, and 2 are sub-grouped. The challenge becomes whether this method of clustering subgroups is meaningful, as these small number of gene combination patterns may not be adequate to determine the cell types, let alone differing functional roles of the three groups of cells. This example shows an overly simplified scenario, but nevertheless, this is the current state of the art in performing "unbiased" single cell subtyping.

We conjecture that one way to address this issue of cell subtype naming (particularly assigning a name suggestive of meaningful biological function) can be done when we utilize prior knowledge known for the context of the experiment. For example, recent studies from the AD research community have been identifying distinct subtypes of microglia such as homeostatic microglia and disease-associated microglia (DAM) [5]. Some even proposes to further subdivide DAM population into Stage 1 DAM (DAM-1), representing a more phagocytic and functional subtype of microglia, and Stage 2 DAM (DAM-2), representing a more dysfunctional subtype and likely contributing to AD pathology [6, 7]. These putative subtypes of microglia were discovered by combining various wet-lab techniques. In addition, naming these subtypes and their functional roles (i.e., if they exhibit a detrimental or beneficial characteristic for the brain) is often controversial [8]. Single cell technology offers a way to deepen our understanding of the biological mechanisms and cross-check regulatory aspects of microglia subtypes, but the current unbiased way of subtyping the cells may not be adequate, particularly, in its inability to utilize prior knowledge into the analysis, and this limitation has motivated us to develop AFA.

The rest of this paper is organized in the following way. Section 2 provides an overview of the computational model underlying AFA. In Sect. 3, we describe what types of computational experiments we performed on eight single-cell transcriptomic datasets of AD related human and mouse microglia cells to demonstrate the effectiveness of AFA and the results obtained from these experiments. Finally, Sect. 4 concludes the paper.

2 Methodology

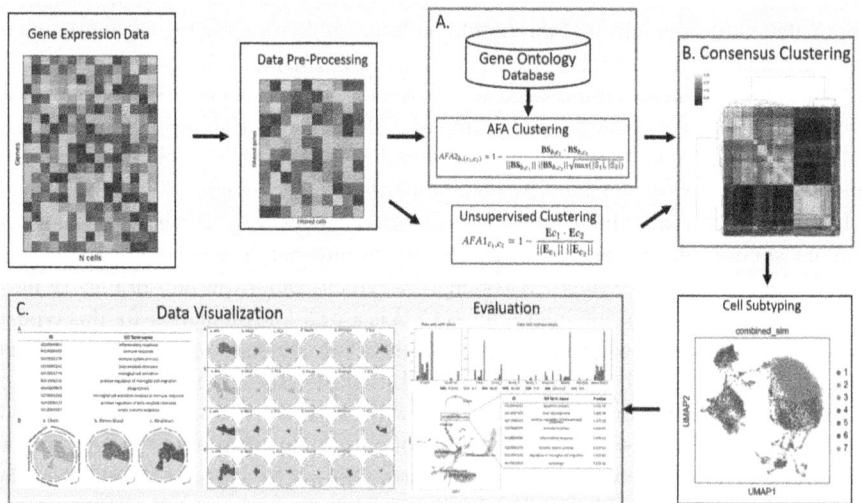

Fig. 2. Overview of the workflow of the Abstract Functional Analysis (AFA).

Figure 2 illustrates the overall process of our proposed framework, Abstract Functional Analysis (AFA). The basic data processing flow of AFA is similar to that of the conventional way of producing single cell subtypes using leiden clustering [9] followed by UMAP/tSNE. As outlined in the figure after data preprocessing, clustering is performed and the identified subtypes are displayed over a two-dimensional map using UMAP/tSNE. The unique aspects of AFA are: (i) introduce Gene Ontology (GO) Biological Process (BP) scores as new features for clustering (Fig. 2A) and (ii) combine both the conventional unsupervised clustering and the consensus clustering obtained using newly added features (i.e., BP scores) (Fig. 2B). AFA also includes clustering assessment measures that are designed to evaluate how the incorporation of BP scores differs or improves from the conventional unbiased clustering with built-in data visualization modules (Fig. 2C). The overall framework of AFA has been implemented in Python using the SCANPY [1], SciPy [10] and Scikitlearn packages [11]. The unique components of AFA are further discussed in the rest of this section.

2.1 Biological Processes of Interest (BPI)

GO Biological Processes (BPs) ([9, 10]) were created as a community resource to help scientists develop computational models of biological systems, and this resource is claimed to be the world's largest source of information on the functions of genes [9]. GO BPs are formed into a hierarchy with multiple inheritance, and each BP is meant to capture annotated genes for the function intended by each corresponding BP term. The recent development of the GO BP curation is to refine and separate the two balancing regulatory aspects of each biological process. For example, for "apoptosis", GO now includes both "positive regulation of execution phase of apoptosis" (GO:1900119) and "negative regulation execution phase of apoptosis" (GO:1900118). In MGI search, 11 genes are annotated for the former BP term, and 10 genes are annotated for the latter. These annotations have been assembled through extensive literature surveys and each annotation is tied to evidence codes. These genes' functional annotations represent critical *prior* knowledge that scientists have been generating over the years, and a key premise behind developing AFA is to use this *prior* knowledge in subtyping cells obtained from single cell omics datasets.

Issues in using this prior knowledge captured as GO BPs for the purpose of AFA are: (i) which BPs should be used for analyzing the omics datasets, and (ii) how reliably can the genes annotated for BPs be used for analysis. These issues are not trivial because gene annotations for BPs are rapidly evolving, and many annotations for a certain gene may be irrelevant to the tissue/cell types of interest in a given experiment or analysis as the BP curation is not tissue or condition specific. In fact, the development of AFA is motivated to overcome these limitations of BPs as a resource to interpret single cell gene expression data.

In AFA, we assume that the user has some ideas about the types of BPs involved in the single cell gene expression experiment. For example, if one were to study AD brain, specifically DAM related microglia cells, they would include the BPs illustrated in Fig. 3A such as Phagocytosis, Lysosome organization, Autophagy, Amyloid beta clearance, Immune response and Inflammatory response, which are commonly documented as DAM related BPs [4–6]. We label these "BPs of Interest (BPI)". We further assume

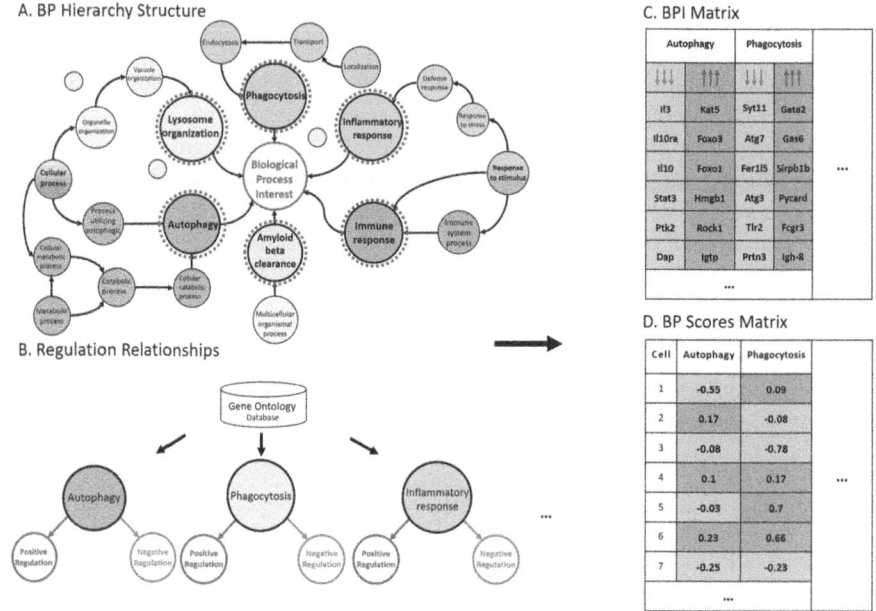

Fig. 3. Flowchart that illustrates the process of constructing a Biological Process of Interest (BPI). A. A BP hierarchy structure is created, and a selection of important biological processes are identified to form the basis of the BPI. B. Each biological process is defined by two states: positive regulation and negative regulation, and two sets of gene lists are constructed using the gene ontology database and thus C. BPI matrix is constructed. D. A BP Scores Matrix is then generated from the gathered data.

that two types of genes are annotated for each BP within the collected BPI: one for positive regulation and one for negative regulation, as capturing this type of balanced regulation patterns is already progressing in GO annotation as illustrated in Fig. 3B. For instance, the AD research community is already aware of the importance of "phagocytosis" as a crucial BP in DAM/microglia cells, and the genes responsible for its positive and negative regulation are known, even though there are currently only five to seven genes in each class. These two groups of genes annotated for Phagocytosis regulation should be captured into the BPI matrix. An example organization for Phagocytosis is given in Fig. 3C (i.e., Syt11, Atg7, ... for Down-regulation and Gata2, Gas6, ... for Up-regulation). The idea behind AFA is to use these known genes to eventually discover additional ones that play a role in either positive or negative regulation of phagocytosis. We note that each BPI can be an assortment of many different types of BPs with one cohesive regulatory function. For example, one might be interested in creating a BPI for inflammation, pain regulation, cellular mobility, etc. Each BPI created should represent a focused analysis of the specific single cell gene expression data. Lastly, BP scores are computed for each single cell based on the input gene expression dataset. For example, Cell IDs 4 and 6 exhibit upregulated patterns for both Autophagy and Phagocytosis (Fig. 3D), and how such BP scores are calculated is explained below.

2.2 Computing Contextual BP Scores

Once the BPI matrix is assembled, the next step is to produce the Contextual BP Scores matrix, as illustrated in Fig. 3D. Let $\mathbb{B}$ be a collection of all BPs that are associated with a specific context (e.g., for the DAM context in our case). That is, for a certain BP $b \in \mathbb{B}$, its Contextual BP Scores (BS) are calculated by Eq. (1). We note that the explanation for this equation has been already detailed in our previous work [11].

$$S_{b,c} = \sum_{g \in \mathbb{S}_b^+} E_{g,c} * w_g - \sum_{g \in \mathbb{S}_b^-} E_{g,c} * w_g \qquad (1)$$

where $E_{g,c}$ is the expression value of the gene g of the cell c, $\mathbb{S}_b^+$ and $\mathbb{S}_b^-$ represent the positive regulation and negative regulation genes annotated for BP b respectively, and w_g is the weight factor of gene g calculated by Eq. (2):

$$w_g = \frac{\sum_{c \in \mathbb{C}} E_{g,c}}{\sum_{g \in \mathbb{G}} \sum_{c \in \mathbb{C}} E_{g,c}} * \eta \qquad (2)$$

where η is a hyper-parameter determining the level of influence as a weight factor that can be chosen for each experiment. Through empirical testing, we selected $\eta = 0.2$ since it generally provides reasonably good statistically significant BPs. $\mathbb{C}$ represents the set of all obtained cells from the data set and $\mathbb{G}$ is the set of all available genes from the data set. All calculations are done before performing the trimmed quantile normalization [12].

2.3 Cell Similarity Measurement

Once the BS matrix is obtained, it is used as inputs to identify sub-types of single cells using BP scores assigned to each cell to discern how two cells are similar or different. We introduce two different distance metrics to measure the relationship between pairs of cells. *AFA1* represents the difference calculated between two cells c_1 and c_2 by utilizing the conventional gene expression data matrix as defined in Eq. 3.

$$AFA1_{c_1,c_2} = 1 - \frac{\mathbf{E}_{c_1} \cdot \mathbf{E}_{c_2}}{\|\mathbf{E}_{c_1}\| \|\mathbf{E}_{c_2}\|} \qquad (3)$$

where $\mathbf{E}_{c_i}$ is the gene expression matrix corresponding to the column of cell $c_i, i \in \{1, 2\}$. *AFA2* is used to quantify the difference between two cells based on their corresponding BP scores. For a given BP b, the *AFA2* between two cells, c_1 and c_2, is calculate by Eq. 4.

$$AFA2_{b,(c_1,c_2)} = 1 - \frac{\mathbf{BS}_{b,c_1} \cdot \mathbf{BS}_{b,c_2}}{\|\mathbf{BS}_{b,c_1}\| \|\mathbf{BS}_{b,c_2}\| \sqrt{\max(|\mathbb{S}_1|, |\mathbb{S}_2|)}} \qquad (4)$$

where $\mathbf{BS}_{b,c_1}$ represents the contextual BP scores. Equation 4 includes a scaling factor $\sqrt{\max(|\mathbb{S}_1|, |\mathbb{S}_2|)}$ which is designed to normalize the impact of different sizes of positive/negative regulation genes annotated for BP b. A Higher value of $AFA2_{b,(c_1,c_2)}$ means that cells c_1 and c_2 are more likely to differ in function in the context of BP b. Before computing the distance matrices using Eqs. (3) and (4), we apply Min-Max normalization to minimize the influence between the two matrices.

2.4 Consensus Clustering

Consensus Adjacency Matrix. After the steps above, two parallel intercellular distance matrices are obtained. The two distance matrices are the normalized by min-max normalization. The normalization step is crucial for eliminating biases that arise from differences in value range thus ensuring unbiased comparisons between the two distance matrices. In AFA, we adapt the consensus clustering strategy introduced in SC3 [13], and it constructs the Consensus Adjacency Matrix using Eq. (5):

$$D_{c_1,c_2} = \hat{\lambda} AFA1 + \left(1 - \hat{\lambda}\right) AFA2 \tag{5}$$

where λ is a hyper-parameter pre-defined by the researchers (default is 0.1), and an optimized $\hat{\lambda}$ is chosen by getting the highest silhouette score [14]. Lastly, we use the Leiden Clustering method [15] to cluster the two distance matrices $AFA1$ and $AFA2$ and the consensus clustering method to merge the labeling results. The Consensus Adjacency Matrix is defined as Eq. 6:

$$M_{C_1,C_2} = \begin{cases} 0 & c_1 \notin LC(c_2) \: or \: c_2 \notin LC(c_1) \\ 1 - D_{c_1,c_2} & c_1 \in LC(c_2), c_2 \in LC(c_1) \end{cases} \tag{6}$$

where M_{C_1,C_2} is the final adjacency matrix between the two cells C_1 and C_2 and $LC(c_1)$ is the Leiden clustering result for cell c. Finally, consensus clustering is performed on the adjacency matrix M.

Refinement by Ward's Hierarchical Clustering. Ward's agglomerative hierarchical clustering approach has been widely used for clustering samples based on a similarity matrix [16]. In this study, we applied Ward's method to obtain clusters by generating a linkage matrix from the adjacency matrix M. Refinement is then performed using the hierarchical clustering defined by the obtained linkage matrix. A scalar threshold is empirically set based on the desired number of clusters, with the default setting being based on the published number of clusters in previous studies.

2.5 CGI: Relevancy Measure Between a Cluster and BPI

One important measure needed in AFA is assessing how relevant a cluster is with respect to the gene regulatory function captured with the input BPI. We introduce a scoring system called the Contextual GO Term Semantic Index (CGI). This method calculates the consistency between two sets of BP terms by computing Mutual Information [17]. One is the input BPI and the other is BPs obtained by applying DAVID [18] analysis on regulatory genes identified from the conventional clustering method (e.g., SCANPY, Seurat, etc.), which is done by performing a Wilcoxon test to identify the marker genes for each cluster. Let the GO term list obtained by applying DAVID be denoted as $GO_\mathbb{D} = \{go_1, go_2, \ldots, go_m\}$. The construction of CGI is based on Eq. (7):

$$CGI = \frac{\sum_{i \in \mathbb{D}} Sim(go_i, GO_{DAM})[-\log(p_i)]}{|\mathbb{D}|} \tag{7}$$

where go_i represents a specific GO term obtained from the DAVID analysis result, denoted by $\mathbb{D}$, and p_i represents the corresponding p-value associated with the term go_i.

The weight factor $-\log(p_i)$ quantifies the significant level of the occurrence of the GO term go_i. Additionally, $Sim(go_i, GO_{DAM})$ measures the semantic similarity between go_i and the set of GO terms $GO_{DAM} = go_1, go_2, \ldots, go_n$, a collection of GO terms associated with the DAM context. This semantic similarity $Sim(go_i, GO_{DAM})$ is defined as follows:

$$Sim(go_i, GO_{DAM}) = \frac{\sum_{j \in DAM} S_{GO}(go_i, go_j)}{|DAM|} \qquad (8)$$

Table 1. Summary of 8 scRNA-Seq Datasets

GSE ID	Datasets ID	Species	Cells	Microglia	Tissue	References
GSE199027	Singh	MM[1]	12,714	12,714	Brains	Singh et al., 2022 [4]
GSE175814	Soreq	HS[2]	21,355	15,093	Anterior hippocampal cortex	Soreq et al., 2023 [21]
GSE214979	Anderson	HS[2]	105,333	3,180	Dorsolateral prefrontal cortex	Anderson et al., 2023 [25]
GSE174367	Morabito	HS[2]	61,473	4,127	Prefrontal cortex	Morabito et al., 2021 [26]
GSE138852	Grubman	HS[2]	13,214	450	Brains	Grubman et al., 2019 [27]
GSE221856	Chen	MM[1]	33,020	16,279	Cortex and Hippocampus	Chen et al., 2023 [28]
GSE98969	Keren-Shaul	MM[1]	41,856	8,620	Whole brain	Keren-Shaul et al., 2017 [29]
GSE102827	Hrvatin	MM[1]	48,266	10,159	Visual Cortex	Hrvatin et al., 2018 [30]

Note: [1] Mus musculus, [2] Homo Sapiens

Wang's measure [19] is used here to calculate the semantic similarity $S_{GO}(go_i, go_j)$. The GO analysis is conducted using the Python library GOATOOLS [20]. A higher value of CGI indicates a stronger relationship between the features derived from the clustering results and the context in which they were identified, for example in our case study, the DAM subtypes.

3 Experiments and Results

3.1 Experimental Datasets

We carried out comparative experiments to test the effectiveness of AFA on a total of eight publicly available AD related scRNA-seq datasets, with four datasets from human brain and four from mouse brain. The details of these eight benchmark datasets are listed in Table 1, which were all obtained from the GEO portal (www.ncbi.nlm.nih.gov/geo/). In Soreq's dataset, we stratified it into three independent samples as each was obtained with different clinical variables [21]. As shown in the table, scRNA-seq experiments were done on different parts of brain, but regardless of the tissue types within the brain (e.g., cortex, vs. hippocampus) the brain's primary immune cell known as microglia is ubiquitous in the entire brain's cell population. We limited our analysis to the microglia subpopulations in each study. When the dataset does not provide the cell type information, we used well-established markers of microglia to isolate a homogeneous microglia population in each dataset. The table shows the total number of microglia cells isolated from the entire population. For example, in Hrvatin's dataset, the whole population has 48,266 cells, but our analysis was limited to 10,159 microglia cells. In addition, to ensure comparability, we normalized the scRNA-seq data using the widely used z-score method.

3.2 Assessing Functional Roles of Identified Subtypes

AFA aims to subgroup cells from the perspective of the input Biological Processes of Interest (BPI). That is, the method calculates how far or closely cells are located with each other when the two types of cell similarity measures *AFA*1 and *AFA*2 are taken into account. Two types of assessments are performed, one using CGI and another using Polar chart.

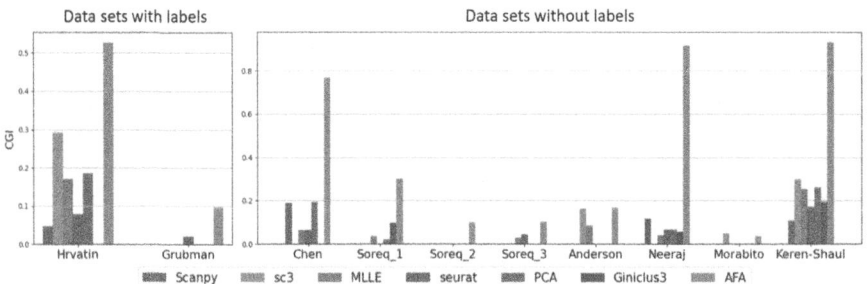

Fig. 4. Methods were evaluated on 10 scRNA-seq datasets, comprising two labeled datasets and eight unlabeled datasets in terms of CGI. Note that those values less than the minimum value of y-axis are missed.

Using CGI. The CGI analysis given in Fig. 4 clearly demonstrates that AFA significantly outperforms other methods on identifying functional roles from clustering tasks—i.e., the higher CGI values, the better. As shown in Fig. 4, we separated the comparison into two groups, one for the Hrvatin's and Grubman's datasets that offered labels for identified subtypes, and one for the other 8 datasets which did not give labels thus requiring us to place subtype labels based on known subtype markers and perform the comparative analysis. For each dataset, the CGI measures are obtained using 6 clustering methods (SCANPY, sc3, MLLE, seurat, PCA and Giniclus3) and these measures are compared to the CGI measure obtained from subtypes determined by AFA. AFA obtained the highest CGI score on most datasets. Hrvatin's dataset may be unique since it contains only two cell types, microglia subtype 1 and microglia subtype 2. Even so, AFA outperforms other methods. For datasets having fewer microglia cells, such as Morabito's and Anderson's datasets, we note that the average CGI scores are relatively low, but this could be due to some low number bias or artifact, so the comparison outcomes may not be considered reliable. Overall, the CGI measure analysis can be summed up into the following statement: If the given single cell gene expression data matrix includes some hidden functional structure, it would be more likely found using AFA than with other conventional "unbiased" clustering methods—as long as the scientists can input a well-constructed BPI (i.e., a prior knowledge) for the analysis.

Using Polar Area Charts. We find that the polar area chart is another effective means to contrast functional identification from sub-grouped cells. Fig. 5A shows how each polar area chart succinctly reveals how 10 different but related GO biological processes are enriched in identified subgroups of cells (i.e., 10 GO BPs from GO:0006955 to GO:0001774 in counterclockwise). Here, the angular axis represents different GO Biological Processes (BPs) with IDs and the radial axis corresponds to the -log(p-value) of the GO terms. The larger colored segment represents higher statistical significance. For example, in Chen's dataset, (GSE221856), GO:0002282 and GO:0002376 are the two highly enriched biological processes.

In Fig. 5B, we show how this polar area chart analysis can reveal a bird's-eye view the AFA's superior performance over five other methods (MLLE, PCA, Seurat, Giniclus3, and SC3) when the same comparative analysis was repeated on the four datasets: Keren-Shaul's (GSE98969), Chen's (GSE221856), Hrvatin's (GSE102827), and Grubman's (GSE138852). It is clear that AFA's polar area charts (the first column) consistently exhibit the highest BP enrichment results across all datasets.

AFA Identifies DAM-1 Subtype and Extends BPI. During our experiments with AFA, we focused on studying two aspects of this new subtyping method: (i) Can this method discover indeed the subtype(s) of cells that is targeted by the given input BPI? and (ii) Does the identified subtype(s) help us further understand the initially given BPI? For example, when the BPI for DAM is used to find subtypes of microglia cells (like DAM cells), will it be able to identify a subgroup of cells that predominantly seem to possess functional signatures of DAM cells? Additionally, can other biological processes, beyond the one initially used for BPI and clustering, be identified from this subgroup of DAM like cells? The evidence for these capabilities of AFA is illustrated in Fig. 6a–c which was obtained by applying AFA to one of Soreq's three stratified datasets (GSM5348376).

Figure 6a shows a UMAP visualization of the AFA clustering performed with its BPI including six biological processes (as illustrated in Fig. 3A). Application of AFA reveals a total of 13 distinct subpopulations (only a subset of these subtypes is annotated in Fig. 3A). To investigate their functional characteristics, we conducted a Wilcoxon test on each of the identified clusters and subsequently performed Gene Ontology (GO) analysis. This GO analysis reveals that cluster ID 13 very likely matches the biological processes of the input BPI. Fig. 6b shows a total of 10 enriched BPs with their p-values less than 7.55E-03. Comparing these BPs with the BPI used to run AFA reveals three interesting findings. First, multiple BPs included in the BPI are found in the top 10 enriched BPs either with precisely the same terms or very closely related terms. The exact same terms are Autophagy for GO:0006914, Inflammatory response for GO:0006954 and Immune response for GO:0006955. The very closely related ones are GO:0002250 (adaptive immune response) and GO:0002376 (immune system process), both of which belonging to the broader BP term "Immune response". Second, more refined BPs than the ones given in BPI are discovered. These include "GO1900223: positive regulation of beta-amyloid clearance" which is a more refined version of Amyloid beta clearance, "GO:0002523: leukocyte migration involved in inflammatory response" which is a more specific process of "Inflammatory response", and "GO:0002282: microglial cell activation involved in immune response" which is a more specific process of "Immune response". Since GO BP terms form a hierarchy, finding a more refined term (i.e., located in a deeper node in the branch) can be an informative discovery. Third, completely new BPs are found. These cases include: "GO:0006915- apoptotic process", and GO:0097242: regulation of microglial cell migration".

These observations strongly suggest that the cluster ID 13 represents or is closely related to DAM stage 1 cells (DAM-1). DAM-1 is defined to exhibit a more functional (beneficial) and phagocytic capabilities [7]. The additional findings of more refined BP terms and new BP terms also align with our literature survey. Microglia cells are known to be a unique and important component of both the innate and adaptive immune responses [22]. Microglia's role in apoptosis is also well established, specifically, TREM2-deficient microglia's increased apoptosis [23]. The connection between microglia and leukocytes in the central nervous system is also well established as both are known to concentrate around amyloid plaques which is a prominent feature of AD [24].

We also performed pseudotemporal cell trajectory analysis to further acquire evidence that the cluster ID 13 is likely a DAM subtype, as illustrated in Fig. 6c. It has been well documented in the literature that homeostatic microglia cells transition to the DAM type [4]. Our trajectory analysis suggests that homeostatic cells (cluster IDs 1 and 12) become DAM-like through sequentially undergoing biological differentiation from transition cells (cluster ID 3) and reaching the DAM-like stage 1 with high phagocytic function (cluster IDs 10 and 13) and eventually becoming DAM-like stage 2 cells (cluster IDs 4 and 11). These findings bolster our assessment that the cluster ID 13 is likely a DAM-1 subtype.

AFA Identifies New Subtypes of Homeostatic Microglia. We also find that AFA can offer a way to place a functional label to a sub-cluster which is difficult to characterize using the conventional marker gene enrichment analysis method. The conventional leiden clustering/UMAP analysis on single cell gene expression intensity data generally fails to separate cells if they are from a homogeneous cell type. The illustration given in Fig. 1a is an example. When the clustering method is forced to produce subtypes, interpreting each subtype (i.e., subgroups of the spade shape clump) is difficult. This type of counter example is abundant. In our other single cell subtyping study involving human knee articular cartilage (AC) cells, the conventional method produces only one big cluster of cells (aka "a hairball") as AC cells are predominantly of one type of cells, chondrocytes.

By applying the AFA, we were able to isolate a meaningful subtype from one big cluster of cells. The results of this experiments illustrated in Fig. 6d–f, using Singh's dataset (GSE199027). Notable in this analysis is that although the visualization includes a large cluster of cells, the method identified cluster ID 2 as homeostatic-transition. The reason for this subtype being part of the largest cluster is because the whole group is made up of microglia cells. However, this sub-group of cells exhibits a strong signature of DAM related biological processes. GO analysis done for this group revealed enrichment in immune-related biological processes, phagocytic functions, lysosomal functions, microglial cell activation, and so on, as shown in Fig. 6d. Interestingly in this GO analysis, beyond the DAM like signatures, it also includes "GO0048667: cell morphogenesis involved in neuron differentiation", possibly suggesting the transitional state of homeostatic microglia cells and justifying the labeling for this subtype "homeostatic-transition". Our subsequent pseudotemporal cell trajectory analysis also suggests its transitional state.

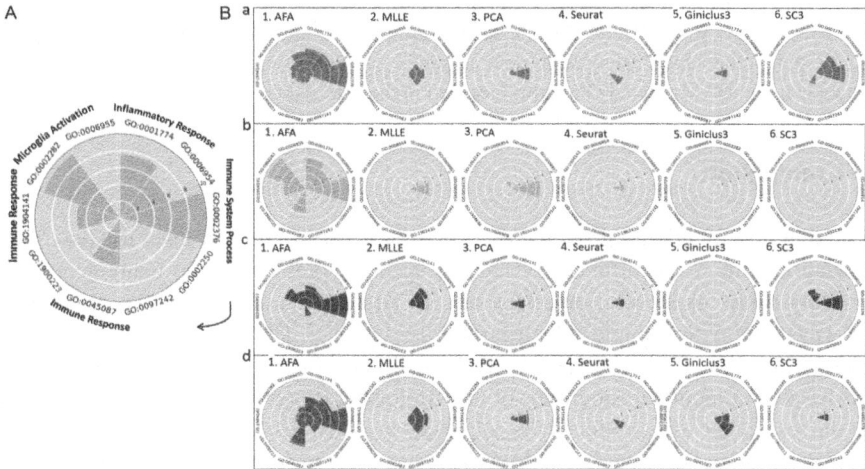

Fig. 5. A: Polar charts were constructed from Chen's data set, with the radial axis corresponding to the significant level of the GO terms, and the angular axis representing different GO term IDs annotated with the names of Biological Processes (BPs). B: Polar charts showcasing the performance of different methods: (a) AFA, (b) MLLE, (c) PCA, (d) Seurat, (e) GiniClus3, and (f) SC3, on four distinct datasets: a) Keren-Shaul, b) Chen, c) Hrvatin, and d) Grubman.

That is, some homeostatic cells (cluster IDs 1, 2 and 3) could also exhibit DAM-like signatures (particularly, cluster ID 2, i.e., homeostatic but DAM-like) through undergoing cellular differentiation possibly from truly homeostatic microglia (cluster IDs 7 and 8). The transitional cells become truly DAM-like stage 1 with its high phagocytosis signature (cluster IDs 4 and 6) and eventually DAM-like stage 2 cells (cluster ID 5). These findings are consistent with our literature survey [4], highlighting the AFA's ability to capture the functional changes during the transitional process.

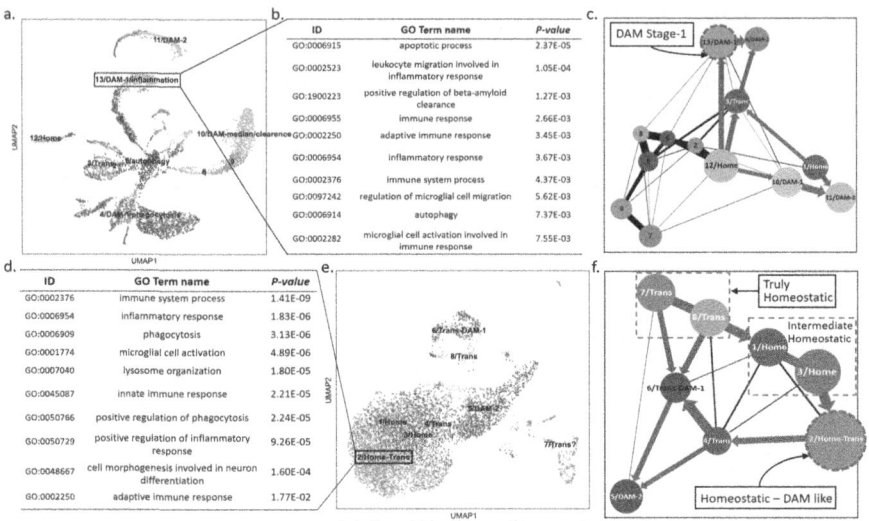

Fig. 6. UMAP visualization results of Singh's and Soreq's first data sets are shown in panels a), b), c), d), e), and f), respectively. The labeled clustering results are depicted in a) and e), while the PAGA graph of the AFA cluster results demonstrating the connectivity of cell clusters can be seen in c) and f). The DAM-related GO terms enriched in the selected clusters are listed in b) and d).

4 Conclusion

Examining the functional grouping of microglia at the single cell level to study their association with AD is difficult for reasons such as complexity (too many genes and too many cells), expression dropouts, etc. To address this challenge, we proposed a new way of subtyping single cells from scRNA-seq data. The method has been labeled Abstract Functional Analysis (AFA), and we showed that it can categorize microglia cells based on functional subgroups of GO Biological Processes. This method assumes that the researcher who produced the single cell gene expression data should have some prior knowledge of the regulatory patterns they expect to see from the gene expression data. For example, some microglia cells should engage in phagocytosis upon encountering Aβ accumulation in AD brain. In that sense, AFA is not an "unbiased" clustering method. Rather, it is a form of semi-supervised clustering method in which the objective is to use what is known *a priori* to uncover unknown patterns. Our experiment with multiple publicly available AD single cell gene expression data sets revealed that AFA can further subdivide microglia cells into distinct functional subtypes which can be labeled in terms of BPs. Our subtype labeling method is different from the conventional "unbiased" clustering method in which scientists attempt to label subgroups—after subtypes are identified—in terms of marker genes that are enriched in cell subgroups. AFA also incorporates the idea of consensus clustering: generating a merged distance matrix by combining the adjacency matrix using the Leiden clustering method. By incorporating functional grouping analysis, our AFA approach offers a promising avenue for exploring the heterogeneity of microglia subtypes and their associations with AD. This framework has the potential to provide valuable insights into the underlying mechanisms of microglia function and their roles in AD. Many issues remain to be solved for further refining AFA, but this method clearly suggests new research opportunities in the direction of semi-supervised clustering in single cell subtyping—using what is known to discover unknown ones.

Acknowledgements. Research reported in this work was supported in part by NIH Grant No. 2RF1AG025493-16. Its contents are solely the responsibility of the authors and do not necessarily represent the views of the NIH.

References

1. Wolf, F.A., Angerer, P., Theis, F.J.: SCANPY: large-scale single-cell gene expression data analysis. Genome Biol. **19**, 1–5 (2018)
2. Butler, A., Hoffman, P., Smibert, P., et al.: Integrating single-cell transcriptomic data across different conditions, technologies, and species. Nat. Biotechnol. **36**, 411–420 (2018)
3. Villani, A.-C., Satija, R., Reynolds, G., et al.: Single-cell RNA-seq reveals new types of human blood dendritic cells, monocytes, and progenitors. Science (1979) **356**, eaah4573 (2017)
4. Singh, N., Benoit, M.R., Zhou, J., et al.: BACE-1 inhibition facilitates the transition from homeostatic microglia to DAM-1. Sci. Adv. **8**, eabo1286 (2022)
5. Deczkowska, A., Keren-Shaul, H., Weiner, A., et al.: Disease-associated microglia: a universal immune sensor of neurodegeneration. Cell **173**, 1073–1081 (2018)
6. Zhou, Y., Song, W.M., Andhey, P.S., et al.: Human and mouse single-nucleus transcriptomics reveal TREM2-dependent and TREM2-independent cellular responses in Alzheimer's disease. Nat. Med. **26**, 131–142 (2020)
7. Huang, Y., Happonen, K.E., Burrola, P.G., et al.: Microglia use TAM receptors to detect and engulf amyloid β plaques. Nat. Immunol. **22**, 586–594 (2021)
8. Aguzzi, A., Barres, B.A., Bennett, M.L.: Microglia: scapegoat, saboteur, or something else? Science (1979) **339**, 156–161 (2013)
9. Consortium GO: The Gene Ontology (GO) database and informatics resource. Nucleic Acids Res. **32**, D258–D261 (2004)
10. Eppig, J.T.: Mouse genome informatics (MGI) resource: genetic, genomic, and biological knowledgebase for the laboratory mouse. ILAR J. **58**, 17–41 (2017)
11. Zhang, C., Joshi, P., Wang, H., et al.: Pola viz reveals microglia polarization at single cell level in Alzheimer's disease. In: 2022 IEEE International Conference on Bioinformatics and Biomedicine (BIBM), pp. 1387–1392 (2022)
12. Wang, H., Joshi, P., Zhang, C., et al.: rCom: a route-based framework inferring cell type communication and regulatory network using single cell data. In: Proceedings of the 13th ACM International Conference on Bioinformatics, Computational Biology and Health Informatics, pp 1–4 (2022)
13. Kiselev, V.Y., Kirschner, K., Schaub, M.T., et al.: SC3: consensus clustering of single-cell RNA-Seq data. Nat. Methods **14**, 483–486 (2017)
14. Rousseeuw, P.J.: Silhouettes: a graphical aid to the interpretation and validation of cluster analysis. J. Comput. Appl. Math. **20**, 53–65 (1987)
15. Traag, V.A., Waltman, L., Van Eck, N.J.: From Louvain to Leiden: guaranteeing well-connected communities. Sci. Rep. **9**, 5233 (2019)
16. Ward, J.H., Jr.: Hierarchical grouping to optimize an objective function. J. Am. Stat. **58**, 236–244 (1963)

17. Ross, B.C.: Mutual information between discrete and continuous data sets. PLoS ONE **9**, e87357 (2014)
18. Dennis, G., Sherman, B.T., Hosack, D.A., et al.: DAVID: database for annotation, visualization, and integrated discovery. Genome Biol. **4**, 1–11 (2003)
19. Wang, J.Z., Du, Z., Payattakool, R., et al.: A new method to measure the semantic similarity of GO terms. Bioinformatics **23**, 1274–1281 (2007)
20. Klopfenstein, D.V., Zhang, L., Pedersen, B.S., et al.: GOATOOLS: a Python library for Gene Ontology analyses. Sci. Rep. **8**, 1–17 (2018)
21. Soreq, L., Bird, H., Mohamed, W., Hardy, J.: Single-cell RNA sequencing analysis of human Alzheimer's disease brain samples reveals neuronal and glial specific cells differential expression. PLoS ONE **18**, e0277630 (2023)
22. Olson, J.K., Miller, S.D.: Microglia initiate central nervous system innate and adaptive immune responses through multiple TLRs. J. Immunol. **173**, 3916–3924 (2004)
23. Mazaheri, F., Snaidero, N., Kleinberger, G., et al.: TREM 2 deficiency impairs chemotaxis and microglial responses to neuronal injury. EMBO Rep. **18**, 1186–1198 (2017)
24. Hansen, D.V., Hanson, J.E., Sheng, M.: Microglia in Alzheimer's disease. J. Cell Biol. **217**, 459–472 (2018)
25. Anderson, A.G., Rogers, B.B., Loupe, J.M., et al.: Single nucleus multiomics identifies ZEB1 and MAFB as candidate regulators of Alzheimer's disease-specific cis-regulatory elements. Cell Genomics **3** (2023)
26. Morabito, S., Miyoshi, E., Michael, N., et al.: Single-nucleus chromatin accessibility and transcriptomic characterization of Alzheimer's disease. Nat. Genet. **53**, 1143–1155 (2021)
27. Grubman, A., Chew, G., Ouyang, J.F., et al.: A single-cell atlas of entorhinal cortex from individuals with Alzheimer's disease reveals cell-type-specific gene expression regulation. Nat. Neurosci. **22**, 2087–2097 (2019)
28. Chen, X., Firulyova, M., , M., et al.: Microglia-mediated T cell infiltration drives neurodegeneration in tauopathy. Nature, 1–10 (2023)
29. Keren-Shaul, H., Spinrad, A., Weiner, A., et al.: A unique microglia type associated with restricting development of Alzheimer's disease. Cell **169**, 1276–1290 (2017)
30. Hrvatin, S., Hochbaum, D.R., Nagy, M.A., et al.: Single-cell analysis of experience-dependent transcriptomic states in the mouse visual cortex. Nat. Neurosci. **21**, 120–129 (2018)

Correction to: Identification of Chimeric RNAs: A Novel Machine Learning Perspective

Paola Bonizzoni, Clelia De Felice, Yuri Pirola, Raffaella Rizzi, Rocco Zaccagnino, and Rosalba Zizza

Correction to:
Chapter 2 in: M. S. Bansal et al. (Eds.): *Computational Advances in Bio and Medical Sciences*, **LNBI 14548, https://doi.org/10.1007/978-3-031-82768-6_2**

In the originally published version of the chapter, first name and last name was incorrectly tagged as "Clelia De" and "Felice" respectively for the second author. The first name and last name tag of the second author has been corrected to "Clelia" and "De Felice" respectively.

The updated version of this chapter can be found at
https://doi.org/10.1007/978-3-031-82768-6_2

Author Index

B
Barzas, Konstantinos 1
Basiri, Mohammad Amin 145
Boddeda, Sriram 132
Bonizzoni, Paola 14
Bumin, Aysegul 27

C
Cao, Guohua 40, 94
Chaturvedi, Ayush 40
Cheng, Chu-Yu 53
Ciccolella, Simone 82
Cordes, Julia 65

D
Daescu, Ovidiu 251
Das, Arghya Kusum 75
Davila, Jaime 65
De Felice, Clelia 14

F
Feng, Wu-chun 40, 94
Fouad, Shereen 1

G
Ganguly, Arnab 75
Gao, Hongchang 119
Gibney, Daniel 75

H
Hara, Karine Marques 107
Hong, Seung-Hyun 262
Huang, Kejun 27

J
Jasa, Gainer 1
Jinna, Nikita 203
Juyal, Akshay 82

K
Kahveci, Tamer 27
Kasparian, Armen 94
Khanmohammadi, Sina 145, 229

L
Landini, Gabriel 1
Li, Bin 119
Liao, Li 178
Lopes, Fabricio Martins 107
Lu, Chung-Chin 53

M
Moussa, Marmar R. 132

N
Nguyen, Sean M. 145

P
Pagolu, Sri Lakshmi Bhavani 154
Pais, Namitha Viona 166
Parekh, Nita 154, 191
Paschoal, Alexandre Rossi 107
Patterson, Murray 82
Paula, Pedro Henrique Mendes de 107
Pinnix, Zandra 203

© The Editor(s) (if applicable) and The Author(s), under exclusive license to Springer Nature Switzerland AG 2025
M. S. Bansal et al. (Eds.): ICCABS 2023, LNBI 14548, pp. 277–278, 2025.
https://doi.org/10.1007/978-3-031-82768-6

Pirola, Yuri 14
Pizzi, Cinzia 238

Q
Qin, Xihan 178

R
Rajasekaran, Sanguthevar 166
Ravishanker, Nalini 166
Razavi, Seyed Majid 229
Rida, Padmashree 203
Rizzi, Raffaella 14

S
Sahoo, Bikram 203
Shi, Xinghua 119
Shin, Dong-Guk 262
Silva, Lucas Otavio Leme 107
Singh, Rahul 218
Soto-Gomez, Mauricio 82
Street, Charles H. 132
Stubby, Emma T. 229
Suba, S. 154
Suseela, Suba 191

T
Tayebi, Zahra 82
Thankachan, Sharma V. 75
Tomasella, Francesco 238

V
Vedova, Gianluca Della 82

W
Wang, Honglin 262
Weinstock, George 166

Y
Yadav, Ankur 251
Yan, Riqiang 262
Yu, Jiadong 218

Z
Zaccagnino, Rocco 14
Zelikovsky, Alex 203
Zelikovsky, Alexander 82
Zhang, Chenyu 262
Zizza, Rosalba 14